Noether's Theorem and Symmetry

Noether's Theorem and Symmetry

Special Issue Editors

P.G.L. Leach
Andronikos Paliathanasis

MDPI • Basel • Beijing • Wuhan • Barcelona • Belgrade

Special Issue Editors
P.G.L. Leach
University of Kwazulu Natal &
Durban University of Technology
Cyprus

Andronikos Paliathanasis
Durban University of Technology
Republic of South Arica

Editorial Office
MDPI
St. Alban-Anlage 66
4052 Basel, Switzerland

This is a reprint of articles from the Special Issue published online in the open access journal *Symmetry* (ISSN 2073-8994) from 2018 to 2019 (available at: https://www.mdpi.com/journal/symmetry/special_issues/Noethers_Theorem_Symmetry).

For citation purposes, cite each article independently as indicated on the article page online and as indicated below:

LastName, A.A.; LastName, B.B.; LastName, C.C. Article Title. *Journal Name* **Year**, *Article Number*, Page Range.

ISBN 978-3-03928-234-0 (Pbk)
ISBN 978-3-03928-235-7 (PDF)

Cover image courtesy of Andronikos Paliathanasis.

© 2020 by the authors. Articles in this book are Open Access and distributed under the Creative Commons Attribution (CC BY) license, which allows users to download, copy and build upon published articles, as long as the author and publisher are properly credited, which ensures maximum dissemination and a wider impact of our publications.
The book as a whole is distributed by MDPI under the terms and conditions of the Creative Commons license CC BY-NC-ND.

Contents

About the Special Issue Editors vii

Preface to "Noether's Theorem and Symmetry" ix

Amlan K Halder, Andronikos Paliathanasis and P.G.L. Leach
Noether's Theorem and Symmetry
Reprinted from: *Symmetry* 2018, 10, 744, doi:10.3390/sym10120744 1

Dmitry S. Kaparulin
Conservation Laws and Stability of Field Theories of Derived Type
Reprinted from: *Symmetry* 2019, 11, 642, doi:10.3390/sym11050642 22

Linyu Peng
Symmetries and Reductions of Integrable Nonlocal Partial Differential Equations
Reprinted from: *Symmetry* 2019, 11, 884, doi:10.3390/sym11070884 42

V. Rosenhaus and Ravi Shankar
Quasi-Noether Systems and Quasi-Lagrangians
Reprinted from: *Symmetry* 2019, 11, 1008, doi:10.3390/sym11081008 53

M. Safdar, A. Qadir, M. Umar Farooq
Comparison of Noether Symmetries and First Integrals of Two-Dimensional Systems of Second Order Ordinary Differential Equations by Real and Complex Methods
Reprinted from: *Symmetry* 2019, 11, 1180, doi:10.3390/sym11091180 73

Marianna Ruggieri and Maria Paola Speciale
Optimal System and New Approximate Solutions of a Generalized Ames's Equation
Reprinted from: *Symmetry* 2019, 11, 1230, doi:10.3390/sym11101230 85

Elena Recio, Tamara M. Garrido, Rafael de la Rosa and María S. Bruzón
Reprinted from: *Symmetry* 2019, 11, 1031, doi:10.3390/sym11081031 96

Almudena P. Márquez and María S. Bruzón
Symmetry Analysis and Conservation Laws of a Generalization of the Kelvin-Voigt Viscoelasticity Equation
Reprinted from: *Symmetry* 2019, 11, 840, doi:10.3390/sym11070840 109

Hassan Azad, Khaleel Anaya, Ahmad Y. Al-Dweik and M. T. Mustafa
Invariant Solutions of the WaveEquation on Static Spherically Symmetric Spacetimes AdmittingG_7 Isometry Algebra
Reprinted from: *Symmetry* 2018, 10, 665, doi:10.3390/sym10120665 118

Sumaira Saleem Akhtar, Tahir Hussain and Ashfaque H. Bokhari
Positive Energy Condition and Conservation Laws in Kantowski-Sachs Spacetime via Noether Symmetries
Reprinted from: *Symmetry* 2018, 10, 712, doi:10.3390/sym10120712 144

Ugur Camci
$F(R, G)$ Cosmology through Noether Symmetry Approach
Reprinted from: *Symmetry* 2018, 10, 719, doi:10.3390/sym10120719 156

About the Special Issue Editors

P.G.L. Leach was born in Melbourne, Australia, just a few days before the bombing of Pearl Harbour. He matriculated from St Patrick's College in 1958 and attended the University of Melbourne to obtain his BSc, BA, and Dip Ed. After teaching at High Schools for some years, he went to the Bendigo Institute of Technology (Victoria, Australia) and resumed his studies at La Trobe University in Melbourne. His supervisor was CJ Eliezer, a former student of PAM Dirac. He graduated with an MSc in 1977, Ph.D. in 1979, and, from the University of Natal, a DSc in 1996. In 1983, he went to the University of the Witwatersrand, Johannesburg, as a Professor of Applied Mathematics, and in 1990, to the University of Natal in Durban. Currently, he is Professor Emeritus at the University of KwaZulu Natal and Extraordinary Research Professor at the Durban University of Technology. Honors: Fellow of the University of Natal (1996); Elected Member of the Academy of Nonlinear Sciences (Russian Republic) (1997); Elected Fellow of the Royal Society of South Africa (1998); Gold Medalist of the South African Mathematical Society (2001); Distinguished Research Award of the South African Mathematical Society (2002). He has some 350 publications.

Andronikos Paliathanasis received his BSc in 2008 from the Physics department of the National and Kapodistrian University of Athens. From the same department, he graduated with an MSc in 2010 and a Ph.D. in 2015. Since then, he has collaborated with the INFN Sez. Di Napoli, in Italy. In November 2015, he was awarded with a postdoctoral fellowship from CONICYT (Chile) in the Universidad Austral de Chile, where he was also an Adjunct Professor and gave various classes in the institute of physics and mathematics. Since June 2017, he has been affiliated with the Institute of System Science of Durban University of Technology, while since July 2019, he has also been a Research Fellow in the same institute. He has more than 100 publications in peer-reviewed journals on the subject of symmetries of differential equations as well as in gravitational physics and cosmology.

Preface to "Noether's Theorem and Symmetry"

This book deals with the two famous theorems of Emmy Noether, published a century ago. The work of Emmy Noether was pioneering for her period and changed the way that we approach physical theories and other applied mathematics theories today. There have been many different readings of her work and various attempts to extend and generalize the original results of 1918. In this book, in a series of articles, we summarize the above attempts to extend Noether's work, and various applications are presented for the modern treatment of Noether's theorems.

P.G.L. Leach, Andronikos Paliathanasis
Special Issue Editors

Article

Noether's Theorem and Symmetry

Amlan K. Halder [1], Andronikos Paliathanasis [2,3,*,†] and Peter G.L. Leach [3,4]

1. Department of Mathematics, Pondicherry University, Kalapet 605014, India; amlan.haldar@yahoo.com
2. Instituto de Ciencias Físicas y Matemáticas, Universidad Austral de Chile, Valdivia 5090000, Chile
3. Institute of Systems Science, Durban University of Technology, PO Box 1334, Durban 4000, South Africa; leachp@ukzn.ac.za
4. School of Mathematical Sciences, University of KwaZulu-Natal, Durban 4000, South Africa
* Correspondence: anpaliat@phys.uoa.gr
† Current address: Instituto de Ciencias Físicas y Matemáticas, Universidad Austral de Chile, Valdivia 5090000, Chile.

Received: 20 November 2018; Accepted: 9 December 2018; Published: 12 December 2018

Abstract: In Noether's original presentation of her celebrated theorem of 1918, allowance was made for the dependence of the coefficient functions of the differential operator, which generated the infinitesimal transformation of the action integral upon the derivatives of the dependent variable(s), the so-called generalized, or dynamical, symmetries. A similar allowance is to be found in the variables of the boundary function, often termed a gauge function by those who have not read the original paper. This generality was lost after texts such as those of Courant and Hilbert or Lovelock and Rund confined attention to point transformations only. In recent decades, this diminution of the power of Noether's theorem has been partly countered, in particular in the review of Sarlet and Cantrijn. In this Special Issue, we emphasize the generality of Noether's theorem in its original form and explore the applicability of even more general coefficient functions by allowing for nonlocal terms. We also look for the application of these more general symmetries to problems in which parameters or parametric functions have a more general dependence on the independent variables.

Keywords: Noether's theorem; action integral; generalized symmetry; first integral; invariant; nonlocal transformation; boundary term; conservation laws; analytic mechanics

1. Introduction

Noether's theorem [1] treats the invariance of the functional of the calculus of variations—the action integral in mechanics—under an infinitesimal transformation. This transformation can be considered as being generated by a differential operator, which in this case is termed a Noether symmetry. The theorem was not developed ab initio by Noether. Not only is it steeped in the philosophy of Lie's approach, but also, it is based on earlier work of more immediate relevance by a number of writers. Hamel [2,3] and Herglotz [4] had already applied the ideas developed in her paper to some specific finite groups. Fokker [5] did the same for specific infinite groups. A then recently-published paper by Kneser [6] discussed the finding of invariants by a similar method. She also acknowledged the contemporary work of Klein [7]. Considering that the paper was presented to the Festschrift in honor of the fiftieth anniversary of Klein's doctorate, this final attribution must have been almost obligatory.

For reasons obscure Noether's theorem has been subsequently subject to downsizing by many authors of textbooks [8–10], which has then given other writers (cf. [11]) the opportunity to 'generalize' the theorem or to demonstrate the superiority of some other method [12,13] to obtain more general results [14–16]. This is possibly due to the simplified form presented in Courant and Hilbert [8]. As Hilbert was present at the presentation by Noether of her theorem to the Festschrift in honor of the fiftieth anniversary of Felix Klein's doctorate, it could be assumed that his description would be accurate. However, Hilbert's sole contribution to the text was his name.

This particularizing tendency has not been uniform, e.g., the review by Sarlet and Cantrijn [17]. According to Noether [1] (pp. 236–237), "In den Transformationen können auch die Ableitungen der u nach den x, also $\partial u/\partial x$, $\partial^2 u/\partial x^2$, ... auftreten", so that the introduction of generalized transformations is made before the statement of the theorem [1] (p. 238). On page 240, after the statement of the theorem, Noether does mention particular results if one restricts the class of transformations admitted and this may be the source of the usage of the restricted treatments mentioned above.

We permit the coefficient functions of the generator of the infinitesimal transformation to be of unspecified dependence subject to any requirement of differentiability.

For the purposes of the clarity of exposition, we develop the theory of the theorem in terms of a first-order Lagrangian in one dependent and one independent variable. The expressions for more complicated situations are given below in a convenient summary format.

2. Noether Symmetries

We consider the action integral:

$$A = \int_{t_0}^{t_1} L(t, q, \dot{q}) \, dt. \tag{1}$$

Under the infinitesimal transformation:

$$\bar{t} = t + \varepsilon \tau, \qquad \bar{q} = q + \varepsilon \eta \tag{2}$$

generated by the differential operator:

$$\Gamma = \tau \partial_t + \eta \partial_q,$$

the action integral (1) becomes:

$$\bar{A} = \int_{\bar{t}_0}^{\bar{t}_1} L(\bar{t}, \bar{q}, \mathring{\bar{q}}) \, d\bar{t}$$

($\mathring{\bar{q}}$ is $d\bar{q}/d\bar{t}$ in a slight abuse of standard notation), which to the first order in the infinitesimal, ε, is:

$$\begin{aligned}\bar{A} &= \int_{t_0}^{t_1} \left[L + \varepsilon \left(\tau \frac{\partial L}{\partial t} + \eta \frac{\partial L}{\partial q} + \zeta \frac{\partial L}{\partial \dot{q}} + \dot{\tau} L \right) \right] dt \\ &\quad + \varepsilon \left[\tau t_1 L(t_1, q_1, \dot{q}_1) - \tau t_0 L(t_0, q_0, \dot{q}_0) \right],\end{aligned} \tag{3}$$

where $\zeta = \dot{\eta} - \dot{q}\dot{\tau}$ and $L(t_0, q_0, \dot{q}_0)$ and $L(t_1, q_1, \dot{q}_1)$ are the values of L at the endpoints t_0 and t_1, respectively.

We demonstrate the origin of the terms outside of the integral with the upper limit. The lower limit is treated analogously.

$$\begin{aligned}\int^{\bar{t}_1} &= \int^{t_1 + \varepsilon \tau(t_1)} \\ &= \int^{t_1} + \int_{t_1}^{t_1 + \varepsilon \tau(t_1)} \\ &= \varepsilon \int^{t_1} + \varepsilon \tau(t_1) L(t_1, q_1, \dot{q}_1)\end{aligned}$$

to the first order in ε. We may rewrite (3) as:

$$\bar{A} = A + \varepsilon \int_{t_0}^{t_1} \left(\tau \frac{\partial L}{\partial t} + \eta \frac{\partial L}{\partial q} + \zeta \frac{\partial L}{\partial \dot{q}} + \dot{\tau} L \right) dt + \varepsilon F,$$

where the number, F, is the value of the second term in brackets in (3). As F depends only on the endpoints, we may write it as:

$$F = -\int_{t_0}^{t_1} \dot{f}\, dt,$$

where the sign is chosen as a matter of later convenience.

The generator, Γ, of the infinitesimal transformation, (2), is a Noether symmetry of (1) if:

$$\bar{A} = A,$$

i.e.,

$$\int_{t_0}^{t_1} \left(\tau \frac{\partial L}{\partial t} + \eta \frac{\partial L}{\partial q} + \zeta \frac{\partial L}{\partial \dot{q}} + \dot{\tau} L - \dot{f} \right) dt = 0$$

from which it follows that:

$$\dot{f} = \tau \frac{\partial L}{\partial t} + \eta \frac{\partial L}{\partial q} + \zeta \frac{\partial L}{\partial \dot{q}} + \dot{\tau} L. \tag{4}$$

Remark 1. *The symmetry is the generator of an infinitesimal transformation, which leaves the action integral invariant, and the existence of the symmetry has nothing to do with the Euler–Lagrange equation of the calculus of variations. The Euler–Lagrange equation follows from the application of Hamilton's principle in which q is given a zero endpoint variation. There is no such restriction on the infinitesimal transformations introduced by Noether.*

3. Noether's Theorem

We now invoke Hamilton's principle for the action integral (1). We observe that the zero-endpoint variation of (1) imposed by Hamilton's principle requires that (1) take a stationary value; not necessarily a minimum! The principle of least action enunciated by Fermat in 1662 as "Nature always acts in the shortest ways" was raised to an even more metaphysical status by Maupertuis [18] (p. 254, p. 267). That the principle applies in classical (Newtonian) mechanics is an accident of the metric! We can only wonder that the quasi-mystical principle has persisted for over two centuries in what are supposed to be rational circles. In the case of a first-order Lagrangian with a positive definite Hessian with respect to $\dot{q}$, Hamilton's principle gives a minimum. This is not necessarily the case otherwise.

The Euler–Lagrange equation:

$$\frac{\partial L}{\partial q} - \frac{d}{dt}\left(\frac{\partial L}{\partial \dot{q}}\right) = 0 \tag{5}$$

follows from the application of Hamilton's principle. We manipulate (4) as follows:

$$\begin{aligned}
0 &= \dot{f} - \tau \frac{\partial L}{\partial t} - \dot{\tau} L - \eta \frac{\partial L}{\partial q} - (\dot{\eta} - \dot{q}\dot{\tau}) \frac{\partial L}{\partial \dot{q}} \\
&= \frac{d}{dt}(f - \tau L) + \tau \left(\dot{q} \frac{\partial L}{\partial q} + \ddot{q} \frac{\partial L}{\partial \dot{q}} \right) + \dot{\tau} \left(\dot{q} \frac{\partial L}{\partial \dot{q}} \right) \\
&\quad - \eta \frac{d}{dt}\left(\frac{\partial L}{\partial \dot{q}}\right) - \dot{\eta} \frac{\partial L}{\partial \dot{q}} \\
&= \frac{d}{dt}\left[f - \tau L - (\eta - \tau \dot{q}) \frac{\partial L}{\partial \dot{q}} \right]
\end{aligned}$$

in the second line of which we have used the Euler–Lagrange Equation (5), to change the coefficient of η. Hence, we have a first integral:

$$I = f - \left[\tau L + (\eta - \dot{q}\tau) \frac{\partial L}{\partial \dot{q}} \right] \tag{6}$$

and an initial statement of Noether's Theorem.

Noether's theorem: If the action integral of a first-order Lagrangian, namely:

$$A = \int_{t_0}^{t_1} L(t, q, \dot{q})\, dt$$

is invariant under the infinitesimal transformation generated by the differential operator:

$$\Gamma = \tau \partial_t + \eta_i \partial_{q_i},$$

there exists a function f such that:

$$\dot{f} = \tau \frac{\partial L}{\partial t} + \eta_i \frac{\partial L}{\partial q_i} + \zeta_i \frac{\partial L}{\partial \dot{q}_i} + \dot{\tau} L, \qquad (7)$$

where $\zeta_i = \dot{\eta}_i - \dot{q}_i \dot{\tau}$, and a first integral given by:

$$I = f - \left[\tau L + (\eta_i - \dot{q}_i \tau) \frac{\partial L}{\partial \dot{q}_i} \right].$$

Γ is called a Noether symmetry of L and I a Noetherian first integral. The symmetry Γ exists independently of the requirement that the variation of the functional be zero. When the extra condition is added, the first integral exists.

We note that there is not a one-to-one correspondence between a Noether symmetry and a Noetherian integral. Once the symmetry is determined, the integral follows with minimal effort. The converse is not so simple because, given the Lagrangian and the integral, the symmetry is the solution of a differential equation with an additional dependent variable, the function f arising from the boundary terms. There can be an infinite number of coefficient functions for a given first integral. The restriction of the symmetry to a point symmetry may reduce the number of symmetries, too effectively, to zero. The ease of determination of a Noetherian integral once the Noether symmetry is known is in contrast to the situation for the determination of first integrals in the case of Lie symmetries of differential equations. The computation of the first integrals associated with a Lie symmetry can be a highly nontrivial matter.

4. Nonlocal Integrals

We recall that the variable dependences of the coefficient functions τ and η were not specified and do not enter into the derivation of the formulae for the coefficient functions or the first integral. Consequently, not only can we have the generalized symmetries of Noether's paper, but we can also have more general forms of symmetry such as nonlocal symmetries [19,20] without a single change in the formalism. Of course, as has been noted for the calculation of first integrals [21] and symmetries in general [22], the realities of computational complexity may force one to impose some constraints on this generality. Once the Euler–Lagrange equation is invoked, there is an automatic constraint on the degree of derivatives in any generalized symmetry.

If one has a standard Lagrangian such as (1), a nonlocal Noether's symmetry will usually produce a nonlocal integral through (6). In that the total time derivative of this function is zero when the Euler–Lagrange equation, (5), is taken into account, it is formally a first integral. However, the utility of such a first integral is at best questionable. Here, Lie and Noether have generically differing outcomes. An exponential nonlocal Lie symmetry can be expected to lead to a local first integral, whereas one could scarcely envisage the same for an exponential nonlocal Noether symmetry.

On the other hand, if the Lagrangian was nonlocal, the combination of nonlocal symmetry and nonlocal Lagrangian could lead to a local first integral. However, we have not constructed a formalism to deal with nonlocal Lagrangians—as opposed to nonlocal symmetries—and so, we cannot simply apply what we have developed above.

The introduction of a nonlocal term into the Lagrangian effectively increases the order of the Lagrangian by one (in the case of a simple integral) and the order of the associated Euler–Lagrange equation by two so that for a Lagrangian regular in $\dot{q}$ instead of a second-order differential equation, we would have a fourth order differential equation in q. To avoid that the Lagrangian would have to be degenerate, i.e., linear, in $\dot{q}$, this cannot, as is well-known, lead to a second-order differential equation. It would appear that nonlocal symmetries in the context of Noether's theorem do not have the same potential as nonlocal Lie symmetries of differential equations.

There is often some confusion of identity between Lie symmetries and Noether symmetries. Although every Noether symmetry is a Lie symmetry of the corresponding Euler–Lagrange equation, we stress that they have different provenances. There is a difference that is more obvious in systems of higher dimension. A Noether symmetry can only give rise to a single first integral because of (3). In an n-dimensional system of second-order ordinary differential equations, a single Lie symmetry gives rise to $(2n-1)$ first integrals [23–26].

5. Extensions: One Independent Variable

The derivation given above applies to a one-dimensional discrete system. The theorem can be extended to continuous systems and systems of higher order. The principle is the same. The mathematics becomes more complicated. We simply quote the relevant results.

For a first-order Lagrangian with n dependent variables:

$$G = \tau \partial_t + \eta_i \partial_{q_i} \tag{8}$$

is a Noether symmetry of the Lagrangian, $L(t, q_i, \dot{q}_i)$, if there exists a function f such that:

$$\dot{f} = \dot{\tau} L + \tau \frac{\partial L}{\partial t} + \eta_i \frac{\partial L}{\partial q_i} + (\dot{\eta}_i - \dot{q}_i \dot{\tau}) \frac{\partial L}{\partial \dot{q}_i} \tag{9}$$

and the corresponding Noetherian first integral is:

$$I = f - \left[\tau L + (\eta_i - \dot{q}_i \tau) \frac{\partial L}{\partial \dot{q}_i} \right] \tag{10}$$

which are the obvious generalizations of (4) and (6), respectively.

In the case of an nth-order Lagrangian in one dependent variable and one independent variable, $L(t, q, \dot{q}, \ldots, q^{(n)})$ with $q^{(n)} = d^n q/dt^n$, the Euler–Lagrange equation is:

$$\sum_{j=0}^{n} (-1)^j \frac{d^j}{dt^j} \left(\frac{\partial L}{\partial q^{(j)}} \right). \tag{11}$$

$\Gamma = \tau \partial_t + \eta \partial_q$ is a Noether symmetry if there exists a function f such that:

$$\dot{f} = \dot{\tau} L + \tau \frac{\partial L}{\partial t} + \sum_{j=0}^{n} (-1)^j \zeta^j \left(\frac{\partial L}{\partial q^{(j)}} \right), \tag{12}$$

where:

$$\zeta^j = \eta^{(j)} - \sum_{k=1}^{j} \binom{j}{k} q^{(j+1-k)} \tau^{(k)}. \tag{13}$$

The expression for the first integral is:

$$I = f - \left[\tau L + \sum_{i=0}^{n-1} \sum_{j=0}^{n-1-i} (-1)^j (\eta - \dot{q}\tau)^{(i)} \frac{d^j}{dt^j} \left(\frac{\partial L}{\partial q^{(i+j+1)}} \right) \right]. \tag{14}$$

In the case of an nth-order Lagrangian in m dependent variables and one independent variable, $L(t, q_k, \dot{q}_k, \ldots, q_k^{(n)})$ with $q_k^{(n)} = d^n q_k / dt^n$, $k = 1, m$, the Euler–Lagrange equation is:

$$\sum_{j=0}^{n} (-1)^j \frac{d^j}{dt^j}\left(\frac{\partial L}{\partial q_k^{(j)}}\right), \quad k = 1, m. \tag{15}$$

$\Gamma = \tau \partial_t + \sum_{k=1}^{m} \eta_k \partial_{q_k}$ is a Noether symmetry if there exists a function f such that:

$$\dot{f} = \dot{\tau} L + \tau \frac{\partial L}{\partial t} + \sum_{k=1}^{m} \sum_{j=0}^{n} (-1)^j \zeta_k^j \left(\frac{\partial L}{\partial q_k^{(j)}}\right), \tag{16}$$

where:

$$\zeta_k^j = \eta_k^{(j)} - \sum_{k=1}^{m} \sum_{i=0}^{j} \binom{j}{i} q_k^{(j+1-i)} \tau^{(i)}. \tag{17}$$

The expression for the first integral is:

$$I = f - \left[\tau L + \sum_{k=1}^{m} \sum_{i=0}^{n-1} \sum_{j=0}^{n-1-i} (-1)^j (\eta_k - \dot{q}_k \tau)^{(i)} \frac{d^j}{dt^j}\left(\frac{\partial L}{\partial q_k^{(i+j+1)}}\right)\right]. \tag{18}$$

The expressions in (14) and (18), although complex enough, conceals an even much greater complexity because each derivative with respect to time is a total derivative and so affects all terms in the Lagrangian and its partial derivatives.

6. Observations

In the case of a first-order Lagrangian with one independent variable, it is well-known [17] that one can achieve a simplification in the calculations of the Noether symmetry in the case that the Lagrangian has a regular Hessian with respect to the $\dot{q}_i$. We suppose that we admit generalized symmetries in which the maximum order of the derivatives present in τ and the η_i is one, i.e., equal to the order of the Lagrangian. Then, the coefficient of each $\ddot{q}_j$ in (9) is separately zero since the Euler–Lagrange equation has not yet been invoked. Thus, we have:

$$\frac{\partial f}{\partial \dot{q}_j} = \frac{\partial \tau}{\partial \dot{q}_j} L + \left(\frac{\partial \eta_i}{\partial \dot{q}_j} - \dot{q}_i \frac{\partial \tau}{\partial \dot{q}_j}\right) \frac{\partial L}{\partial \dot{q}_i}. \tag{19}$$

We differentiate (10) with respect to $\dot{q}_j$ to obtain:

$$\frac{\partial I}{\partial \dot{q}_j} = \frac{\partial f}{\partial \dot{q}_j} - \left[\frac{\partial \tau}{\partial \dot{q}_j} L + \tau \frac{\partial L}{\partial \dot{q}_j} + \left(\frac{\partial \eta_i}{\partial \dot{q}_j} - \delta_{ij}\tau - \dot{q}_i \frac{\partial \tau}{\partial \dot{q}_j}\right) \frac{\partial L}{\partial \dot{q}_i} + (\eta_i - \dot{q}_i \tau) \frac{\partial^2 L}{\partial \dot{q}_i \partial \dot{q}_j}\right], \tag{20}$$

where δ_{ij} is the usual Kronecker delta, which, when we take (19) into account, gives:

$$\frac{\partial I}{\partial \dot{q}_j} = -(\eta_i - \dot{q}_i \tau) \frac{\partial^2 L}{\partial \dot{q}_i \partial \dot{q}_j}. \tag{21}$$

Consequently, if the Lagrangian is regular with respect to the $\dot{q}_i$, we have:

$$(\eta_i - \dot{q}_i \tau) = -g_{ij} \frac{\partial I}{\partial \dot{q}_j}, \tag{22}$$

where:

$$g_{ik}\frac{\partial^2 L}{\partial \dot{q}_k}\partial \dot{q}_j = \delta_{ij}.$$

The relations (21) and (22) reveal two useful pieces of information. The first is that the derivative dependence of the first integral is determined by the nature of the generalized symmetry (modulo the derivative dependence in the Lagrangian). The second is that there is a certain freedom of choice in the structure of the functions τ and η_i in the symmetry. Provided generalized symmetries are admitted, there is no loss of generality in putting one of the coefficient functions equal to zero. An attractive candidate is τ as it appears the most frequently. The choice should be made before the derivative dependence of the coefficient functions is assumed. We observe that in the case of a 'natural' Lagrangian, i.e., one quadratic in the derivatives, the first integrals can only be linear or quadratic in the derivatives if the symmetry is assumed to be point.

7. Examples

The free particle:

We consider the simple example of the free particle for which:

$$L = \tfrac{1}{2}y'^2.$$

Equation (7) is:

$$(\eta' - y'\xi')y' + \tfrac{1}{2}y'^2 = f'. \tag{23}$$

If we assume that Γ is a Noether point symmetry, (23) gives the following determining equations:

$$y'^3 : \quad -\tfrac{1}{2}\frac{\partial \xi}{\partial y} = 0$$

$$y'^2 \quad : \quad \frac{\partial \eta}{\partial y} - \tfrac{1}{2}\frac{\partial \xi}{\partial x} = 0$$

$$y'^1 \quad : \quad \frac{\partial \eta}{\partial x} - \frac{\partial f}{\partial y} = 0$$

$$y'^0 \quad : \quad \frac{\partial f}{\partial x} = 0$$

from which it is evident that:

$$\xi = a(x)$$
$$\eta = \tfrac{1}{2}a'y + b(x)$$
$$f = \tfrac{1}{4}a''y^2 + b'y + c(x)$$
$$0 = \tfrac{1}{4}a'''y^2 + b''y + c'.$$

Hence:

$$a = A_0 + A_1 x + A_2 x^2$$
$$b = B_0 + B_1 x$$
$$c = C_0.$$

Because c is simply an additive constant, it is ignored. There are five Noether point symmetries, which is the maximum for a one-dimensional system [27]. They and their associated first integrals are:

$$\Gamma_1 = \partial_y \qquad I_1 = -y'$$
$$\Gamma_2 = x\partial_y \qquad I_2 = y - xy'$$
$$\Gamma_3 = \partial_x \qquad I_3 = \tfrac{1}{2}y'^2$$
$$\Gamma_4 = x\partial_x + \tfrac{1}{2}y\partial_y \qquad I_4 = -\tfrac{1}{2}y'(y - xy')$$
$$\Gamma_5 = x^2\partial_x + xy\partial_y \qquad I_5 = \tfrac{1}{2}(y - xy')^2.$$

The corresponding Lie algebra is isomorphic to $A_{5,40}$ [28]. The algebra is structured as $2A_1 \oplus_s sl(2, R)$, which is a proper subalgebra of the Lie algebra for the differential equation for the free particle, namely $sl(3, R)$, which is structured as $2A_1 \oplus_s \{sl(2, R) \oplus_s A_1\} \oplus_s 2A_1$. The missing symmetries are the homogeneity symmetry and the two non-Cartan symmetries. The absence of the homogeneity symmetry emphasizes the distinction between the Lie and Noether symmetries.

Noether symmetries of a higher-order Lagrangian:

Suppose that $L = \tfrac{1}{2}y''^2$. The condition for a Noether point symmetry is that:

$$\zeta_2 \frac{\partial L}{\partial y''} + \xi' L = f', \tag{24}$$

where $\zeta_2 = \eta'' - 2y''\xi' - y'\xi''$, so that (24) becomes:

$$(\eta'' - 2y''\xi' - y'\xi'')y'' + \tfrac{1}{2}\xi' y''^2 = f'. \tag{25}$$

Assume a point transformation, i.e., $\xi = \xi(x, y)$ and $\eta = \eta(x, y)$. Then:

$$\left[\frac{\partial^2 \eta}{\partial x^2} + 2y' \frac{\partial^2 \eta}{\partial x \partial y} + y'^2 \frac{\partial^2 \eta}{\partial y^2} + y'' \frac{\partial \eta}{\partial y} - 2y'' \left(\frac{\partial \xi}{\partial x} + y' \frac{\partial \xi}{\partial y} \right) \right.$$
$$\left. -y' \left(\frac{\partial^2 \xi}{\partial x^2} + 2y' \frac{\partial^2 \xi}{\partial x \partial y} + y'^2 \frac{\partial^2 \xi}{\partial y^2} + y'' \frac{\partial \xi}{\partial y} \right) \right] y'' + \tfrac{1}{2} \left(\frac{\partial \xi}{\partial x} + y' \frac{\partial \xi}{\partial y} \right) y''^2$$
$$= \frac{\partial f}{\partial x} + y' \frac{\partial f}{\partial y} + y'' \frac{\partial f}{\partial y'}.$$

from the coefficient of $y'y''^2$, videlicet:

$$-\tfrac{5}{2} \frac{\partial \xi}{\partial y} = 0,$$

we obtain:

$$\xi = a(x).$$

The coefficient of y''^2,

$$\frac{\partial \eta}{\partial y} - \tfrac{3}{2} \frac{\partial \xi}{\partial x} = 0,$$

results in:

$$\eta = \tfrac{3}{2} a' y + b(x)$$

and the coefficient of y'',

$$\frac{\partial^2 \eta}{\partial x^2} + 2y' \frac{\partial^2 \eta}{\partial x \partial y} + y'^2 \frac{\partial^2 \eta}{\partial y^2} - y' \frac{\partial^2 \xi}{\partial x^2} = \frac{\partial f}{\partial y'},$$

gives f as:

$$f = a'' y'^2 + (\tfrac{3}{2} a''' y + b'') y' + c(x, y).$$

The remaining terms give:

$$y' \frac{\partial f}{\partial y} + \frac{\partial f}{\partial x} = 0,$$

i.e.,
$$y'\left[\frac{3}{2}a'''y' + \frac{\partial c}{\partial y}\right] + a'''y'^2 + \left(\frac{3}{2}a^{iv}y + b'''\right)y' + \frac{\partial c}{\partial x} = 0.$$

The coefficient of y'^2 is $\frac{5}{2}a''' = 0$, from which it follows that:
$$a = A_0 + A_1 x + A_2 x^2.$$

The coefficient of y' is:
$$\frac{\partial c}{\partial y} + \frac{3}{2}a^{iv}y + b''' = 0$$

and so:
$$c = -b'''y + d(x).$$

The remaining terms give:
$$\frac{\partial c}{\partial x} = 0,$$

i.e.,
$$-b^{iv}y + d' = 0$$

from which d is a constant (and can therefore be ignored) and:
$$b = B_0 + B_1 x + B_2 x^2 + B_3 x^3.$$

There are seven Noetherian point symmetries for $L = \frac{1}{2}y''^2$. They and the associated "gauge functions" are:

$B_0:$	$\Gamma_1 = \partial_y$	$f_1 = 0$
$B_1:$	$\Gamma_2 = x\partial_y$	$f_2 = 0$
$B_2:$	$\Gamma_3 = x^2\partial_y$	$f_3 = 2xy'$
$B_3:$	$\Gamma_4 = x^3\partial_y$	$f_4 = 6xy' - 6y$
$A_0:$	$\Gamma_5 = \partial_x$	$f_5 = 0$
$A_1:$	$\Gamma_6 = x\partial_x + \frac{3}{2}y\partial_y$	$f_6 = 0$
$A_2:$	$\Gamma_7 = x^2\partial_x + 3xy\partial_y$	$f_7 = 2y'^2.$

The Euler–Lagrange equation for $L = \frac{1}{2}y''^2$ is $y^{(iv)} = 0$, which has Lie point symmetries the same as the Noether point symmetries plus $\Gamma_8 = y\partial_y$. Note that there is a contrast here in comparison with the five Noether point symmetries of $L = \frac{1}{2}y'^2$ and the eight Lie point symmetries of $y'' = 0$. The additional Lie symmetries are $y\partial_y$ as above for $y^{iv} = 0$ and the two non-Cartan symmetries, $X_1 = y\partial_x$ and $X_2 = xy\partial_x + y^2\partial_y$.

For $L = \frac{1}{2}y''^2$, the associated first integrals have the structure:
$$I = f - \frac{1}{2}\xi y''^2 + (\eta - y'\xi)y''' - (\eta' - y''\xi - y'\xi')y''$$

and are:
$$\begin{aligned}
I_1 &= y''' \\
I_2 &= xy''' - y'' \\
I_3 &= x^2 y''' - 2xy'' + 2xy' \\
I_4 &= x^3 y'''^2 y'' + 6xy' - 6y \\
I_5 &= -y'y''' + \frac{1}{2}y''^2 \\
I_6 &= -xy'y + \frac{1}{2}xy''^2 - \frac{1}{2}y'y'' + \frac{3}{2}yy''' \\
I_7 &= x(3y - xy')y''' - (3y - xy' - \frac{1}{2}x^2 y'')y'' + 2y'^2.
\end{aligned}$$

Note that I_1–I_4 associated with Γ_1–Γ_4, respectively, are also integrals obtained by the Lie method. However, each Noether symmetry produces just one first integral, whereas each Lie symmetry has three first integrals associated with it.

In this example, only point Noether symmetries have been considered. One may also determine symmetries that depend on derivatives, effectively up to the third order when one is calculating first integrals of the Euler–Lagrange equation.

Omission of the gauge function:

In some statements of Noether's theorem, the so-called gauge function, f, is taken to be zero. In the derivation given here, f comes from the contribution of the boundary terms produced by the infinitesimal transformation in t and so is not a gauge function in the usual meaning of the term. However, it does function as one since it is independent of the trajectory in the extended configuration space and depends only on the evaluation of functions at the boundary (end points in a one-degree-of-freedom case) and can conveniently be termed one especially in light of Boyer's theorem [29].

Consider the example $L = \frac{1}{2} y'^2$ without f. The equation for the symmetries,

$$f' = \zeta \frac{\partial L}{\partial x} + \eta \frac{\partial L}{\partial y} + (\eta' - y'\zeta') \frac{\partial L}{\partial y'} + \zeta' L,$$

becomes:

$$0 = \left(\frac{\partial \eta}{\partial x} + y' \frac{\partial \eta}{\partial y} - y' \frac{\partial \zeta}{\partial x} - y'^2 \frac{\partial \zeta}{\partial y} \right) y' + \frac{1}{2} y'^2 \left(\frac{\partial \zeta}{\partial x} + y' \frac{\partial \zeta}{\partial y} \right).$$

We solve this in the normal way: the coefficients of y'^3, y'^2 and of y' give in turn:

$$\zeta = a(x)$$
$$\eta = \tfrac{1}{2} a' y + b(x)$$
$$\tfrac{1}{2} a'' y + b' = 0$$

which hold provided that:

$$a = A_0 + A_1 x \qquad b = B_0,$$

i.e., only three symmetries are obtained instead of the five when the gauge function is present.

It makes no sense to omit the gauge function when the infinitesimal transformation is restricted to be point and only in the dependent variables.

A higher-dimensional system:

We determine the Noether point symmetries and their associated first integrals for:

$$L = \tfrac{1}{2}(\dot{x}^2 + \dot{y}^2)$$

(which is the standard Lagrangian for the free particle in two dimensions). The determining equation is:

$$\frac{\partial f}{\partial t} + \dot{x} \frac{\partial f}{\partial x} + \dot{y} \frac{\partial f}{\partial y} = \left(\frac{\partial \eta}{\partial t} + \dot{x} \frac{\partial \eta}{\partial x} + \dot{y} \frac{\partial \eta}{\partial y} - \dot{x} \left(\frac{\partial \zeta}{\partial t} + \dot{x} \frac{\partial \zeta}{\partial x} + \dot{y} \frac{\partial \zeta}{\partial y} \right) \right) \dot{x}$$
$$+ \left(\frac{\partial \zeta}{\partial t} + \dot{x} \frac{\partial \zeta}{\partial x} + \dot{y} \frac{\partial \zeta}{\partial y} - \dot{y} \left(\frac{\partial \zeta}{\partial t} + \dot{x} \frac{\partial \zeta}{\partial x} + \dot{y} \frac{\partial \zeta}{\partial y} \right) \right) \dot{y},$$

where $\eta_1 = \eta$ and $\eta_2 = \zeta$.

We separate by powers of $\dot{x}$ and $\dot{y}$. Firstly taking the third-order terms, we have:

$$\dot{x}^3: \quad -\frac{\partial \xi}{\partial x} = 0$$
$$\dot{x}^2\dot{y}: \quad -\frac{\partial \xi}{\partial y} = 0$$
$$\dot{x}\dot{y}^2: \quad -y\frac{\partial \xi}{\partial x} = 0$$
$$\dot{y}^3: \quad -\frac{\partial \xi}{\partial y} = 0$$

which implies $\xi = a(t)$. We now consider the second-order terms: the coefficient of $\dot{x}^2$ gives η as $\eta = \dot{a}x + b(y,t)$; that of $\dot{x}\dot{y}$ gives ζ as:

$$\zeta = -\frac{\partial b}{\partial y}x + c(y,t);$$

and that of $\dot{y}^2$ gives $c = \dot{a}y + d(t)$ and $b = e(t)y + g(t)$. Thus far, we have:

$$\xi = a(t) \qquad \eta = \dot{a}x + ey + g \qquad \zeta = -ex + \dot{a}y + d.$$

The coefficient of $\dot{x}$ gives f as:

$$f = \tfrac{1}{2}\ddot{a}x^2 + \dot{e}xy + \dot{g}x + K(y,t).$$

The coefficient of $\dot{y}$ requires that:

$$\dot{e}x + \frac{\partial K}{\partial y} = -\dot{e}x + \ddot{a}y + \dot{d}$$

which implies:

$$\dot{e} = 0 \qquad K = \tfrac{1}{2}\ddot{a}y^2 + \dot{d}y + h(t).$$

The remaining term requires that:

$$\tfrac{1}{2}\dddot{a}x^2 + \ddot{g}x + \tfrac{1}{2}\dddot{a}y^2 + \ddot{d}y + \dot{h} = 0$$

whence:

$$a = A_0 + A_1 t + A_2 t^2$$
$$g = G_0 + G_1 t$$
$$d = D_0 + D_1 t$$
$$h = H_0$$

(we ignore H_0, as it is an additive constant to f).

The coefficient functions are:

$$\xi = A_0 + A_1 t + A_2 t^2$$
$$\eta = (A_1 + 2A_2 t)x + E_0 y + G_0 + G_1 t$$
$$\zeta = -E_0 x + (A_1 + 2A_2 t)y + D_0 + D_1 t$$

and the gauge function is:

$$f = A_2 x^2 + G_1 x + A_2 y^2 + D_1 y.$$

We obtain three symmetries from a, namely:

$$\begin{aligned}\Gamma_1 &= \partial_t \\ \Gamma_2 &= t\partial_t + x\partial_x + y\partial_y \\ \Gamma_3 &= t^2\partial_t + 2t\left(x\partial_x + y\partial_y\right)\end{aligned}$$

which form $sl(2,R)$, one from e,

$$\Gamma_4 = y\partial_x - x\partial_y$$

which is $so(2)$, and four from g and d, namely:

$$\begin{aligned}\Gamma_5 &= \partial_x \\ \Gamma_6 &= t\partial_x \\ \Gamma_7 &= \partial_y \\ \Gamma_8 &= t\partial_y.\end{aligned}$$

The last four are the "solution" symmetries and form the Lie algebra $4A_1$.

8. More than One Independent Variable: Preliminaries

8.1. Euler–Lagrange Equation

Noether's original formulation of her theorem was in the context of Lagrangians for functions of several independent variables. We have deliberately separated the case of one independent variable from the general discussion to be able to present the essential ideas in as simple a form as possible. The discussion of the case of several independent variables is inherently more complex simply from a notational point of view, although there is no real increase in conceptual difficulty.

We commence with the simplest instance of a Lagrangian of this class, which is $L(t, x, u, u_t, u_x)$, i.e., one dependent variable, u, and two dependent variables, t and x. We recall the derivation of the Euler–Lagrange equation for $u(t, x)$ consequent upon the application of Hamilton's principle. In the action integral:

$$A = \int_\Omega L\left(t, x, u, u_t, u_x\right) dxdt \tag{26}$$

we introduce an infinitesimal variation of the dependent variable,

$$\bar{u} = u + \varepsilon v(t, x), \tag{27}$$

where ε is the infinitesimal parameter, $v(t, x)$ is continuously differentiable in both independent variables and is required to be zero on the boundary, $\partial\Omega$, of the domain of integration, Ω, which in this introductory case is some region in the (t, x) plane. Otherwise, v is an arbitrary function. We have:

$$\bar{A} = \int_\Omega L\left(t, x, \bar{u}, \bar{u}_t, \bar{u}_x\right) dxdt \tag{28}$$

and we require the action integral to take a stationary value, i.e., $\delta A = \bar{A} - A$ be zero. Now:

$$\begin{aligned}
\delta A &= \int_\Omega \left[L\left(t,x,\bar{u},\bar{u}_t,\bar{u}_x\right) - L\left(t,x,u,u_t,u_x\right)\right] dxdt \\
&= \varepsilon \int_\Omega \left[\frac{\partial L}{\partial u}v + \frac{\partial L}{\partial u_t}\frac{v}{t} + \frac{L}{u_x}\frac{v}{x}\right] dxdt + O(\varepsilon^2) \\
&= \varepsilon \int_\Omega \left[\frac{L}{u} - \frac{\partial}{\partial t}\left(\frac{L}{u_t}\right) - \frac{\partial}{\partial x}\left(\frac{\partial L}{\partial \bar{u}_x}\right)\right] vdtdx \\
&\quad + \int_{\partial\Omega} \left[\frac{\partial L}{\partial u_t}dt + \frac{\partial L}{\partial u_x}dx\right]v + O(\varepsilon^2).
\end{aligned} \qquad (29)$$

The second integral in (29) is the sum of four integrals, two along each of the intervals (t_1, t_2) and (x_1, x_2) with $x = x_1$ and $x = x_2$ for the two integrals with respect to t and $t = t_1$ and $t = t_2$ for the two integrals with respect to x. Because v is zero on the boundary, this term must be zero. As v is otherwise arbitrary, the expression within the brackets in the first integral must be zero, and so, we have the Euler–Lagrange equation:

$$\frac{\partial L}{\partial u} - \frac{\partial}{\partial t}\left(\frac{\partial L}{\partial u_t}\right) - \frac{\partial}{\partial x}\left(\frac{\partial L}{\partial \bar{u}_x}\right) = 0. \qquad (30)$$

A conservation law for (30) is a vector-valued function, **f**, of t, x, u and the partial derivatives of u, which is divergence free, i.e.,

$$\begin{aligned}
\text{div.}\mathbf{f} &= \frac{\partial f^1}{\partial t} + \frac{\partial f^2}{\partial x} \\
&= 0.
\end{aligned} \qquad (31)$$

In (31), the operators ∂_t and ∂_x are operators of total differentiation with respect to t and x, respectively, and henceforth, we denote these operators by D_t and D_x. The standard symbol for partial differentiation, ∂_A, indicates differentiation solely with respect to A. In this notation, (30) and (31) become respectively:

$$\frac{\partial L}{\partial u} - D_t \frac{\partial L}{\partial u_t} - D_x \frac{\partial L}{\partial u_x} = 0 \text{ and} \qquad (32)$$

$$\text{div.}\mathbf{f} = D_t f^1 + D_x f^2 = 0. \qquad (33)$$

Naturally, there is no distinction between D_t, $\partial u/\partial t$ and u_t, likewise for the derivatives with respect to x.

8.2. Noether's Theorem for $L(t, x, u, u_t, u_x)$

We introduce into the action integral an infinitesimal transformation,

$$\bar{t} = t + \varepsilon\tau \quad \bar{x} = x + \varepsilon\xi \quad \bar{u} = u + \varepsilon\eta \qquad (34)$$

generated by the differential operator:

$$\Gamma = \tau\partial_t + \xi\partial_x + \eta\partial_u, \qquad (35)$$

which, because the Lagrangian depends on u_t and u_x, we extend once to give:

$$\Gamma^{[1]} = \Gamma + (D_t\eta - u_t D_t\tau - u_t D_t\xi)\partial_{u_t} + (D_x\eta - u_x D_t\tau - u_x D_x\xi)\partial_{u_x}. \qquad (36)$$

The coefficient functions τ, ξ and η may depend on the derivatives of u, as well as t, x, and u.

The change in the action due to the infinitesimal transformation is given by:

$$\begin{aligned}\delta A &= \bar{A} - A \\ &= \int_{\bar{\Omega}} L\left(\bar{t}, \bar{x}, \bar{u}, \bar{u}_{\bar{t}}, \bar{u}_{\bar{x}} d\bar{t}\right) d\bar{x} - \int_{\Omega} L\left(t, x, u, u_t, u_x\right) dt dx, \end{aligned} \qquad (37)$$

where $\bar{\Omega}$ is the transformed domain. We recall that Noether's theorem comes in two parts. In the first part, with which we presently deal, the discussion is about the action integral and not the variational principle. Consequently, there is no reason why the domain over which integration takes place should be the same before and after the transformation. Equally, there is no reason to require that the coefficient functions vanish on the boundary of Ω. To make progress in the analysis of (37), we must reconcile the variables and domains of integration. For the variables of integration, we have:

$$\begin{aligned} d\bar{t} d\bar{x} &= \frac{\partial \left(\bar{t}, \bar{x}\right)}{\partial \left(t, x\right)} dt dx \\ &= \begin{vmatrix} D_t \bar{t} & D_t \bar{x} \\ D_x \bar{t} & D_x \bar{x} \end{vmatrix} dt dx \\ &= \begin{vmatrix} 1 + \varepsilon D_t \tau & \varepsilon D_t \xi \\ \varepsilon D_x \tau & 1 + \varepsilon D_x \xi \end{vmatrix} dt dx \\ &= \left[1 + \varepsilon \left(D_t \tau + D_x \xi\right) + O\left(\varepsilon^2\right)\right] dt dx. \end{aligned} \qquad (38)$$

For the domain, we have simply that:

$$\bar{\Omega} = \Omega + \delta \Omega \qquad (39)$$

which, as the transformation is infinitesimal, in general means the evaluation of the surface integral and in this two-dimensional case the evaluation of the line integral along the boundary of the original domain. Although this domain is arbitrary, it is fixed for the variational principle we are using. We can use the divergence theorem to express this in terms of the volume integral over the original domain of the divergence of some vector-valued function. Combining these considerations with (37) and (38) and expanding the integrand of the first integral in (37) as a Taylor series, we can write the condition that the action integral be invariant under the infinitesimal transformation as:

$$\begin{aligned} 0 &= \int_{\Omega} \left\{ L + \varepsilon \left[\tau \frac{\partial L}{\partial t} + \xi \frac{\partial L}{\partial x} + \eta \frac{\partial L}{\partial u} + \left(D_t \eta - u_t D_t \tau - u_t D_t \xi\right) \frac{\partial L}{\partial u_t} \right. \right. \\ &\quad \left. \left. + \left(D_x \eta - u_x D_t \tau - u_x D_x \xi\right) \frac{\partial L}{\partial u_x} \right] - \varepsilon \operatorname{div}.\mathbf{F} \right\} \left[1 + \varepsilon \left(D_t \tau + D_x \xi\right)\right] dt dx \\ &\quad - \int_{\Omega} L dt dx + O\left(\varepsilon^2\right), \end{aligned} \qquad (40)$$

where $\mathbf{F}$ represents the contribution from the boundary term. If we require that this be true for any domain in which the Lagrangian is validly defined, the first-order term in (40) gives the condition for the Lagrangian to possess a Noether symmetry, videlicet:

$$\begin{aligned} \operatorname{div}.\mathbf{F} &= \left(D_t \tau + D_x \xi\right) L + \tau \frac{\partial L}{\partial t} + \xi \frac{\partial L}{\partial x} + \eta \frac{\partial L}{\partial u} \\ &\quad + \left(D_t \eta - u_t D_t \tau - u_x D_t \xi\right) \frac{\partial L}{\partial u_t} + \left(D_x \eta - u_t D_x \tau - u_x D_x \xi\right) \frac{\partial L}{\partial u_x}. \end{aligned} \qquad (41)$$

The rest is just a matter of computation! There does not appear to be code that enables one to solve (41) for a given Lagrangian even for point symmetries. One is advised [30] (p. 273) to calculate the (generalized) symmetries of the corresponding Euler–Lagrange equation and then test whether there exists an **F** such that each of these symmetries in turn satisfies (41).

There is a theorem in Olver [31] (p. 326) that the set of generalized symmetries of the Euler–Lagrange equation contains the set of generalized Noether symmetries of the Lagrangian. A purist could well prefer to be able to solve (41) directly. Alan Head, the distinguished Australian scientist, who wrote one of the more successful codes for differential equations in 1978, considered the effort involved to write the requisite code twenty years later excessive when the indirect route was available (A K Head, private communication, December, 1997).

A conservation law corresponding to a Noether symmetry "derived" from (41) is obtained when the Euler–Lagrange equation is taken into account. We rewrite the right side of (41) and have:

$$
\begin{aligned}
\text{div.}\mathbf{F} &= D_t\left[\tau L + \eta \frac{\partial L}{\partial u_t}\right] + D_x\left[\xi L + \eta \frac{\partial L}{\partial u_x}\right] + \tau \frac{\partial L}{\partial t} + \xi \frac{\partial L}{\partial x} + \eta \frac{\partial L}{\partial u} \\
&\quad - (u_t D_t \tau + u_x D_t \xi)\frac{\partial L}{\partial u_t} - (u_t D_x \tau + u_x D_x \xi)\frac{\partial L}{\partial u_x} \\
&\quad - \tau D_t L - \xi D_x L - \eta D_t \frac{\partial L}{\partial u_t} - \eta D_x \frac{\partial L}{\partial u_x} \\
&= D_t\left[\tau L + \eta \frac{\partial L}{\partial u_t}\right] + D_x\left[\xi L + \eta \frac{\partial L}{\partial u_x}\right] + \tau \frac{\partial L}{\partial t} + \xi \frac{\partial L}{\partial x} - (u_t D_t \tau + u_x D_t \xi)\frac{\partial L}{\partial u_t} \\
&\quad - (u_t D_x \tau + u_x D_x \xi)\frac{\partial L}{\partial u_x} - \tau D_t L - \xi D_x L \\
&= D_t\left[\tau L + (\eta - u_t \tau - u_x \xi)\frac{\partial L}{\partial u_t}\right] + D_x\left[\xi L + (\eta - u_t \tau - u_x \xi)\frac{\partial L}{\partial u_x}\right] \\
&\quad + \tau\left[\frac{\partial L}{\partial t} - D_t L + D_t\left(u_t \frac{\partial L}{\partial u_t}\right) + D_x\left(u_t \frac{\partial L}{\partial u_x}\right)\right] \\
&\quad + \xi\left[\frac{\partial L}{\partial x} - D_x L + D_t\left(u_x \frac{\partial L}{\partial u_t}\right) + D_x\left(u_x \frac{\partial L}{\partial u_x}\right)\right] \\
&= D_t\left[\tau L + (\eta - u_t \tau - u_x \xi)\frac{\partial L}{\partial u_t}\right] + D_x\left[\xi L + (\eta - u_t \tau - u_x \xi)\frac{\partial L}{\partial u_x}\right] \quad (42)
\end{aligned}
$$

when the Euler–Lagrange equation is taken into account. Hence, there is the vector of the conservation law:

$$
\mathbf{I} = \mathbf{F} - \left[\tau L + (\eta - u_t \tau - u_x \xi)\frac{\partial L}{\partial u_t}\right]\mathbf{e}_t - \left[\xi L + (\eta - u_t \tau - u_x \xi)\frac{\partial L}{\partial u_x}\right]\mathbf{e}_x, \quad (43)
$$

where $\mathbf{e}_t$ and $\mathbf{e}_x$ are the unit vectors in the (t, x) plane.

We consider the simple example of the Lagrangian:

$$
L = \tfrac{1}{12}(u_x)^4 + \tfrac{1}{2}(u_t)^2.
$$

The condition for the existence of a Noether symmetry, (41), becomes:

$$
\text{div.}\mathbf{F} = (D_t \tau + D_x \xi)\left(\tfrac{1}{12}(u_x)^4 + \tfrac{1}{2}(u_t)^2\right) + (D_t \eta - u_t D_t \tau - u_x D_t \xi) u_t \\
+ (D_x \eta - u_t D_x \tau - u_x D_x \xi)\tfrac{1}{3}u_x^3. \quad (44)
$$

The Lagrangian has the Euler–Lagrange equation:

$$
u_x^2 u_{xx} + u_{tt} = 0. \quad (45)
$$

The Lie point symmetries of (45) are:

$$\begin{aligned}
\Gamma_1 &= \partial_t & \Gamma_4 &= t\partial_u \\
\Gamma_2 &= \partial_x & \Gamma_5 &= t\partial_t - u\partial_u \\
\Gamma_3 &= \partial_u & \Gamma_6 &= x\partial_x + 2u\partial_u.
\end{aligned} \qquad (46)$$

The Lie point symmetries Γ_1–Γ_4 give a zero vector $\mathbf{F}$ except for Γ_4, which gives $(u, 0)$. The symmetries Γ_5 and Γ_6 give nonlocal vectors and so nonlocal conservation laws, which could be interpreted as meaning that they are not Noether symmetries for the given Lagrangian. The four local conservation laws are:

$$\begin{aligned}
I_1 &= \left(\tfrac{1}{2}u_t^2 - u_x^4, \tfrac{1}{12}u_t u_x^3\right) \\
I_2 &= \left(u_t u_x, \tfrac{1}{4}u_x^4 - \tfrac{1}{2}u_t^2\right) \\
I_3 &= \left(u_t, \tfrac{1}{3}u_x^3\right) \\
I_4 &= \left(u - tu_t, -\tfrac{1}{3}tu_x^3\right).
\end{aligned}$$

9. The General Euler–Lagrange Equation

In the case of a pth-order Lagrangian in m dependent variables, u_i, $i = 1, m$, and n independent variables, x_j, $j = 1, n$, the Lagrangian, $L(x, u, u_1, \ldots, u_p)$, under an infinitesimal transformation:

$$\bar{u}^i(x) = u^i(x) + \varepsilon v^i(x),$$

where ε is the parameter of smallness and $v(x)$ is $k - 1$-times differentiable and zero on the boundary $\partial\Omega$ of the domain of integration Ω of the action integral,

$$A = \int_\Omega L\left(x, u, u_1, \ldots, u_p\right) dx, \qquad (47)$$

becomes:

$$\begin{aligned}
\bar{L} &= L\left(\bar{x}, \bar{u}, \bar{u}_1, \ldots, \bar{u}_p\right) \\
&= L\left(x, u, u_1, \ldots, u_p\right) + \varepsilon v^i_{j_1, j_2, \ldots, j_k} \frac{\partial L}{\partial u^i_{j_1, j_2, \ldots, j_k}} + O\left(\varepsilon^2\right)
\end{aligned} \qquad (48)$$

in which summation over repeated indices is implied and $i = 1, m$, $j = 1, n$, and $k = 0, p$. The variation in the action integral is:

$$\begin{aligned}
\delta A &= \int_\Omega [\bar{L} - L]\, dx \\
&= \varepsilon \int_\Omega v^i_{j_1, j_2, \ldots, j_k} \frac{\partial L}{\partial u^i_{j_1, j_2, \ldots, j_k}} dx + O\left(\varepsilon^2\right).
\end{aligned} \qquad (49)$$

We consider one set of terms in (49) with summation only over j_k.

$$\begin{aligned}
&\int_\Omega v^i_{j_1, j_2, \ldots, j_k} \frac{\partial L}{\partial u^i_{j_1, j_2, \ldots, j_k}} dx \\
&= \int_\Omega \left\{ D_{j_k}\left[v^i_{j_1, j_2, \ldots, j_{k-1}} \frac{\partial L}{\partial u^i_{j_1, j_2, \ldots, j_k}}\right] - v^i_{j_1, j_2, \ldots, j_{k-1}} D_{j_k}\left[\frac{\partial L}{\partial u^i_{j_1, j_2, \ldots, j_k}}\right] \right\} dx \\
&= \int_{\partial\Omega} D_{j_k}\left[v^i_{j_1, j_2, \ldots, j_{k-1}} \frac{\partial L}{\partial u^i_{j_1, j_2, \ldots, j_k}}\right] n_{j_k}\, d\sigma - \int_\Omega v^i_{j_1, j_2, \ldots, j_{k-1}} D_{j_k}\left[\frac{\partial L}{\partial u^i_{j_1, j_2, \ldots, j_k}}\right] dx \\
&= \int_\Omega v^i_{j_1, j_2, \ldots, j_{k-1}} D_{j_k}\left[\frac{\partial L}{\partial u^i_{j_1, j_2, \ldots, j_k}}\right] dx,
\end{aligned} \qquad (50)$$

where on the right side in passing from the first line to the second line, we have made use of the divergence theorem and from the second to the third the requirement that v and its derivatives up to the $(p-1)^{\text{th}}$ be zero on the boundary. If we apply this stratagem repeatedly to (50), we eventually obtain that:

$$\int_\Omega v^i_{j_1,j_2,\ldots,j_k} \frac{\partial L}{\partial u^i_{j_1,j_2,\ldots,j_k}} \, dx = (-1)^k \int_\Omega v^i D_{j_1} D_{j_2} \ldots D_{j_k} \frac{\partial L}{\partial u^i_{j_1,j_2,\ldots,j_k}} \, dx. \tag{51}$$

We substitute (51) into (49) to give:

$$\delta A = \varepsilon (-1)^k \int_\Omega v^i D_{j_1} D_{j_2} \ldots D_{j_k} \frac{\partial L}{\partial u^i_{j_1,j_2,\ldots,j_k)}} \, dx + O\left(\varepsilon^2\right). \tag{52}$$

Hamilton's principle requires that δA be zero for a zero-boundary variation. As the functions $v^i(x)$ are arbitrary subject to the differentiability condition, the integrand in (52) must be zero for each value of the index i, and so, we obtain the m Euler–Lagrange equations:

$$(-1)^k D_{j_1} D_{j_2} \ldots D_{j_k} \frac{\partial L}{\partial u^i_{j_1,j_2,\ldots,j_k}} = 0, \quad i = 1, m, \tag{53}$$

with the summation on j being from one to n and on k from zero to p.

10. Noether's Theorem: Original Formulation

Under the infinitesimal transformation:

$$\bar{x}^j = x^j + \varepsilon \xi^j \qquad \bar{u}^i = u^i + \varepsilon \eta^i \tag{54}$$

of both independent and dependent variables generated by the differential operator:

$$\Gamma = \xi_j \partial_{x_j} + \eta_i \partial_{u_i}, \tag{55}$$

in which summation on i and j from 1–m and from 1–n, respectively, is again implied, the action integral,

$$A = \int_\Omega L(x, u, u_1, \ldots, u_p) \, dx, \tag{56}$$

becomes:

$$\begin{aligned}
\bar{A} &= \int_{\bar\Omega} L(\bar{x}, \bar{u}, \bar{u}_1, \ldots, \bar{u}_p) \, d\bar{x} \\
&= \int_{\Omega+\delta\Omega} L\left(x + \varepsilon \xi, u + \varepsilon \eta, u_1 + \varepsilon \eta^{(1)}, \ldots, u_p + \varepsilon \eta^{(p)}\right) J(\bar{x}, x) \, dx.
\end{aligned} \tag{57}$$

The notation $\delta\Omega$ indicates the infinitesimal change in the domain of integration Ω induced by the infinitesimal transformation of the independent variables.

The notation $\eta^{(j)}$ is a shorthand notation for the jth extension of Γ. For the j_1^{th} derivative of u^i, we have specifically:

$$\eta^{i(1)}_{j_1} = D_{j_1} \eta^i - u^i_l D_{j_1} \xi^l \tag{58}$$

and for higher derivatives, we can use the recursive definition:

$$\eta^{i(k)}_{j_1 j_2 \ldots j_k} = D_k \eta^{i(k-1)}_{j_1 j_2 \ldots j_{k-1}} - u^i_{j_1 j_2 \ldots j_k l} D_{j_k} \xi^l \tag{59}$$

in which the terms in parentheses are not to be taken as summation indices.

The Jacobian of the transformation may be written as:

$$\begin{aligned} J(\tilde{x}, x) &= \left| \frac{\partial \tilde{x}^i}{\partial x^j} \right| \\ &= \left| \delta_{ij} + \varepsilon D_j \xi^i + O\left(\varepsilon^2\right) \right| \\ &= 1 + \varepsilon D_j \xi^j + O\left(\varepsilon^2\right). \end{aligned} \qquad (60)$$

We now can write (57) as:

$$\tilde{A} = \int_{\Omega} \left\{ L + \varepsilon \left[\varepsilon L D_j \xi^i + \xi^j \frac{\partial L}{\partial x_j} + \eta^{i(k)}_{j_1 j_2 \dots j_k} \frac{\partial L}{\partial u^i_{j_1 j_2 \dots j_k}} \right] \right\} dx + \int_{\delta \Omega} L dx + O\left(\varepsilon^2\right). \qquad (61)$$

Because the transformation is infinitesimal, to the first order in the infinitesimal parameter, ε, the integral over $\delta\Omega$ can be written as:

$$\begin{aligned} \int_{\delta\Omega} L dx &= \varepsilon \int_{\partial\Omega} L d\sigma \\ &= -\int_{\Omega} D_j F^j dx, \end{aligned} \qquad (62)$$

where **F** is an as yet arbitrary function. The requirement that the action integral be invariant under the infinitesimal transformation now gives:

$$D_j F_j = L D_j \xi^j + \xi^j \frac{\partial L}{\partial x_j} + \eta^{i(k)}_{j_1 j_2 \dots j_k} \frac{\partial L}{\partial u^i_{j_1 j_2 \dots j_k}}. \qquad (63)$$

This is the condition for the existence of a Noether symmetry for the Lagrangian. We recall that the variational principle was not used in the derivation of (63), and so, the Noether symmetry exists for all possible curves in the phase space and not only the trajectory for which the action integral takes a stationary value.

To obtain a conservation law corresponding to a given Noether symmetry, we manipulate (63) taking cognizance of the Euler–Lagrange equations. As:

$$\xi^i D_j L = \xi^j \left(\frac{\partial L}{\partial x_j} + D_j u^i_{j_1 j_2 \dots j_k} \frac{\partial L}{\partial u^i_{j_1 j_2 \dots j_k}} \right), \qquad (64)$$

we may write (63) as:

$$\begin{aligned} D_j \left[F_j - L \xi^j \right] &= \left[\eta^{i(k)}_{j_1 j_2 \dots j_k} - \xi^j D_j u^i_{j_1 j_2 \dots j_k} \right] \frac{\partial L}{\partial u^i_{j_1 j_2 \dots j_k}} \\ &= \left(\eta^i - \xi^j D_j u^i \right) \frac{\partial L}{\partial u^i} + \sum_{k=1}^{p} \left[\eta^{i(k)}_{j_1 j_2 \dots j_k} - \xi^j D_j u^i_{j_1 j_2 \dots j_k} \right] \frac{\partial L}{\partial u^i_{j_1 j_2 \dots j_k}} \\ &= -\left(\eta^i - \xi^j D_j u^i \right) (-1)^k D_{j_1} D_{j_2} \dots D_{j_k} \frac{\partial L}{\partial u^i_{j_1 j_2 \dots j_k}} \\ &\quad + \sum_{k=1}^{p} \left[\eta^{i(k)}_{j_1 j_2 \dots j_k} - \xi^j D_j u^i_{j_1 j_2 \dots j_k} \right] \frac{\partial L}{\partial u^i_{j_1 j_2 \dots j_k}} \end{aligned} \qquad (65)$$

in the second line of which we have separated the first term from the summation and used the Euler–Lagrange equation. We may rewrite the first term as:

$$D_{j_k}\left[\left(\eta^i - \xi^j D_j u^i\right)(-1)^k D_{j_1} D_{j_2} \ldots D_{j_{k-1}} \frac{\partial L}{\partial u^i_{j_1 j_2 \ldots j_k}}\right]$$
$$-\left(D_{j_k}\eta^i - \left(D_{j_k}\xi^j\right) D_{j_k} u^i - \xi^j D_{jj_k} u^i\right)(-1)^k D_{j_1} D_{j_2} \ldots D_{j_{k-1}} \frac{\partial L}{\partial u^i_{j_1 j_2 \ldots j_k}}.$$

The first term, being a divergence, can be moved to the left side (after replacing the repeated index j_k with j). We observe that the second term may be written as (58):

$$-\left(\eta^{i(1)}_{j_k} - \xi^j u^i_{jj_k}\right)(-1)^k D_{j_1} D_{j_2} \ldots D_{j_{k-1}} \frac{\partial L}{\partial u^i_{j_1 j_2 \ldots j_k}}$$
$$\Leftrightarrow -D_{j_{k-1}}\left[\left(\eta^{i(1)}_{j_k} - \xi^j u^i_{jj_k}\right)(-1)^k D_{j_1} D_{j_2} \ldots D_{j_{k-2}} \frac{\partial L}{\partial u^i_{j_1 j_2 \ldots j_k}}\right]$$
$$+ \left[D_{j_{k-1}}\left(\eta^{i(1)}_{j_k} - \xi^j u^i_{jj_k}\right)\right](-1)^k D_{j_1} D_{j_2} \ldots D_{j_{k-2}} \frac{\partial L}{\partial u^i_{j_1 j_2 \ldots j_k}}$$
$$\Leftrightarrow -D_{j_{k-1}}\left[\left(\eta^{i(1)}_{j_k} - \xi^j u^i_{jj_k}\right)(-1)^k D_{j_1} D_{j_2} \ldots D_{j_{k-2}} \frac{\partial L}{\partial u^i_{j_1 j_2 \ldots j_k}}\right]$$
$$+ \left(\eta^{i(2)}_{j_{k-1} j_k} - \xi^j u^i_{jj_{k-1} j_k}\right)(-1)^k D_{j_1} D_{j_2} \ldots D_{j_{k-2}} \frac{\partial L}{\partial u^i_{j_1 j_2 \ldots j_k}} \quad (66)$$

in which we see the same process repeated. Eventually, all terms can be included with the divergence, and we have the conservation law:

$$D_j\left\{F_j - L\xi^j - \left(\eta^{i(l)}_{jk\ldots j_{k-l+1}} - \xi^m u^i_{mjk\ldots j_{k-l+1}}\right)(-1)^k D_{j_1} D_{j_2} \ldots D_{j_{k-l}} \frac{\partial L}{\partial u^i_{j_1 j_2 \ldots j_k}}\right\} = 0. \quad (67)$$

The relations (63) and (67) constitute Noether's theorem for Hamilton's principle.

11. Noether's Theorem: Simpler Form

The original statement of Noether's theorem was in terms of infinitesimal transformations depending on dependent and independent variables and the derivatives of the former. Thus, the theorem was stated in terms of generalized symmetries ab initio. The complexity of the calculations for even a system of a moderate number of variables and derivatives of only low order in the coefficient functions is difficult to comprehend and the thought of hand calculations depressing. We have already mentioned that one is advised to calculate generalized Lie symmetries for the corresponding Euler–Lagrange equation using some package and then to check whether there exists an **F** such that (63) is satisfied for the Lie symmetries obtained. Even this can be a nontrivial task. Fortunately, there exists a theoretical simplification, presented by Boyer in 1967 [29], which reduces the amount of computation considerably. The basic result is that under the set of generalized symmetries:

$$\Gamma = \xi^i \partial_{x^i} + \eta^i \partial_{u^i},$$

where the ξ^i and η^i are functions of u, x and the derivatives of u with respect to x, and:

$$\Gamma = \bar{\eta}^i \partial_{u^i}, \quad \bar{\eta}^i = \eta^i - u^i_j \xi^j$$

one obtains the same results [32,33].

This enables (63) and (67) to be written without the coefficient functions ξ^i. This is a direct generalization of the result for a first-order Lagrangian in one independent variable. One simply must ensure that generality is not lost by allowing for a sufficient generality in the dependence of the η^i upon the derivatives of the dependent variables. The only caveat one should bear in mind is that the physical or geometric interpretation of a symmetry may be impaired if the symmetry is given in a form that is not its natural form. This does raise the question of what is the "natural" form of a symmetry. It does not provide the beginnings of an answer. It would appear that the natural form is often determined in the eye of the beholder, cf. [34].

The proof of the existence of equivalence classes of generalized transformation depends on the fact that two transformations can produce the same effect on a function.

12. Conclusions

In this review article, we perform a detailed discussion on the formulation of Noether's theorems and on its various generalizations. More specifically, we discuss that in the original presentation of Noether's work [1], the dependence of the coefficient functions of the infinitesimal transformation can be upon the derivatives of the dependent variables. Consequently, a series of generalizations on Noether' theorem, like hidden symmetries, generalized symmetries, etc., are all included in the original work of Noether. That specific point and that the boundary function on the action integral can include higher-order derivatives of the dependent variables were the main subjects of discussion for this work. Our aim was to recover for the audience that generality that has been lost after texts, for instance Courant, Hilbert, Rund, and many others, where they identify as Noether symmetries only the point transformations. The discussion has been performed for ordinary and partial differential equations, while the corresponding conservation laws/flows are given in each case.

Author Contributions: The authors contributed equivalently.

Funding: APacknowledges the financial support of FONDECYT Grant No. 3160121. PGLL Thanks the Durban University of Technology, the University of KwaZulu-Natal, and the National Research Foundation of South Africa for support. AKH expresses grateful thanks to UGC(India), NFSC, Award No. F1-17.1/201718/ RGNF-2017-18-SC-ORI-39488, for financial support.

Conflicts of Interest: The authors declare no conflict of interest. The funders had no role in the design of the study; in the collection, analyses, or interpretation of data, in the writing of the manuscript; nor in the decision to publish the results.

References and Notes

1. Noether, E. Invariante Variationsprobleme. *Königlich Gesellschaft der Wissenschaften Göttingen Nachrichten Mathematik-Physik Klasse* **1918**, *2*, 235–267.
2. Hamel, G. Ueber die Grundlagen der Mechanik. *Math. Ann.* **1908**, *66*, 350–397. [CrossRef]
3. Hamel, G. Ueber ein Prinzig der Befreiung bei Lagrange. *Jahresbericht der Deutschen Mathematiker-Vereinigung* **1917**, *25*, 60–65.
4. Herglotz, G. Über den vom Standpunkt des Relativitätsprinzips aus als "starr" zu bezeichnenden Körper. *Annalen der Physik* **1910**, *336*, 393–415. [CrossRef]
5. Fokker. *Verslag die Amsterdamer Akad*; Holland, 1917.
6. Kneser, A. Kleinste Wirkung und Galileische Relativität. *Math. Z.* **1918**, *2*, 326–349. [CrossRef]
7. Klein, F. *Königlich Gesellschaft der Wissenschaften Göttingen Nachrichten Mathematik-Physik Klasse*; Germany, 1918; p. 2.
8. Courant, R.; Hilbert, D. *Methods of Mathematical Physics*; Wiley Interscience: New York, NY, USA, 1953.
9. Lovelock, D.; Rund, H. *Tensors, Differential Forms and Variational Principles*; Wiley: New York, NY, USA, 1975.
10. Logan, J.D. *Invariant Variational Principles*; Academic: New York, NY, USA, 1977.
11. Stan, M. On the Noether's Theorem in Hamiltonian formalism. *Sci. Bull. Politech. Univ. Buchar. Ser. A Appl. Math. Phys.* **1994**, *56*, 93–98.

12. Anco, S.C. Generalization of Noether's Theorem in Modern Form to Non-variational Partial Differential Equations. In *Recent Progress and Modern Challenges in Applied Mathematics, Modeling and Computational Science*; Melnik, R., Makarov, R., Belair, J., Eds.; Fields Institute Communications; Springer: New York, NY, USA, 2017; Volume 79.
13. Hojman, S.A. A new conservation law constructed without using either Lagrangians or Hamiltonians. *J. Phys. A Math. Gen.* **1992**, *25*, L291–L295. [CrossRef]
14. González-Gascón, F. Geometric foundations of a new conservation law discovered by Hojman. *J. Phys. A Math. Gen.* **1994**, *27*, L59–L60. [CrossRef]
15. Crampin, M. Hidden symmetries and Killing tensors. *Rep. Math. Phys.* **1984**, *20*, 31–40. [CrossRef]
16. Crampin, M.; Mestdag, T. The Cartan form for constrained Lagrangian systems and the nonholonomic Noether theorem. *Int. J. Geom. Methods Mod. Phys.* **2011**, *8*, 897–923. [CrossRef]
17. Sarlet, W.; Cantrijn, F. Generalizations of Noether's Theorem in Classical Mechanics. *SIAM Rev.* **1981**, *23*, 467–494. [CrossRef]
18. Dugas, R. *A History of Mechanics*; Maddox, J.R., Translated; Dover: New York, NY, USA, 1988.
19. Govinder, K.S.; Leach, P.G.L. Paradigms of Ordinary Differential Equations. *S. Afr. J. Sci.* **1995**, *91*, 306–311.
20. Pillay, T.; Leach, P.G.L. Comment on a theorem of Hojman and its generalizations. *J. Phys. A Math. Gen.* **1996**, *29*, 6999–7002. [CrossRef]
21. Leach, P.G.L. Applications of the Lie theory of extended groups in Hamiltonian mechanics: The oscillator and the Kepler problem. *J. Aust. Math. Soc.* **1981**, *23*, 173–186. [CrossRef]
22. Govinder, K.S.; Leach, P.G.L.; Maharaj, S.D. Integrability analysis of a conformal equation arising in relativity. *Int. J. Theor. Phys.* **1995**, *34*, 625–639. [CrossRef]
23. Leach, P.G.L.; Govinder, K.S.; Abraham-Shrauner, B. Symmetries of First Integrals and Their Associated Differential Equations. *J. Math. Anal. Appl.* **1999**, *235*, 58–83. [CrossRef]
24. Leach, P.G.L.; Warne, R.R.; Caister, N.; Naicker, V.; Euler, N. Symmetries, integrals and solutions of ordinary differential equations of maximal symmetry. *Proc. Math. Sci.* **2010**, *120*, 123. [CrossRef]
25. Leach, P.G.L.; Mahomed, F.M. Maximal subalgebra associated with a first integral of a system possessing sl(3,R) algebra. *J. Math. Phys.* **1988**, *29*, 1807. [CrossRef]
26. Paliathanasis, A.; Leach, P.G.L. Nonlinear Ordinary Differential Equations: A discussion on Symmetries and Singularities. *Int. J. Geom. Meth. Mod. Phys.* **2016**, *13*, 1630009. [CrossRef]
27. Mahomed, F.M.; Kara, A.H.; Leach, P.G.L. Lie and Noether counting theorems for one-dimensional systems. *J. Math. Anal. Appl.* **1993**, *178*, 116–129. [CrossRef]
28. Patera, J.; Winternitz, P. Algebras of real three- and four-dimensional Lie algebras. *J. Math. Phys.* **1977**, *18*, 1449–1455. [CrossRef]
29. Boyer, T.H. Continuous symmetries and conserved currents. *Ann. Phys.* **1967**, *42*, 445–466. [CrossRef]
30. Bluman, G.W.; Kumei, S. *Symmetries and Differential Equations*; Springer: New York, NY, USA, 1989.
31. Olver, P.J. *Applications of Lie Groups to Differential Equations*; Springer: Berlin, Germany, 1986.
32. Katzin, G.H.; Levine, J. Characteristic functional structure of infinitesimal symmetry mappings of classical dynamical systems. I. Velocity-dependent mappings of second-order differential equations. *J. Math. Phys.* **1985**, *26*, 3080. [CrossRef]
33. Katzin, G.H.; Levine, J. Characteristic functional structure of infinitesimal symmetry mappings of classical dynamical systems. II. Mappings of first-order differential equations. *J. Math. Phys.* **1985**, *26*, 3100. [CrossRef]
34. Moyo, S.; Leach, P.G.L. Exceptional properties of second and third order ordinary differential equations of maximal symmetry. *J. Math. Anal. Appl.* **2000**, *252*, 840–863. [CrossRef]

© 2018 by the authors. Licensee MDPI, Basel, Switzerland. This article is an open access article distributed under the terms and conditions of the Creative Commons Attribution (CC BY) license (http://creativecommons.org/licenses/by/4.0/).

Article

Conservation Laws and Stability of Field Theories of Derived Type

Dmitry S. Kaparulin

Faculty of Physics, Tomsk State University, Tomsk 634050, Russia; dsc@phys.tsu.ru

Received: 4 April 2019; Accepted: 5 May 2019; Published: 7 May 2019

Abstract: We consider the issue of correspondence between symmetries and conserved quantities in the class of linear relativistic higher-derivative theories of derived type. In this class of models the wave operator is a polynomial in another formally self-adjoint operator, while each isometry of space-time gives rise to the series of symmetries of action functional. If the wave operator is given by n-th-order polynomial then this series includes n independent entries, which can be explicitly constructed. The Noether theorem is then used to construct an n-parameter set of second-rank conserved tensors. The canonical energy-momentum tensor is included in the series, while the other entries define independent integrals of motion. The Lagrange anchor concept is applied to connect the general conserved tensor in the series with the original space-time translation symmetry. This result is interpreted as existence of multiple energy-momentum tensors in the class of derived systems. To study stability we seek for bounded-conserved quantities that are connected with the time translations. We observe that the derived theory is stable if its wave operator is defined by a polynomial with simple and real roots. The general constructions are illustrated by the examples of the Pais–Uhlenbeck oscillator, higher-derivative scalar field, and extended Chern–Simons theory.

Keywords: Noether's theorem; generalized symmetry; energy-momentum tensor; Lagrange anchor

1. Introduction

Once the Noether theorem [1] is applied to the higher-derivative theories, the models whose Lagrangians involve second and higher-derivatives in time, the canonical energy usually appears to be unbounded. This is often interpreted as instability of higher-derivative dynamics [2]. The presence of classical trajectories with runaway behavior and the absence of a well-defined vacuum state with the lowest energy at the quantum level are considered as typical indicators of stability problems. The recent research [3–5] demonstrates that the models with unbounded classical energy are not necessarily unstable. Various ideas were applied to the study of stability of higher-derivative theories, including the non-Hermitian quantum mechanics [6–8], alternative Hamiltonian formulations [9–12], adiabatic invariants [13], and special boundary conditions [14]. For constrained systems, the energy can be bounded on-shell due to constraints [15–17]. The f(R)-gravity [18–20] is the most studied model of such a type.

In [21], it has been observed that the higher-derivative dynamics can be stabilized by another bounded-conserved quantity. Even though the Noether theorem associates this bounded quantity with a certain higher symmetry, the Lagrange anchor can be used to connect the additional integral of motion with the time translation. The Lagrange anchor was first introduced to quantize non-variational models in [22]. Later, it was observed that it also connects symmetries and conserved quantities in both Lagrangian and non-Lagrangian theories [23]. In the first-order formalism, the Lagrange anchor defines the Poisson bracket [24]. The bounded-conserved quantity, which is connected with the time translation symmetry, serves as Hamiltonian with respect to this Poisson bracket. In this way, the stability of dynamics is retained at both classical and quantum levels.

In [25], an interesting class of higher-derivative theories of derived type was introduced. In these models the wave operator is a constant coefficient polynomial (characteristic polynomial) in another formally self-adjoint operator of lower order. The Pais–Uhlenbeck theory [26], Podolsky electrodynamics [27], conformal gravities in four and six dimensions [28,29], and extended Chern–Simons [30] are particular examples of theories of derived type. The general result of [25] is the following: the derived theory with the n-th-order characteristic polynomial admits n-parameter series of conserved quantities, which can be bounded or unbounded. The bounded-conserved quantities can stabilize the classical dynamics of the theory. It was also claimed (without detailed analysis) that all of the conserved quantities in the series are connected with the time translation symmetry by a Lagrange anchor.

In the present article, we study the stability of higher-derivative dynamics from the viewpoint of a more general correspondence between symmetries and conservation laws, which is established by the Lagrange anchor. We show that the derived model with the n-th-order characteristic polynomial admits an n-parameter series of Lagrange anchors, which connects the general conserved quantity with time translation symmetry. To get this result we reformulate the correspondence between symmetries and conserved quantities in terms of algebra of polynomials and apply the Bezout Lemma. To study stability we address the issue of correspondence between the time translations and bounded-conserved quantities. This problem is special because the bounded-conserved quantities are not general representatives of the conserved quantity series. We conclude that in non-singular theories a bounded-conserved quantity can be connected with the time translation if all of the roots of the characteristic polynomial are real and simple. As for gauge models, a more accurate analysis is performed.

The rest of the paper is organized as follows. In Section 2, we recall some basic facts about symmetries, characteristics, and conservation laws in linear systems. In Section 3, we introduce the Lagrange anchor and establish a correspondence between symmetries and conservation laws. The class of derived models is introduced in Section 4. Section 5 studies the issue of the relationship between the bounded quantities and time translations. We also obtain the stability conditions of higher-derivative theories of derived type in Section 5. Section 6 illustrates general constructions in the theories of the Pais–Uhlenbeck oscillator, higher-derivative scalar, and extended Chern–Simons. The conclusion summarizes the results.

2. Symmetries, Characteristics, and Conserved Quantities of Linear Systems

Given the d-dimensional Minkowski space (We use the mostly minus convention for the space-time metric throughout this paper) with the local coordinates x^μ, $\mu = 0, 1, \ldots, d-1$, we consider the set of fields $\varphi^i(x)$. The multi-index i includes all the tensor, spinor, and isotopic indices, which label the fields. We assume the existence of an appropriate constant metric which can be used to raise and lower the multi-indices. This gives rise to the inner product of fields,

$$\langle \varphi, \psi \rangle = \varphi_i(x)\psi^i(x) \tag{1}$$

(no integration over space-time). Zero boundary conditions are assumed for the fields at infinity. In this setting, the most general linear theory reads as PDE,

$$T_i(\varphi) = M_{ij}(\partial)\varphi^j = 0, \tag{2}$$

where summation over repeated index is implied, and M is the matrix differential operator. By the matrix differential operator, we mean the matrix whose entries are polynomials in the formal variable $\partial_\mu = \partial/\partial x^\mu$.

The formal adjoint of the matrix differential operator M is defined as follows:

$$M^\dagger_{ij}(\partial) = M_{ji}(-\partial). \tag{3}$$

If the wave operator is formally self-adjoint, the corresponding Equation (1) are variational with the action functional

$$S[\varphi(x)] = \frac{1}{2}\int \langle \varphi, M\varphi \rangle d^d x. \tag{4}$$

In this class of models, M is often called the wave operator. The following identity relates the operator and its adjoint:

$$\frac{1}{2}\left(\langle \varphi, M\psi \rangle - \langle M^\dagger \varphi, \psi \rangle\right) = \partial_\mu \left(j_M(\varphi, \psi)\right)^\mu. \tag{5}$$

Here, ψ and φ are test functions, and the right-hand side is a divergence of some vector. The index M in j_M labels the operator M, which is involved in the left-hand side of Equation (5).

We allow gauge freedom for the model (2). If the wave operator M has right null-vectors in the class of matrix differential operators, i.e.,

$$M(\partial)R_\alpha(\partial) \equiv 0, \tag{6}$$

with α being some multi-index, Equation (2) are invariant with respect to the gauge transformation

$$\delta_\varepsilon \varphi^i = R_\alpha^i(\partial)\varepsilon^\alpha. \tag{7}$$

Here, $\varepsilon^\alpha = \varepsilon^\alpha(x)$ are functions of space-time coordinates, and summation over the repeated index α is assumed. The matrix differential operators R_α are called gauge generators.

The wave operator can have left null-vectors in the class of matrix differential operators. All such null-vectors determine gauge identities between equations of motion of the following form:

$$Z_A(\partial)T(\varphi) \equiv 0, \tag{8}$$

where the multi-index A labels gauge identities. For linear variational theories, the identity generators are adjoint of gauge generators,

$$Z_A(\partial) = R_\alpha^\dagger(\partial), \quad A \equiv \alpha. \tag{9}$$

The statement about the relation between gauge symmetries and gauge identities is often called the second Noether theorem, and equality (9) is a particular form of it. In non-variational theories, the gauge symmetries and identity generators are unrelated to each other.

The matrix differential operator L is called the symmetry (The notion of symmetry can be introduced in various ways. In this article we give non-rigorous explanations about symmetries, which are sufficient for our consideration. For systematical introduction into the subject we refer to the book [31]) of linear theory (2) if it is interchangeable with M in the following sense:

$$M(\partial)L(\partial) = Q(\partial)M(\partial), \tag{10}$$

with Q being some matrix differential operator. Symmetry induces a linear transformation of the fields that preserves the mass shell (2):

$$\delta_\xi \varphi = \xi L(\partial)\varphi, \quad \delta_\xi T(\varphi) = \xi L(\partial)T(\varphi) \approx 0. \tag{11}$$

Here, the constant ξ is the infinitesimal transformation parameter, and the sign $\approx$ means equality modulo Equation (2). The symmetries of linear theory form an associative algebra with respect to the composition of the operators [31]. Equation (10) has a lot of trivial solutions of the form

$$L_{\text{triv}}(\partial) = U(\partial)M(\partial) + R_\alpha(\partial)U^\alpha(\partial), \tag{12}$$

where U and U^α are some matrix differential operators. These symmetries are present in every theory and do not contain any valuable information about the dynamics of the model. In what follows, we systematically ignore them.

We term the symmetry variational if the transformation (11) preserves the action functional (4). The defining condition for the variational symmetry reads

$$L^\dagger(\partial)M(\partial) + M(\partial)L(\partial) = 0, \tag{13}$$

which is relation (10) for $Q = L^\dagger$. Trivial variational symmetries have the form

$$L^{\text{var}}_{\text{triv}}(\partial) = G(\partial)M(\partial) + R_\alpha(\partial)U^\alpha(\partial), \quad G^\dagger = -G. \tag{14}$$

The vector-valued function of the fields and their derivatives j is called a conserved current of the model (4) if its divergence vanishes on the mass shell, i.e.,

$$\partial_\mu j^\mu(\varphi, \partial\varphi, \partial^2\varphi, \ldots) = \langle N(\varphi), T(\varphi)\rangle \tag{15}$$

for some characteristics, N being the function of fields and its derivatives. The conserved current j defines the integral of motion:

$$J = \int j^0(\varphi) d^{d-1}x, \tag{16}$$

where the integration is held over the space coordinates. By construction, J is a constant for all the solutions to the classical Equation (2).

The above definition of conserved current and characteristic has natural ambiguities. Two conserved currents are considered equivalent if their difference is the divergence of some anti-symmetric tensor on-shell,

$$j'^\mu(\varphi) - j^\mu(\varphi) \approx \partial_\nu \Sigma^{\nu\mu}(\varphi), \quad \Sigma^{\nu\mu}(\varphi) = -\Sigma^{\mu\nu}(\varphi). \tag{17}$$

Equivalent conserved currents define one and the same integral of motion (16). As for characteristics, the corresponding condition reads

$$N'(\varphi) - N(\varphi) \approx U^\alpha(\varphi)R_\alpha. \tag{18}$$

Equivalent characteristics determine equivalent conserved currents. This establishes a one-to-one correspondence between the equivalence classes of characteristics and conservation laws.

In linear theories, the quadratic conserved currents are the most relevant. Once the conserved current is bilinear in fields, the characteristic is linear. It can be written in the form

$$N(\varphi) = N(\partial)\varphi, \tag{19}$$

where N is a matrix differential operator, which we call the operator of characteristics. The defining condition for the operator of characteristic reads

$$N^\dagger(\partial)M(\partial) + M^\dagger(\partial)N(\partial) = 0. \tag{20}$$

The formula can be applied to both variational and non-variational theories. Trivial operators of characteristics read

$$N_{\text{triv}}(\partial) = G(\partial)M(\partial) + \sum_A (Z^\dagger)_A(\partial)U^A(\partial), \quad G^\dagger = -G. \tag{21}$$

In the class of variational models' conditions (20) and (21) take the form of (13) and (14). This brings us to the identification of variational symmetries and characteristics,

$$L(\partial) \equiv N(\partial). \tag{22}$$

The relationship between the variational symmetries and conserved currents is obtained from (20) and (21) using the integration by parts by means of Equation (5),

$$j = j_M(\varphi, N\varphi), \tag{23}$$

(the on-shell vanishing terms are omitted). In fact, (22) and (23) represent one of the possible formulations of the classical Noether theorem [1].

As we can see from the above, the classical Noether theorem essentially uses two distinct facts: the relationship between characteristics and conserved currents (23), and the connection between symmetries and characteristics (22). The first of these facts holds true for the variational and non-variational models, while the second uses the presence of action functional. We introduce Formula (22) because it allows us to extend the Noether theorem beyond the scope of variational dynamics. A crucial ingredient of generalization of the Noether theorem is the Lagrange anchor.

3. Lagrange Anchor and Generalization of the Noether Theorem

The matrix differential operator V is called a Lagrange anchor (For the general definition of Lagrange anchor see [22]) of the linear theory (2) if the following relation is satisfied:

$$V^\dagger(\partial)M^\dagger(\partial) - M(\partial)V(\partial) = 0. \tag{24}$$

The defining equation for the Lagrange anchor has a lot of trivial solutions of the form

$$V_{\text{triv}}(\partial) = G(\partial)M(\partial) + \sum_\alpha R_\alpha(\partial)U^\alpha(\partial), \quad G^\dagger = G, \tag{25}$$

where G is a self-adjoint operator, and U^α are arbitrary operators. The trivial Lagrange anchors do not contain valuable information about the dynamics of the theory, and they are also useless for the connection of symmetries and conserved quantities. We systematically ignore them.

The Lagrange anchor is called transitive if it has zero kernel in the class of matrix differential operators,

$$V(\partial)K(\partial) = 0 \Leftrightarrow K(\partial) = 0. \tag{26}$$

We mostly consider transitive Lagrange anchors. The variational models admit the canonical Lagrange anchor, $V = id$, which is transitive. In non-variational theories, some examples of transitive Lagrange anchors can be found in [32,33].

The Lagrange anchor connects symmetries and conserved quantities. If N is the characteristic of a conserved quantity, and V is a Lagrange anchor, the corresponding symmetry reads

$$L(\partial) = V(\partial)N(\partial). \tag{27}$$

conditions (20) and (24) ensure that the right-hand side of this expression is symmetry in the sense of condition (10). Once the variational theory is equipped with the canonical Lagrange anchor, the identity map connects characteristics and symmetries,

$$L(\partial) = N(\partial), \quad V = id. \tag{28}$$

This is the standard Noether correspondence between symmetries and conserved quantities (22), (23). If V is non-canonical, the symmetries and conserved quantities are connected in a non-canonical way.

Relation (27) can be lifted at the level of classes of equivalence of non-trivial symmetries and conserved quantities. To ensure this fact, we should prove that the trivial characteristics are mapped into trivial symmetries. The following relations are used:

$$V(\partial)N_{\text{triv}}(\partial) = V(\partial)G(\partial)M(\partial) + \sum_\alpha V(\partial)(Z^\dagger)_A(\partial)U^A(\partial) =$$
$$= (V(\partial)G(\partial))M(\partial) + \sum_\alpha R_\alpha(\partial)U^\alpha(\partial) = L_{\text{triv}}(\partial). \tag{29}$$

For the sake of simplicity, we assume that the set of gauge generators is complete. In this case, the identity $M(\partial)V(\partial)Z_A(\partial)U^A(\partial) = 0$ implies $V(\partial)Z_A(\partial)U^A(\partial) = R_\alpha(\partial)U^\alpha(\partial)$ for some U^α, summation over repeated indices is implied.

The classes of equivalence of characteristics and conserved currents are in one-to-one correspondence. This means that Formula (27) connects the classes of equivalence of conservation laws and symmetries, like the Noether theorem. We call relation (27) the generalization of the Noether theorem. As the requirement of the existence of the Lagrange anchor is less restrictive than the presence of the least-action principle, the generalization of the Noether theorem can be applied to connect symmetries and conserved quantities in non-variational theories.

The matrix differential operator L is called a proper symmetry with respect to the Lagrange anchor V if $L(\partial) = V(\partial)P(\partial)$ for some P, and

$$V(\partial)(P^\dagger(\partial)M(\partial) + M^\dagger(\partial)P(\partial)) = 0. \tag{30}$$

the proper symmetries are symmetries in the usual sense. Indeed, by relations (24) and (27), we get

$$M(\partial)V(\partial)P(\partial) = (V^\dagger(\partial)M^\dagger(\partial) - M(\partial)V(\partial))P(\partial) + M(\partial)V(\partial)P(\partial) =$$
$$= V^\dagger(\partial)(M^\dagger(\partial)P(\partial) + P^\dagger(\partial)M(\partial)) - V^\dagger(\partial)P^\dagger(\partial)M = V^\dagger(\partial)P^\dagger(\partial)M(\partial). \tag{31}$$

In the class of variational models equipped with the canonical Lagrange anchor, the proper symmetries are just variational ones. If the Lagrange anchor is transitive, the proper symmetries can be connected with integrals of motion. Applying (26) to (30), we get that P is the characteristic,

$$P^\dagger(\partial)M(\partial) + M^\dagger(\partial)P(\partial) = 0. \tag{32}$$

The characteristic defines the conserved current by Formula (16). This establishes a correspondence between proper symmetries and conserved quantities. In the class of variational models equipped with the canonical Lagrange anchor, relation (28) establishes the Noether correspondence between the variational symmetries and integrals of motion. In the class of non-variational theories, the proper symmetries constitute a special subset in the space of all symmetries, which can be connected with conserved quantities.

The connection between symmetries and conserved currents is not necessarily unique for a given system of equations, because multiple Lagrange anchors can be admissible by the model. If this takes place, one and the same conserved current can be associated with several different symmetries. Alternatively, one and the same symmetry can come from different pairs of Lagrange anchors and conserved currents. In the rest of the paper we mostly deal with the variational theories of derived type, which are known to admit multiple Lagrange anchors. In particular, we show that the isometries of space-time can be connected with the series of conserved tensors in this class of models.

4. Higher-Derivative Theories of Derived Type

Given a set of fields φ^i, we introduce the variational primary model without higher-derivatives

$$W_{ij}(\partial)\varphi^j = 0, \quad W^\dagger = W. \tag{33}$$

The model (2) falls into the class of theories of derived type if its wave operator is a constant coefficient polynomial in W,

$$M(\alpha;W) = \sum_{p=0}^{n} \alpha_p W^p, \quad \alpha_n \neq 0. \tag{34}$$

The constants $\alpha_0, \ldots, \alpha_n$ distinguish different representatives in the class. The equations of motion (33) come from the least-action principle for the functional

$$S[\varphi(x)] = \frac{1}{2}\int \langle \varphi, W\varphi \rangle d^d x. \tag{35}$$

Each derived theory is defined by a primary operator W, and a characteristic polynomial in the complex variable z,

$$M(\alpha;z) = \sum_{p=0}^{n} \alpha_p z^p. \tag{36}$$

The symmetries, characteristics, and Lagrange anchors in the class of derived systems can be systematically obtained from the corresponding quantities of the primary model. Even though the derived quantities do not cover all symmetries, characteristics, and Lagrange anchors, they contain valuable information about dynamics. In the present paper, we use derived conserved quantities to study the stability of higher-derivative models.

Let us explain the details of the construction. Suppose that the primary theory admits the series of variational symmetries with the operator $L(\xi)$,

$$L(\xi) = \sum_{a=1}^{s} \xi^a L_a, \quad [L_a, W] = 0, \quad L_a = -(L_a)^\dagger, \tag{37}$$

where the indices $a = 1, \ldots, s$ label the generators of symmetry series, and ξ's are the transformation parameters, being real constants. In this case, the Noether theorem (22), (23) associates the conserved currents j_a, $a = 1, \ldots, s$ with these symmetries. The conserved currents are linearly independent if the symmetry generators are linearly independent (modulo trivial symmetries). In what follows in this section, we assume the linear independence of the generators L_a (37). The operators L_a (37) can form a Lie algebra, at least in some models. The Poincare symmetry in the relativistic theories is one of the relevant examples of such kind. We do not specify the structure of the algebra of symmetries because we are mostly interested in the correspondence between the linear spaces of symmetries and conserved quantities.

A single variational symmetry (37) defines the n-parameter series of variational symmetries of the action functional (4), (34),

$$N(\beta, \xi; W) = L(\xi)N(\beta; W), \quad N(\beta; W) \equiv \sum_{p=0}^{n-1} \beta_p W^p, \tag{38}$$

where β_p, $p = 0, \ldots, n-1$, are constant parameters. For $p = 0$, this series includes the symmetries (37) of the primary model (33). The other entries in (38) are higher-derivative operators whose origin is a consequence of the derived structure (34) of the equations of motion (2). The symmetries in the series (38) are usually independent, even though it is not a theorem. The general argument is that if all the powers of the primary operator are independent initial data and L_a's are transitive operators, the characteristic (38) cannot be zero on-shell. The exceptions are possible in the class of gauge theories, where some entries of the series (38) can trivialize on the mass-shell with account of gauge symmetries.

The Noether theorem (22), (23) associates the following set of conserved currents with the variational symmetry (38):

$$j(\beta, \xi) = \sum_{p=0}^{n-1} \beta_p j^p(\xi),$$
$$j^p(\xi) = \sum_{q=0}^{n} \alpha_q \left(\sum_{s=1}^{q-p} j_W(L(\xi) W^{p+s-1} \varphi, W^{q-s} \varphi) + \sum_{s=1}^{p-q} j_W(L(\xi) W^{p-s} \varphi, W^{q+s-1} \varphi) \right). \tag{39}$$

The derivation of this formula uses integration by parts by means of Equation (5). All the on-shell vanishing contributions are omitted in (39). In this set, j^0's represents the canonical conserved quantities of the derived model (4), (34) that are connected with the original symmetry (37) of the primary model (33). The other conserved currents, j^p, $p = 1, \ldots, n-1$, come from the higher-symmetries in the set (38). The conditions of linear independence of conserved currents are the same as that of symmetries.

The Lagrange anchors in the class of derived theories are just polynomials in the primary operator,

$$V(\gamma; W) = \sum_{p=0}^{n-1} \gamma_p W^p. \tag{40}$$

The higher-powers of W do not define new non-trivial Lagrange anchors, see Equation (25). The canonical Lagrange anchor is included in this series for

$$\beta_0 = 1, \quad \beta_1 = \beta_2 = \ldots = \beta_{n-1} = 0. \tag{41}$$

The Lagrange anchor (40) connects characteristic (38) with symmetry of the derived theory (34) by the rule (27). The resulting symmetry is the original one if

$$V(\gamma; W) N(\beta, \xi; W) = L(\xi) + \text{trivial symmetry}. \tag{42}$$

We are interested in the problem of connecting of conserved quantities with the original symmetry because it is relevant for studying stability.

Let us first consider the theories without gauge symmetries. In this case, the relevant trivial symmetry in the right-hand side of (42) has the form

$$L_{\text{triv}} = K(\delta; W) M(\alpha; W) L(\xi), \quad K(\delta; W) = \sum_{p=0}^{n-2} \delta_p W^p, \tag{43}$$

with K being some polynomial in W. To meet (42), it is sufficient to impose the condition

$$V(\gamma; W) N(\beta; W) = \text{id} + K(\delta; W) M(\alpha; W). \tag{44}$$

If all the powers of the characteristic operator are independent, the formula is equivalent to the relation between the characteristic polynomials of the involved quantities

$$V(\gamma; z) N(\beta; z) - K(\delta; z) M(\alpha; z) = 1. \tag{45}$$

The Bezout Lemma states that the problem has a unique solution for V and K if, and only if, N and M have no common roots. Once two general polynomials have no common roots, almost all the conserved quantities in the series are related to the original symmetry.

In the class of gauge models, some additional symmetry can trivialize on-shell. The relevant trivial symmetry in the right-hand side of (42) can be chosen in the form

$$L_{\text{triv}} = \left[K(\delta; W) M(\alpha; W) + \sum_{k=1}^{s} K_s(W) M_s(W) \right] L(\xi), \qquad (46)$$

where M_s are additional generators of trivial symmetries. In principle, the wave operator and M_s can be dependent in some models. We should ignore M in the formulas below in these circumstances. The analogue of relations (44), (45) reads

$$V(\gamma; W) N(\beta; W) - K(\delta; W) M(\alpha; W) - \sum_{k=1}^{s} K_s(W) M_s(W) = \text{id}; \qquad (47)$$

$$V(\gamma; z) N(\beta; z) - K(\delta; z) M(\alpha; z) - \sum_{k=1}^{s} K_s(z) M_s(z) = 1. \qquad (48)$$

These equations are consistent if N and at least one characteristic polynomial of the trivial gauge symmetry generators M, M_s, have no common roots. This condition is less restrictive than in the case of non-gauge models.

5. Stability of Higher-derivative Dynamics

In this section, we address the problem of establishing a relationship between the bounded-conserved quantities and the time translation symmetry. This problem needs more accurate investigation because bounded-conserved quantities are not necessarily general representatives of conserved current series, while general conserved currents may define unbounded-conserved charges. We proceed in two steps. First, the bounded-conserved quantities in the series are identified. After that, we address the issue of connecting bounded-conserved quantities with the space-time translation. Throughout the section we assume that $L(\xi) = \xi^\mu \partial_\mu$.

Let us discuss the structure of conserved quantities in the series (39). We introduce the factorization of the wave operator (34) in the form

$$M(\alpha; W) = \alpha_n \prod_{k=1}^{r} (W - \lambda_k)^{m_k}, \qquad (49)$$

where the numbers λ_k and m_k label different roots and their multiplicities. The integer r is the total number of different roots. We assume that all the quantities λ_p are real numbers. We do not consider complex roots because the corresponding derived theory is always unstable.

Using factorization (49), we define the new dynamical variables absorbing the derivatives of the original field by the receipt of Pais and Uhlenbeck [26]:

$$\varphi_k = \Lambda_k(\partial)\varphi, \quad \Lambda_k(\partial) = \alpha_n \prod_{i=1, i \neq k}^{r} (W - \lambda_i)^{m_i}, \quad k = 1, \ldots, r. \qquad (50)$$

We term the new quantities as components. By construction, the original field φ can be expressed on-shell as the linear combination:

$$\varphi = \sum_{k=1}^{r} C_k(\zeta; W) \varphi_k, \quad C_k(\zeta; W) = \sum_{k=0}^{m_k - 1} \zeta_k W^k. \qquad (51)$$

The following system of equations is valid for the components:

$$(W - \lambda_k)^{m_k} \varphi_k = 0, \quad k = 1, \ldots, r. \tag{52}$$

It ensures that the components, which are associated with different roots of characteristic equation, are independent degrees of freedom. The one-to-one correspondence between the solutions of systems (2) and (52) is established by Formulas (50) and (51).

The system of equations for components comes from the action functional

$$S = \sum_{k=1}^{r} S_k[\varphi_k(x)], \quad S_k[\varphi_k(x)] = \frac{1}{2} \int \langle \varphi_k, (W - \lambda_k)^{m_k} \varphi_k \rangle d^d x. \tag{53}$$

Once $L(\xi)$ is a set of symmetries of the primary model, the following transformations are variational symmetries:

$$\delta_\xi \varphi_k = (W - \lambda_k)^l L(\xi) \varphi_k, \quad k = 1, \ldots, r, \quad l = 0, \ldots, m_k - 1, \tag{54}$$

with no sum in k. There are n independent symmetries in this set. In terms of the original field set, these transformations correspond to the derived symmetries (38). The following conserved currents are associated with them:

$$j^{k;m_k - l - 1}(\xi) = \sum_{s=1}^{m_k - l} j_{W - \lambda_k} (L(\xi)(W - \lambda_k)^{l+s-1} \varphi_k, (W - \lambda_k)^{m_k - s} \varphi_k), \tag{55}$$
$$k = 1, \ldots, r, l = 0, \ldots, m_k - 1.$$

In this set, the leading representatives $j^{k;m_k-1}$ are the canonical energy-momentum currents of the components, while the others are independent quantities. It is obvious that (55) are linear combinations of conserved tensors (39).

Once ξ is an arbitrary constant vector, the conserved quantities are tensors. The second-rank conserved tensors $(T^{k;l})^{\mu\nu}$ are defined by the rule

$$(j^{k;l}(\xi))^\mu = \xi_\nu (T^{k;l})^{\mu\nu}. \tag{56}$$

As for the structure of conserved tensors (56), the following observation is relevant. The leading representatives in the conserved tensor series,

$$(T^{k;0})^{00} = (j_{W - \lambda_k} (\partial^0 (W - \lambda_k)^{m_k - 1} \varphi_k, (W - \lambda_k)^{m_k - 1} \varphi_k))^0, \tag{57}$$

have bounded from below 00-component if the canonical energy of the primary model is bounded on-shell

$$(T^{k;0})^{00} \geq 0 \Leftrightarrow T^{00}_{W - \lambda_k}(\varphi) \equiv (j_{W - \lambda_k}(\partial^0 \varphi, \varphi))^0 \geq 0, \quad \forall \varphi. \tag{58}$$

These conditions can be satisfied in many cases. For example, it is sufficient to assume

$$T^{00}_W(\varphi) \geq 0, \quad T^{00}_{\lambda_k}(\varphi) = -\lambda_k \langle \varphi, \varphi \rangle \geq 0. \tag{59}$$

The other contributions in (55) are not bounded from below because they are linear in the variable

$$(W - \lambda_k)^{m_k - 1} \varphi_k. \tag{60}$$

Once this quantity is an initial data of the model, the conserved tensors (56) cannot have bounded 00-component for $l > 0$.

The observations above can be summarized as follows. If the primary theory is stable, the leading terms in the conserved tensor set (56) have bounded 00-components, while the other contributions are unbounded unless they vanish due to gauge symmetries or constraints. The bounded series of conserved currents in unconstrained theory has the form

$$(j^+)^\mu(\beta,\xi) = \sum_{k=1}^r \beta_k \xi_v (T^{k;0})^{\mu v}. \qquad (61)$$

In the case of multiple roots of the characteristic equation, this subseries is special because it involves only some initial data (60). Such conserved quantities cannot ensure the classical stability of the model. This brings us to the conclusion that non-singular theories of derived type with multiple roots should be unstable. This observation is supported by the models of the Pais–Uhlenbeck oscillator and the higher-derivative scalar field, where the models with resonance are unstable.

Let us now show that theories with multiple roots cannot be stable at the quantum level. This amounts to the fact that the bounded-conserved quantity cannot be connected with the time translation symmetry. The characteristic for the bounded-conserved tensor reads

$$N^+(\beta,\xi;W) = \sum_{k=1}^r L(\xi)\beta_k (W-\lambda_k)^{m_k-1}\Lambda_k(\lambda;W), \qquad (62)$$

where Λ's were introduced in (50). Given the Lagrange anchor (40), the corresponding symmetry is (42). This expression defines the space-time translation if

$$L(\xi)V(\gamma;W)N^+(\beta;W) = L(\xi) + \text{trvial symmetry}. \qquad (63)$$

Assuming that no gauge symmetries are present in the model, and $L(\xi)$ is a transitive operator (26), we conclude that

$$V(\gamma;W)N^+(\beta;W) - K(\delta;W)M(\alpha;W) = \text{id}, \qquad (64)$$

where K is some polynomial. Once all the powers of W are linearly independent, this equation is equivalent to the following relation between characteristic polynomials of conserved quantities:

$$V(\gamma;z)N^+(\beta;z) - K(\delta;z)M(\alpha;z) = 1. \qquad (65)$$

By the Bezout Lemma, this condition is consistent for fixed N^+ and M if, and only if, these polynomials have no common roots. On the other hand, each multiple root is common for M and N^+. Hence, the bounded-conserved quantity can only be connected with space-time translations if all the roots of the characteristic equation are simple. The Pais–Uhlenbeck oscillator and higher-derivative scalar field models again serve as demonstrations for this observation.

The case of gauge theories needs more accurate consideration. There are two important points: unbounded contributions in (55) can trivialize on the mass-shell, and the relation between symmetries and conserved quantities is more relaxed. This allows us to connect a bounded-conserved tensor with the time translation symmetry even if the characteristic polynomial has multiple roots. We illustrate this possibility in the extended Chern–Simons theory in Section 6.3.

6. Examples

6.1. Fourth-order Pais–Uhlenbeck Oscillator

The Pais–Uhlenbeck oscillator of the fourth-order is a theory of a single dynamical variable $x(t)$ with the action functional

$$S[x(t)] = \frac{1}{2}\int \left(x\left(\frac{d^2}{dt^2} + \omega_1^2\right)\left(\frac{d^2}{dt^2} + \omega_2^2\right)x \right)dt, \qquad (66)$$

where the frequencies ω_1, ω_2 are parameters of the model. The Euler–Lagrange equation for the action functional is of derived type,

$$\frac{\delta S}{\delta x} = \left(\frac{d^2}{dt^2} + \omega_1^2\right)\left(\frac{d^2}{dt^2} + \omega_2^2\right)x = 0, \tag{67}$$

with the primary operator being the second time derivative. The squares of frequencies determine the roots of the characteristic polynomial (36), which can be equal or different. The order of the characteristic polynomial equals two.

If the frequencies of the Pais–Uhlenbeck oscillator are different, two integrals of motion (56) are admissible by the model,

$$J^k = \frac{1}{2(\omega_1^2 + \omega_2^2)}\left((\ddot{x} + \omega_k^2 x)^2 + (\omega_1^2 + \omega_2^2 - \omega_k^2)(\dot{x} + \omega_k^2 x)^2\right), \quad k = 1, 2. \tag{68}$$

These conserved quantities are obviously bounded from below. The bounded subseries of conserved quantities (61) is the linear combination of these expressions,

$$J^+ = \sum_{k=1}^{2} \beta_k J^k, \quad \beta_k > 0. \tag{69}$$

The characteristic for the bounded-conserved quantity reads

$$N^+\left(\beta, \frac{d^2}{dt^2}\right) = \sum_{k=1}^{2} \frac{\beta_k}{\omega_1^2 + \omega_2^2}\left(\frac{d^2}{dt^2} + \omega_k^2\right)\frac{d}{dt}. \tag{70}$$

The Lagrange anchor,

$$V\left(\frac{d^2}{dt^2}\right) = \frac{(\omega_1^2 + \omega_2^2)}{(\omega_2^2 - \omega_1^2)^2}\left(\frac{\beta_1 + \beta_2}{\beta_1 \beta_2}\frac{d^2}{dt^2} + \frac{\beta_1 \omega_2^2 + \beta_2 \omega_1^2}{\beta_1 \beta_2}\right), \tag{71}$$

connects the conserved quantity (69) with the time translation symmetry whenever the product $\beta_1 \beta_2$ is nonzero. This gives an alternative proof of the stability of the Pais–Uhlenbeck oscillator with different frequencies.

In the case of resonance $\omega_1 = \omega_2 = \omega$, the Pais–Uhlenbeck oscillator has two conserved quantities, only one of which is bounded from below,

$$J^1 = \frac{1}{2\omega^2}\left((\ddot{x} + \omega^2 x)^2 + (\dot{x} + \omega^2 x)^2\right), \quad J^2 = \frac{1}{2\omega^2}(2\dot{x}\ddot{x} - \ddot{x}^2) + x^2 + \frac{1}{2}\omega^2 x^2. \tag{72}$$

The series of bounded-conserved quantities (61) includes a single entry J^1. The characteristic (62) for this conserved quantity reads

$$N^+\left(\frac{d^2}{dt^2}\right) = \frac{1}{\omega^2}\left(\frac{d^2}{dt^2} + \omega^2\right)\frac{d}{dt}. \tag{73}$$

It has the common root ω^2 with the characteristic polynomial of the wave operator (67). This means that the bounded integral of motion cannot be connected with the time translation symmetry in the model with resonance.

The results of this subsection show that the fourth-order Pais–Uhlenbeck oscillator is stable if its frequencies are different. This result confirms the general observation about the connection of structure of the roots of characteristic polynomial and its stability.

6.2. Higher-derivative Scalar Field

Consider the theory of real scalar field $\varphi(x)$ on d-dimensional Minkowski space with the action functional

$$S[\varphi(x)] = \frac{1}{2} \int \varphi \left(\prod_{p=1}^{n} \left(\frac{\partial^2}{\partial x^\mu \partial x_\mu} + \mu_p^2 \right) \right) \varphi d^d x. \tag{74}$$

In this formula, the real numbers μ_p determine the spectrum of masses in the theory. The equations of motion belong to derived type, with the primary operator being d'Alembertian,

$$\frac{\delta S}{\delta \varphi} = \prod_{p=1}^{n} \left(\frac{\partial^2}{\partial x^\mu \partial x_\mu} + \mu_p^2 \right) \varphi = 0, \quad W = \frac{\partial^2}{\partial x^\mu \partial x_\mu}. \tag{75}$$

The primary model is the theory of free mass-less scalar field,

$$\frac{\partial^2}{\partial x^\mu \partial x_\mu} \varphi = 0. \tag{76}$$

This theory is invariant under the Poincare symmetries, including the space-time translations.

The structure of conserved quantities in the model depends on the values of the roots of the characteristic polynomial. Once all roots are different, all the dynamical degrees of freedom are scalars with different masses. The lower order formulation (52) for the model reads

$$\left(\frac{\partial^2}{\partial x^\mu \partial x_\mu} + \mu_p^2 \right) \varphi_p = 0, \quad \varphi_p = \prod_{q=1, q \neq p}^{n} \left(\frac{\partial^2}{\partial x^\mu \partial x_\mu} + \mu_q^2 \right) \varphi, \quad p = 1, \ldots, n. \tag{77}$$

The conserved tensors (56) are just energies of the components,

$$(T^p)^{\mu\nu} = \partial^\mu \varphi_p \partial^\nu \varphi_p - \frac{1}{2} \eta^{\mu\nu} (\partial^\rho \varphi_p \partial_\rho \varphi_p - \mu_p^2 \varphi_p^2), \quad p = 1, \ldots, n. \tag{78}$$

It is clear that all of these quantities have bounded from below 00-component. The subseries of bounded-conserved quantities (61) have the form

$$(j^+)^\mu (\beta; \xi) = \sum_{p=1}^{n} \beta_p \xi_\nu (T^p)^{\mu\nu}, \quad \beta_p > 0. \tag{79}$$

By the general theorem above, all of these quantities can be connected with the space-time translations by the appropriate Lagrange anchor. We derive the explicit expression for such a Lagrange anchor in Appendix A (For the fourth-order theory (case $n = 2$) such a Lagrange anchor has been first introduced in [21]).

If multiple roots are admissible for the characteristic polynomial, two or more conserved quantities are related with one and the same root. Below, we give expressions for additional integrals of motion in the simplest option, where only one root has multiplicity two, and all the other roots are simple. Without loss of generality we assume that

$$\mu_{n-1} = \mu_n = \mu. \tag{80}$$

In this case, the system (52) reads

$$\left(\frac{\partial^2}{\partial x^\mu \partial x_\mu} + \mu_p^2 \right) \varphi_p = 0, \quad \varphi_p = \prod_{q=1, q \neq p}^{n} \left(\frac{\partial^2}{\partial x^\mu \partial x_\mu} + \mu_q^2 \right) \varphi, \quad p = 1, \ldots, n-2, \tag{81}$$

$$\left(\frac{\partial^2}{\partial x^\mu \partial x_\mu} + \mu^2\right)^2 \varphi_{n-1} = 0, \quad \varphi_{n-1} = \prod_{q=1}^{n-2}\left(\frac{\partial^2}{\partial x^\mu \partial x_\mu} + \mu_q^2\right)\varphi. \tag{82}$$

The components φ_p, $p = 1, \ldots, n-2$ are usual scalar fields, while φ_{n-1} obeys higher-derivative equations. For simple roots the conserved tensors have the form (78), we do not repeat the expressions for them. Two conserved tensors are associated with the multiple root,

$$\begin{aligned}(T^{n-1;0})^{\mu\nu} &= \tfrac{1}{\mu^2}\big(\partial^\mu \widetilde{\varphi}_{n-1}\partial^\nu \widetilde{\varphi}_{n-1} - \tfrac{1}{2}\eta^{\mu\nu}(\partial^\rho \widetilde{\varphi}_{n-1}\partial_\rho \widetilde{\varphi}_{n-1} - \mu^2 \widetilde{\varphi}_{n-1}^2)\big), \widetilde{\varphi}_{n-1} \equiv (\partial_\mu\partial^\mu + \mu^2)\varphi_{n-1};\\ (T^{n-1;1})^{\mu\nu} &= \tfrac{1}{\mu^2}\big(\partial^\mu \varphi_{n-1}\partial^\nu\partial^\rho \varphi_{n-1} - \partial^\mu \partial_\rho \varphi_{n-1}\partial^\nu\partial^\rho \varphi_{n-1} + 2\mu^2 \partial^\mu \varphi_{n-1}\partial^\nu \varphi_{n-1} \\ &\quad -\tfrac{1}{2}\eta^{\mu\nu}(-\partial^\rho \partial^\tau \varphi_{n-1}\partial_\rho \partial_\tau \varphi_{n-1} + \mu^2 \partial^\rho \varphi_{n-1}\partial_\rho \varphi_{n-1} - \mu^4 \varphi_{n-1}^2)\big).\end{aligned} \tag{83}$$

The bounded-conserved quantity reads

$$(j^+)^\mu(\beta;\xi) = \sum_{p=1}^{n-1} \beta_p \xi_\nu (T^{p;0})^{\mu\nu}, \quad \beta_p > 0. \tag{84}$$

The characteristic for the bounded-conserved tensor has the form

$$N^+\left(\beta, \xi; \frac{\partial^2}{\partial x^\mu \partial x_\mu}\right) = \left(\frac{\partial^2}{\partial x^\mu \partial x_\mu} + \mu^2\right)\left(\sum_{p=1}^{n-1}\beta_p \xi_\mu \partial^\mu \prod_{q=1, q\neq p}^{n-1}\left(\frac{\partial^2}{\partial x^\mu \partial x_\mu} + \mu_q^2\right)\right). \tag{85}$$

Its characteristic polynomial,

$$N^+(\beta; z) = (z + \mu^2)\sum_{p=1}^{n-1}\left(\beta_p \prod_{q=1, q\neq p}^{n-1}(z + \mu_q^2)\right), \tag{86}$$

has the common root $-\mu^2$ with the characteristic polynomial of the theory. In doing so, the bounded-conserved quantity cannot be connected with the time translations. This demonstrates that the free higher-derivative scalar field theory with different masses is stable, while the multiple roots indicate the instability of the model.

6.3. Extended Chern–Simons Model

Consider the theory of the vector field $A = A_\mu(x)dx^\mu$ on 3d Minkowski space with the action functional

$$S[A(x)] = \frac{1}{2}\int\left(A, \sum_{p=1}^{n}\alpha_p(*d)^p A\right)d^3x, \tag{87}$$

where the round brackets denote the standard inner product of differential forms, $*$ is the Hodge dual, and d is the de-Rham differential. The parameters of the model are the constants $\alpha_1, \ldots, \alpha_n$. The action of the Chern–Simons operator on the vector field is determined by the relation

$$(*dA)_\mu dx^\mu = \varepsilon_{\mu\nu\rho}dx^\mu \partial^\nu A^\rho. \tag{88}$$

Here, ε is the 3d Levi-Civita symbol, with $\varepsilon_{012} = 1$. The Euler–Lagrange equations for the model have the form

$$\frac{\delta S}{\delta A} = \sum_{p=1}^{n}\alpha_p(*d)^p A = 0. \tag{89}$$

The primary operator of the theory is the Chern–Simons one, see Equation (88). The primary theory for the model is the usual abelian Chern–Simons theory.

The primary operator (88) is Poincare-invariant, so that the space-time translations are symmetries of the model. The series of derived symmetries reads

$$N(\beta,\xi;*d) = \sum_{p=0}^{n-1} \xi^\mu \beta_p(*d)^p \partial_\mu. \tag{90}$$

The corresponding set of conserved tensors has the form

$$T^{\mu\nu}(\beta) = \frac{1}{2}\sum_{p,q=0}^{n-1} C_{p,q}(\alpha,\beta)(F^{(p)\mu}F^{(q)\nu} + F^{(p)\mu}F^{(q)\nu} - \eta^{\mu\nu}\eta_{\rho\sigma}F^{(p)\rho}F^{(q)\sigma}), \tag{91}$$

all of the space-time indices are raised and lowered by the Minkowski metric, and the following notation is used:

$$F^{(p)} = (*d)^p A, \quad p = 0,\ldots,n-1. \tag{92}$$

The quantity $C(\alpha,\beta)$ (91) is the Bezout matrix of the characteristic polynomial of the model and the characteristic polynomial of the symmetry. It is defined by the generating relation,

$$C_{p,q}(\alpha,\beta) = \frac{\partial^{p+q}}{\partial^p z \partial^q u}\left(\frac{M(z)N(u)-M(u)N(z)}{z-u}\right)\bigg|_{z=u=0},$$
$$M(z) \equiv \sum_{p=1}^{n}\alpha_p z^p, \quad N(z) \equiv \sum_{p=0}^{n-1}\beta_p z^{p+1}. \tag{93}$$

where z and u are two independent variables. The individual conserved tensors in the set (91) can be found by the following receipt:

$$(T^p)^{\mu\nu} = \frac{\partial T^{\mu\nu}(\beta)}{\partial \beta_p}, \quad p = 0,\ldots,n-1. \tag{94}$$

In this set, the quantities $(T^p)^{\mu\nu}$, $p = 0,\ldots,n-2$, are independent, while

$$(T^{n-1})^{\mu\nu} = -\sum_{p=0}^{n-2}\frac{\alpha_p}{\alpha_n}(T^p)^{\mu\nu}. \tag{95}$$

We mention this fact to illustrate possible dependence among conserved quantities in the class of gauge models. We also notice that the set of the conserved currents of the extended Chern–Simons model (87) was introduced in the work [25], while the compact form (91)–(94) is proposed in [34].

The structure of the conserved currents can be studied along the lines of the previous section. The components (50) are introduced by the standard rule,

$$A_k = \Lambda_k(\partial)A, \quad \Lambda_k(\partial) = \alpha_n \prod_{i=1,i\neq k}^{r}(*d-\lambda_i)^{m_i}, \quad k = 1,\ldots,r. \tag{96}$$

The corresponding conserved quantities (55) are determined by Formulas (91)–(94) with

$$M = (z-\lambda_k)^{m_k}, \quad Q = (z-\lambda_k)^l, \quad k = 1,\ldots,r, \quad l = 0,\ldots,m_k-1. \tag{97}$$

These conserved currents can be found for each particular value of the roots. Without loss of generality we assume that the zero root corresponds to $k = 1$, while all the other numbers λ_k are non-zero.

As for the structure of the conserved current, we mention that the contribution with the highest derivative has the form

$$(T^{k;l})^{\mu\nu} = \tfrac{1}{2}(-\lambda)^{m_k}(F_k^{(m_k-1)\mu}F_k^{(l)\nu} + F_k^{(m_k-1)\nu}F_k^{(l)\mu} - \eta^{\mu\nu}\eta_{\rho\sigma}F_k^{(m_k-1)\rho}F_k^{(l)\sigma}) + \ldots, \qquad (98)$$
$$F_k^{(l)} = (*d - \lambda_k)^l \Lambda_k A, \quad k=1,\ldots,r, \quad l=0,\ldots,m_k-1.$$

The dots denote all the contributions with lower derivatives. These conserved quantities cannot be bounded unless $l = m_k - 1$. For $l = m_k - 1$, expressions (98) are exact with no dotted terms, and the corresponding conserved quantities are bounded.

As is seen from Formula (98), the leading representatives $T^{k;0}$ in the conserved-quantity series are bounded. For simple zero roots the corresponding conserved quantity is the Chern–Simons energy, which is trivial. The additional conserved currents are associated with multiple roots. In the case of non-zero root, all these conserved quantities are independent and unbounded. In the case of zero roots, the additional quantity has a bounded 00-component (It is the canonical energy of 3d electrodynamics, which is known to be bounded), while all the other independent entries are independent and unbounded. This means that all of the additional conserved quantities in the set (56) are unbounded if the multiple zero root has a multiplicity greater than two. The subseries (61) of conserved quantities with the bounded 00-component has the form

$$(j^+)^\mu(\beta;\xi) = \sum_{k=2}^{r} \beta_{k;0}\xi_\nu(T^{k;0})^{\mu\nu} + \beta_{1;1}\xi_\nu(T^{1;1})^{\mu\nu}. \qquad (99)$$

The characteristic of the bounded-conserved tensor series reads

$$N^+(\beta;\xi) = -\sum_{k=2}^{r} sgn(\lambda_k)\beta_{k;0}\Lambda_k(*d)L(\xi) + \beta_{1;1}(*d)\Lambda_1(*d)L(\xi). \qquad (100)$$

This is not the general representative of a characteristic series (90), because it involves fewer entries.

Let us discuss the stability of the model. The series (40) of the Lagrange anchors for the extended Chern–Simons model has the following form:

$$V(\gamma;*d) = \sum_{p=0}^{n-1} \gamma_p (*d)^p, \qquad (101)$$

with $\gamma_0, \ldots, \gamma_{n-1}$ being real numbers. All the entries of the series are non-trivial. The Lagrange anchor (101) takes the characteristic (100) into a symmetry by the rule (27). This symmetry is the space-time translation if the condition (42) is satisfied. The general trivial symmetry in the considered case reads

$$L_{triv}(\xi;*d) = L_\xi K(\delta;*d)M^-(\alpha;*d), \quad M^-(\alpha;*d) = \sum_{p=1}^{n} \alpha_p(*d)^{p-1}, \qquad (102)$$

where K is an arbitrary polynomial in $*d$. To ensure trivialization of this symmetry, one can use the Cartan formula for the Lie derivative,

$$L_\xi K(\delta;*d)M^-(\delta;*d) = (i_\xi d + di_\xi)K(\delta;*d)M^-(\delta;*d) = \qquad (103)$$
$$= i_\xi * K(\delta;*d)M(\delta;*d) + di_\xi K(\delta;*d)M^-(\delta;*d).$$

Substituting (100), (101), and (102) into (47), we get the equation

$$V(\gamma;*d)N^+(\beta;*d) - K(\delta;*d)M^-(\alpha;*d) = id, \qquad (104)$$

where the constants γ, δ are unknown. In terms of characteristic polynomials of the Lagrange anchor and symmetry, the following equation is relevant:

$$V(\gamma;z)N^+(\beta;z) - K(\delta;z)M^-(\alpha;z) = 1. \tag{105}$$

It is consistent if N^+ and M^- have no common roots. The following restriction on the multiplicity of roots is implied:

$$m_1 = 1,2, \quad m_k = 0,1, \quad k = 2,\ldots,r. \tag{106}$$

In this case, all the non-zero roots should be simple and real, and the zero root should have a multiplicity of one or two. This requirement is less restrictive than the absence of a common root between N^+ and the characteristic polynomial of the derived theory, which has the place in case of non-gauge systems.

7. Conclusions

In this article, we have studied the stability of the class of relativistic higher-derivative theories of derived type from the viewpoint of a more general correspondence between symmetries and conserved quantities, which is established by the Lagrange anchor. Assuming that the wave operator of the linear model is the n-th-order polynomial in the lower-order operator, we have obtained the following results. First, we observed that n-parameter series of second-rank conserved tensors and Lagrange anchors are admissible by the derived model. The canonical energy, which is unbounded, is included into the series in all of the instances. The other integral of motion can be bounded or unbounded depending on the structure of the roots of the characteristic polynomial. The general conserved tensor in the series can be connected with the space-time symmetries by an appropriate Lagrange anchor. Second, we studied the stability of higher-derivative dynamics from the viewpoint of correspondence between the time translation symmetry and bounded-conserved quantities. It has been observed that this relationship can be established if all of the roots of the characteristic polynomial are real and simple. For multiple roots, bounded-conserved quantities are admissible, but they are unrelated with the time translation.

Our stability analysis is applicable to free models. The real issue is the stability of higher-derivative dynamics at the non-linear level. This subject has been studied in many works and we cite recent papers [3–5,35,36] and references therein. The method of proper deformation [37] has been proposed to systematically deform the equations of motion and conserved quantities. Once this method uses the Lagrange anchor construction it can be used to deform bounded-conserved tensors, which are found in this paper. In doing so, the stability of linear higher-derivative models, which are studied in the present paper, can be extended at the non-linear level.

Funding: This research was funded by the state task of Ministry of Science and Higher Education of Russian Federation, grant number 3.9594.2017/8.9.

Acknowledgments: I thank S.L. Lyakhovich, A.A. Sharapov and V.A. Abakumova for valuable discussions on the subject of this research. I benefited from the valuable comments of two anonymous referees that helped me to improve the initial version of the manuscript. I also thank my family who supported me at all stages of this work.

Conflicts of Interest: The funders had no role in the design of the study, in the collection, analyses, or interpretation of data, in the writing of the manuscript, or in the decision to publish the results.

Appendix A

In this appendix, we demonstrate that the general bounded-conserved tensor (79) in the theory of the higher-derivative scalar field with different masses (74) can be connected with the space-time translations by the appropriate Lagrange anchor. To solve the problem, we find the characteristic polynomials $V(\gamma,z)$ and $K(\delta,z)$ that satisfy condition (45). In this case, the Lagrange anchor is defined by the Formula (40), with W being the d'Alembertian.

At first, we identify all the ingredients in (45), including the characteristic polynomials of the wave operator and characteristic,

$$M(\mu;z) = \prod_{p=1}^{n}(z+\mu_p^2), \quad N^+(\beta;z) = \sum_{p=1}^{n}\beta_p\Lambda_p(z), \quad V(\gamma;z) = \sum_{p=1}^{n}\gamma_p\Lambda_p(z),$$
$$K(\delta;z) = \prod_{p=0}^{n-2}\delta_p z^p, \quad \Lambda_p(z) \equiv \prod_{q=1, q\neq p}^{n}(z+\mu_q^2). \tag{A1}$$

Here, the expressions $M(\mu;z)$ and $N(\beta;z)$ are fixed by the model parameters and the choice of the selected representative in the conserved tensor series, while γ, δ are unknown constants. The quantities $\Lambda_p(z)$ denote characteristic polynomials of operators (50). By construction, $\Lambda_p(z)$ form the basis in the space of polynomials of order $n-1$ in the variable z. In this setting, the chosen ansatz for $V(\gamma;z)$ is a different parameterization of Lagrange anchor series (40).

The defining Equation (45), (A1) for the characteristic polynomial $V(\gamma;z)$ has the form

$$\sum_{p=1}^{n}\sum_{q=1}^{n}\beta_p\gamma_q\Lambda_p(z)\Lambda_q(z) - K(\delta;z)M(\mu;z) = 1. \tag{A2}$$

From here, the polynomial $K(\delta,z)$ is immediately found,

$$K(\delta;z) = \sum_{p=1}^{n}\sum_{q=1}^{n}\beta_p\gamma_q\Lambda_{pq}(z), \quad \Lambda_{pq}(z) = \prod_{r\neq p,q}(z+\mu_r^2). \tag{A3}$$

As $K(\delta,z)$ is the fraction of $V(\gamma;z)\cdot N(\beta,z)$ and $M(\mu;z)$, no restrictions on the parameters γ in the Lagrange anchor appear at this step. After that we estimate left- and right-hand sides of the relation (A2) for

$$z = -\mu_p^2, \quad p = 1,\ldots,n, \tag{A4}$$

being the roots of characteristic polynomial. We arrive to the following system of equations:

$$\beta_p\gamma_p\Lambda_p^2(-\mu_p^2) = 1, \quad p = 1,\ldots,n. \tag{A5}$$

From here, the parameters γ are determined,

$$\gamma_p = \frac{1}{\beta_p\Lambda_p^2(-\mu_p^2)}. \tag{A6}$$

This solution is well-defined since $\beta_p > 0$, and $\Lambda_p(-\mu_p^2)$'s are non-zero. Moreover, we observe that

$$\gamma_p > 0. \tag{A7}$$

The Lagrange anchor, being defined by the characteristic polynomial $V(\gamma;z)$ (A1), has the form

$$V(\gamma;z) = \sum_{p=1}^{n}\frac{1}{\beta_p\Lambda_p^2(-\mu_p^2)}\Lambda_p\left(\frac{\partial^2}{\partial x^\mu \partial x_\mu}\right). \tag{A8}$$

This Lagrange anchor is non-canonical because the coefficient at the highest power of the primary operator is positive.

References

1. Kosmann-Schwarzbach, Y. *The Noether theorems: Invariance and conservation laws in the twentieth century*; Springer: New York, NY, USA, 2011; pp. 1–199.
2. Woodard, R.P. The theorem of Ostrogradski. *Scholarpedia* **2015**, *10*, 32243. [CrossRef]
3. Tomboulis, E.T. Renormalization and unitarity in higher-derivative and nonlocal gravity theories. *Mod. Phys. Lett. A* **2015**, *30*, 1540005. [CrossRef]
4. Pavsic, M. Pais–Uhlenbeck oscillator and negative energies. *Int. J. Geom. Meth. Mod. Phys.* **2016**, *13*, 1630015. [CrossRef]
5. Smilga, A.V. Classical and Quantum Dynamics of Higher-Derivative Systems. *Int. J. Mod. Phys. A* **2017**, *32*, 1730025. [CrossRef]
6. Bender, C.M.; Mannheim, P.D. No-ghost theorem for the fourth-order derivative Pais–Uhlenbeck oscillator model. *Phys. Rev. Lett.* **2008**, *100*, 110402. [CrossRef]
7. Bender, C.M. Giving up the ghost. *J. Phys. A Math. Theor.* **2008**, *41*, 304018. [CrossRef]
8. Mostafazadeh, A. A Hamiltonian formulation of the Pais–Uhlenbeck oscillator that yields a stable and unitary quantum system. *Phys. Lett. A* **2010**, *375*, 93–98. [CrossRef]
9. Bolonek-Lason, K.; Kosinski, P. Hamiltonian structures for Pais–Uhlenbeck oscillator. *Acta Phys. Polon. B* **2005**, *36*, 2115–2131.
10. Damaskinsky, E.V.; Sokolov, M.A. Remarks on quantization of Pais–Uhlenbeck oscillators. *J. Phys. A Math. Gen.* **2006**, *39*, 10499. [CrossRef]
11. Andrzejewski, K.; Bolonek-Lason, K.; Gonera, J.; Maslanka, P. Canonical formalism and quantization of perturbative sector of higher-derivative theories. *Phys. Rev. A* **2007**, *76*, 032110. [CrossRef]
12. Masterov, I. An alternative Hamiltonian formulation for the Pais–Uhlenbeck oscillator. *Nucl. Phys. B* **2016**, *902*, 95–114. [CrossRef]
13. Boulanger, N.; Buisseret, F.; Dierick, F.; White, O. Higher-derivative harmonic oscillators: Stability of classical dynamics and adiabatic invariants. *Eur. Phys. J. C* **2019**, *79*, 60. [CrossRef]
14. Maldacena, J. Einstein Gravity from Conformal Gravity. Available online: https://arxiv.org/abs/1105.5632 (accessed on 6 May 2019).
15. Bergshoeff, E.A.; Kovacevic, M.; Rosseel, J.; Townsend, P.K.; Yin, Y. A spin-4 analog of 3D massive gravity. *Class. Quant. Grav.* **2011**, *28*, 245007. [CrossRef]
16. Chen, T.; Fasiello, M.; Lim, E.A.; Tolley, A.J. Higher-derivative theories with constraints: exorcising Ostrogradskis ghost. *J. Cosmol. Astropart. Phys.* **2013**, *1302*, 42. [CrossRef]
17. Nitta, M.; Yokokura, R. Topological couplings in higher-derivative extensions of supersymmetric three-form gauge theories. Available online: https://arxiv.org/abs/1810.12678 (accessed on 6 May 2019).
18. Strominger, A. Positive energy theorem for $R+R^2$ gravity. *Phys. Rev. D* **1984**, *30*, 2257–2259. [CrossRef]
19. Sotiriou, T.P.; Faraoni, V. f(R) theories of gravity. *Rev. Mod. Phys.* **2010**, *82*, 451–497. [CrossRef]
20. De Felice, A.; Tsujikawa, S. f(R) theories. *Living Rev. Rel.* **2010**, *13*, 3. [CrossRef]
21. Kaparulin, D.S.; Lyakhovich, S.L.; Sharapov, A.A. Classical and quantum stability of higher-derivative dynamics. *Eur. Phys. J. C* **2014**, *74*, 3072. [CrossRef]
22. Kazinski, P.O.; Lyakhovich, S.L.; Sharapov, A.A. Lagrange structure and quantization. *J. High Energy Phys.* **2005**, *507*, 76. [CrossRef]
23. Kaparulin, D.S.; Lyakhovich, S.L.; Sharapov, A.A. Rigid symmetries and conservation laws in non-Lagrangian field theory. *J. Math. Phys.* **2010**, *51*, 082902. [CrossRef]
24. Kaparulin, D.S.; Lyakhovich, S.L.; Sharapov, A.A. BRST analysis of general mechanical systems. *J. Geom. Phys.* **2013**, *74*, 164–184. [CrossRef]
25. Kaparulin, D.S.; Karataeva, I.Y.; Lyakhovich, S.L. Higher-derivative extensions of *3d* Chern–Simons models: conservation laws and stability. *Eur. Phys. J. C* **2015**, *75*, 552. [CrossRef]
26. Pais, A.; Uhlenbeck, G.E. On field theories with non-localized action. *Phys. Rev.* **1950**, *79*, 145–165. [CrossRef]
27. Podolsky, B.; Schwed, P. Review of a generalized electrodynamics. *Rev. Mod. Phys.* **1948**, *20*, 40–50. [CrossRef]
28. Lu, H.; Pang, Y.; Pope, C.N. Conformal Gravity and Extensions of Critical Gravity. *Phys. Rev. D* **2011**, *84*, 064001. [CrossRef]
29. Lu, H.; Pang, Y.; Pope, C.N. Black holes in six-dimensional conformal gravity. *Phys. Rev. D* **2013**, *87*, 104013. [CrossRef]

30. Deser, S.; Jackiw, R. Higher-derivative Chern–Simons extensions. *Phys. Lett. B* **1999**, *451*, 73–76. [CrossRef]
31. Fushchich, W.I.; Nikitin, A.G. *Symmetries of Equations of Quantum Mechanics*; Allerton Press Inc.: New York, NY, USA, 1994; pp. 1–480.
32. Lyakhovich, S.L.; Sharapov, A.A. Quantization of Donaldson-Yhlenbeck-Yau theory. *Phys. Lett B* **2007**, *656*, 265–271. [CrossRef]
33. Kaparulin, D.S.; Lyakhovich, S.L.; Sharapov, A.A. Lagrange Anchor and Characteristic Symmetries of Free Massless Fields. *Symmetry Integr. Geom. Methods Appl.* **2012**, *8*, 1–18. [CrossRef]
34. Abakumova, V.A.; Kaparulin, D.S.; Lyakhovich, S.L. Stable interactions in higher-derivative theories of derived type. *Phys. Rev. D* **2019**, *99*, 045020. [CrossRef]
35. Smilga, A.V. Benign vs malicious ghosts in higher-derivative theories. *Nucl. Phys. B* **2005**, *706*, 598–614. [CrossRef]
36. Avendao-Camacho, M.; Vallejo, J.A.; Vorobiev, Y. A perturbation theory approach to the stability of the Pais–Uhlenbeck oscillator. *J. Math. Phys.* **2017**, *58*, 093501. [CrossRef]
37. Kaparulin, D.S.; Lyakhovich, S.L.; Sharapov, A.A. Stable interactions via proper deformations. *J. Phys. A Math. Theor.* **2016**, *49*, 155204. [CrossRef]

© 2019 by the author. Licensee MDPI, Basel, Switzerland. This article is an open access article distributed under the terms and conditions of the Creative Commons Attribution (CC BY) license (http://creativecommons.org/licenses/by/4.0/).

Article

Symmetries and Reductions of Integrable Nonlocal Partial Differential Equations

Linyu Peng

Waseda Institute for Advanced Study, Waseda University, Tokyo 169-8050, Japan; l.peng@aoni.waseda.jp

Received: 8 June 2019; Accepted: 2 July 2019; Published: 5 July 2019

Abstract: In this paper, symmetry analysis is extended to study nonlocal differential equations. In particular, two integrable nonlocal equations are investigated, the nonlocal nonlinear Schrödinger equation and the nonlocal modified Korteweg–de Vries equation. Based on general theory, Lie point symmetries are obtained and used to reduce these equations to nonlocal and local ordinary differential equations, separately; namely, one symmetry may allow reductions to both nonlocal and local equations, depending on how the invariant variables are chosen. For the nonlocal modified Korteweg–de Vries equation, analogously to the local situation, all reduced local equations are integrable. We also define complex transformations to connect nonlocal differential equations and differential-difference equations.

Keywords: continuous symmetry; symmetry reduction; integrable nonlocal partial differential equations

1. Introduction

Symmetry has proved to be fundamentally important in understanding the solutions of differential equations (see, e.g., [1–5]). It also reveals the integrability of partial differential equations (PDEs); for instance, the Ablowitz–Ramani–Segur conjecture stated that every ordinary differential equation (ODE) obtained by an exact reduction of an integrable evolution equation solvable by inverse scattering transforms is of the P-type, i.e., ODEs without movable critical points [6]. In this paper, powerful symmetry techniques are extended to study nonlocal differential equations with space and/or time reflections. This can not only provide insights for obtaining analytic solutions, but also reveal the integrability of the nonlocal equations. After writing down a general theory in Section 2, two integrable nonlocal differential equations—the nonlocal nonlinear Schrödinger (NLS) equation [7] and the nonlocal modified Korteweg–de Vries (mKdV) equation [8]—are separately investigated as illustrative examples. The results are immediately applicable to the many nonlocal differential equations proposed in the recent literature (see, e.g., [9–13]).

The nonlocal NLS equation

$$i q_t(x,t) + q_{xx}(x,t) + 2q^2(x,t)q^*(-x,t) = 0, \qquad (1)$$

was derived by Ablowitz and Musslimani [7] by reduction of the AKNS system. The nonlocal NLS equation admits a great number of good properties that the classical NLS equation possesses, such as being PT-symmetric, admitting a Lax-pair and infinitely many conservation laws, and being solvable using inverse scattering transforms. Integrable nonlocal systems have recently received a great amount of attention with many newly-proposed models (e.g., the nonlocal vector NLS equation [13], a multi-dimensional extension of the nonlocal NLS equation [10], the nonlocal sine-Gordon equation, the nonlinear derivative NLS equation and related systems [9], the nonlocal mKdV equation [8], Alice–Bob physics [11], and the nonlocal Sasa–Satsuma equation [12], to mention only a few). Solutions of these systems have also been explored by many scholars; see, for example, [8,12,14–19].

One issue, as Ablowitz and Musslimani have noticed [7,9], is that reductions of nonlocal equations amount to nonlocal ODEs; for example, nonlocal Painlevé-type equations. In this paper, we show alternative ways which allow us to avoid such an inconvenience. We, first, classify all Lie point symmetries of the nonlocal NLS Equation (1) and the nonlocal mKdV equation [8]

$$u_t(x,t) + u(x,t)u(-x,-t)u_x(x,t) + u_{xxx}(x,t) = 0. \tag{2}$$

Then, possible symmetry reductions are conducted for both equations. We find that one may reduce a nonlocal differential equation to both nonlocal and local ODEs by choosing the invariant variables in different ways. In other words, we are able to kill all nonlocal terms in the reduced ODEs by choosing the invariant variables in a proper manner. In particular, for the nonlocal mKdV equation, all reduced local ODEs are integrable. These results are included in Sections 3 and 4. In Section 5, simple transformations are defined to connect nonlocal differential equations with differential-difference equations (DDEs).

2. The Linearized Symmetry Condition for Nonlocal Differential Equations

We, first, introduce the multi-index notations needed for the symmetry techniques of local differential equations (see, e.g., [5]), which will be extended to nonlocal differential equations.

Let $x = (x^1, x^2, \ldots, x^m) \in \mathbb{R}^m$ be the independent variables and let $u = (u^1, u^2, \ldots, u^n) \in \mathbb{R}^n$ be the dependent variables. Note that, in many occasions, people also tend to use (x,t) to denote independent variables as the space x and time t; this convention will occur in the next sections but, for now, we are happy without distinguishing one another. Partial derivatives of u^α are written in the multi-index form u_J^α, where $J = (j_1, j_2, \ldots, j_m)$ with each index j_i a non-negative integer, denoting the number of derivatives with respect to x^i; namely,

$$u_J^\alpha = \frac{\partial^{|J|} u^\alpha}{\partial (x^1)^{j_1} \partial (x^2)^{j_2} \ldots \partial (x^m)^{j_m}}, \tag{3}$$

where $|J| = j_1 + j_2 + \cdots + j_m$. Consider a one-parameter group of Lie point transformations as follows:

$$\tilde{x} = \tilde{x}(\varepsilon; x, u), \quad \tilde{u} = \tilde{u}(\varepsilon; x, u), \tag{4}$$

subject to $\tilde{x}|_{\varepsilon=e} = x$, $\tilde{u}|_{\varepsilon=e} = u$. Here, $\varepsilon = e$ is the identity element of the one-parameter group. Define the total derivative with respect to x^i as

$$D_i = \frac{\partial}{\partial x^i} + \sum_{\alpha, J} u_{J+1_i}^\alpha \frac{\partial}{\partial u_J^\alpha}, \tag{5}$$

where 1_i is the m-tuple with only one non-zero entry 1 in the i-th position. The notation $\partial_{x^i} = \frac{\partial}{\partial x^i}$ and so forth will also be used. The corresponding infinitesimal generator is

$$\mathbf{v} = \xi^i(x,u)\partial_{x^i} + \phi^\alpha(x,u)\partial_{u^\alpha}, \tag{6}$$

where

$$\xi^i = \frac{d\tilde{x}^i}{d\varepsilon}\bigg|_{\varepsilon=e}, \quad \phi^\alpha = \frac{d\tilde{u}^\alpha}{d\varepsilon}\bigg|_{\varepsilon=e}. \tag{7}$$

Note that the Einstein summation convention is used here (and elsewhere, if necessary). The prolonged generator pr $\mathbf{v}$ can be written in terms of u, ξ, ϕ, and their derivatives, as:

$$\mathrm{pr}\,\mathbf{v} = \mathbf{v} + \sum_{\alpha, |J| \geq 1} \phi_J^\alpha(x, [u]) \partial_{u_J^\alpha}, \tag{8}$$

where $[u]$ is shorthand for u and finitely many of its partial derivatives, and the coefficients are recursively given by

$$\phi^\alpha_{J+1_i}(x,[u]) = D_i \phi^\alpha_J(x,[u]) - (D_i \xi^j(x,u)) u^\alpha_{J+1_j}. \tag{9}$$

It is often more convenient to equivalently write prolonged generators in terms of the so-called characteristics of symmetries $Q^\alpha = \phi^\alpha - \xi^i u^\alpha_{1_i}$; that is

$$\operatorname{pr} \mathbf{v} = \xi^i D_i + \sum_{\alpha,J} (D_J Q^\alpha) \partial_{u^\alpha_J}. \tag{10}$$

Here, we use the shorthand notation $D_J = D_1^{j_1} D_2^{j_2} \cdots D_m^{j_m}$ for $J = (j_1, j_2, \ldots, j_m)$. The invariance of a system of local differential equations

$$\{F_k(x,[u]) = 0\}_{k=1}^l, \tag{11}$$

corresponding to the transformations (4), leads to the linearized symmetry condition

$$\operatorname{pr} \mathbf{v}(F_k(x,[u])) = 0, \text{ whenever } \{F_k(x,[u]) = 0\}_{k=1}^l, \tag{12}$$

where $\mathbf{v}$ is the infinitesimal generator (6).

To extend the above analysis to the nonlocal equations of our interest, we define the following reflections for $i = 1, 2, \ldots, m$,

$$\operatorname{Ref}^i : (x^1, \ldots, x^i, \ldots, x^m) \mapsto (x^1, \ldots, -x^i, \ldots, x^m) \tag{13}$$

and

$$\operatorname{Ref}^i : f(x^1, \ldots, x^i, \ldots, x^m) \mapsto f(x^1, \ldots, -x^i, \ldots, x^m), \tag{14}$$

for a function f defined on proper domains. In particular,

$$\operatorname{Ref}^i u^\alpha(x^1, \ldots, x^i, \ldots, x^m) = u^\alpha(x^1, \ldots, -x^i, \ldots, x^m), \quad \alpha = 1, 2, \ldots, n. \tag{15}$$

A system of nonlocal differential equations is, then, given by

$$\mathcal{A} = \left\{ F_k \left(x, [u], [\operatorname{Ref}^i u], [\operatorname{Ref}^i \circ \operatorname{Ref}^j u]_{i<j}, \ldots, [u(-x)] \right) = 0 \right\}_{k=1}^l. \tag{16}$$

For simplicity, we will sometimes omit the arguments if they are local variables x. Let us consider transformations of the form (4) with the infinitesimal generator (6). Now, the prolongation formula involving the reflections becomes

$$\operatorname{pr}_{\operatorname{Ref}} \mathbf{v} = \mathbf{v} + \sum_{\alpha, |J| \geq 1} \phi^\alpha_J \frac{\partial}{\partial u^\alpha_J}$$

$$+ \sum_{i, \alpha, |J| \geq 1} \left(\operatorname{Ref}^i \phi^\alpha_J \right) \frac{\partial}{\partial \left(\operatorname{Ref}^i u^\alpha_J \right)} + \sum_{i<j, \alpha, |J| \geq 1} \left(\operatorname{Ref}^i \circ \operatorname{Ref}^j \phi^\alpha_J \right) \frac{\partial}{\partial \left(\operatorname{Ref}^i \circ \operatorname{Ref}^j u^\alpha_J \right)} \tag{17}$$

$$+ \cdots + \sum_{\alpha, |J| \geq 1} \phi^\alpha_J(-x, [u(-x)]) \frac{\partial}{\partial u^\alpha_J(-x)},$$

where the functions $\phi^\alpha_J = \phi^\alpha_J(x,[u])$ are again defined by (9). Invariance of the nonlocal system (16) with respect to the transformations (4) is equivalent to the linearized symmetry condition that

$$\operatorname{pr}_{\operatorname{Ref}} \mathbf{v} \left(F_k \left(x, [u], [\operatorname{Ref}^i u], [\operatorname{Ref}^i \circ \operatorname{Ref}^j u]_{i<j}, \ldots, [u(-x)] \right) \right) = 0, \text{ whenever } \mathcal{A} \text{ holds}, \tag{18}$$

which are the first order terms about ε in the Taylor expansions of the nonlocal system (16) evaluated at the new variables $\tilde{x}, \tilde{u}$, and so forth.

In the next two sections, we will apply this general theory to two integrable nonlocal differential equations: The nonlocal NLS equation and the nonlocal mKdV equation.

3. The Nonlocal NLS Equation

An integrable nonlocal NLS equation was proposed by Ablowitz and Musslimani [7]:

$$i\,q_t(x,t) + q_{xx}(x,t) + 2q^2(x,t)q^*(-x,t) = 0, \tag{19}$$

where $*$ denotes the complex conjugate and $q(x,t)$ is a complex-valued function of real variables x and t. They showed that it possesses a Lax pair and infinitely many conservation laws, and is solvable by the inverse scattering transform. We study its continuous symmetries in this section.

3.1. Lie Point Symmetries

As $q(x,t)$ is complex-valued, two alternative approaches may be used to calculate its continuous symmetries. Under the co-ordinate $(x,t,q(x,t),q^*(x,t))$, we consider the following local transformations

$$\begin{aligned}
x &\mapsto x + \varepsilon \xi\,(x,t,q(x,t),q^*(x,t)) + O(\varepsilon^2),\\
t &\mapsto t + \varepsilon \tau\,(x,t,q(x,t),q^*(x,t)) + O(\varepsilon^2),\\
q(x,t) &\mapsto q(x,t) + \varepsilon \phi\,(x,t,q(x,t),q^*(x,t)) + O(\varepsilon^2).
\end{aligned} \tag{20}$$

Again, we omit the arguments if they are local variables (x,t). The corresponding infinitesimal generator is

$$\mathbf{v} = \xi\,(x,t,q,q^*)\,\partial_x + \tau\,(x,t,q,q^*)\,\partial_t + \phi\,(x,t,q,q^*)\,\partial_q. \tag{21}$$

From Section 2, we know that the prolongation formula for an infinitesimal generator $\xi(x,t,u)\partial_x + \tau(x,t,u)\partial_t + \phi(x,t,u)\partial_u$ of local differential equations is (see, also, [5,20])

$$\xi D_x + \tau D_t + Q\partial_u + (D_x Q)\partial_{u_x} + (D_t Q)\partial_{u_t} + \cdots + \left(D_x^k D_t^l Q\right)\partial_{\left(D_x^k D_t^l u\right)} + \cdots, \tag{22}$$

where the characteristic function Q is $Q = \phi - \xi u_x - \tau u_t$, D_x^k denotes k times of total derivatives with respect to x and D_t^l denotes l times of total derivatives with respect to t. For the nonlocal NLS equation, we then adopt the following prolongation formula:

$$\begin{aligned}
\mathrm{pr}_{\mathrm{Ref}}\,\mathbf{v} = \mathbf{v} &+ \phi^*\,(-x,t,q(-x,t),q^*(-x,t))\,\partial_{q^*(-x,t)} + (D_t\phi - (D_t\xi)q_x - (D_t\tau)q_t)\,\partial_{q_t}\\
&+ \left(D_x^2\phi - (D_x^2\xi)q_x - 2(D_x\xi)q_{xx} - (D_x^2\tau)q_t - 2(D_x\tau)q_{tx}\right)\partial_{q_{xx}} + \cdots.
\end{aligned} \tag{23}$$

It is generalised from the prolongation formula (22) for symmetries of local differential equations with real variables, adding the conjugate terms and their prolongations.

A vector field $\mathbf{v}$ generates a group of symmetries for the nonlocal NLS equation if the following linearized symmetry condition

$$\mathrm{pr}_{\mathrm{Ref}}\,\mathbf{v}\left(i\,q_t + q_{xx} + 2q^2 q^*(-x,t)\right) = 0 \tag{24}$$

holds identically for all solutions of the nonlocal NLS Equation (19). We, first, expand the left hand side of (24) and obtain

$$\begin{aligned}
i\,(D_t\phi - (D_t\xi)q_x - (D_t\tau)q_t) + D_x^2\phi - (D_x^2\xi)q_x - 2(D_x\xi)q_{xx} - (D_x^2\tau)q_t - 2(D_x\tau)q_{tx}\\
+ 4qq^*(-x,t)\phi + 2q^2\phi^*\,(-x,t,q(-x,t),q^*(-x,t)) = 0,
\end{aligned} \tag{25}$$

restricted to solutions of the nonlocal NLS equation. We, then, substitute $q_t = i\left(q_{xx} + 2q^2q^*(-x,t)\right)$ and $q_t^* = -i\left(q_{xx} + 2q^2q^*(-x,t)\right)^*$, leading to a polynomial for q_x, q_x^*, q_{xx}, q_{xx}^*, and so forth, which equals zero identically. It is necessary and sufficient for the coefficients of the polynomial to vanish, amounting to a system of PDEs for ξ, τ, and ϕ, as follows:

$$D_x\tau = 0,\ \xi_q = 0,\ \xi_{q^*} = 0,\ \phi_{q^*} = 0,\ \phi_{qq} = 0,\ \tau_t - 2\xi_x = 0,\ i\xi_t + \xi_{xx} - 2\phi_{xq} = 0,$$
$$i\phi_t + \phi_{xx} - 2(\phi_q - \tau_t)q^2q^*(-x,t) + 4qq^*(-x,t)\phi + 2q^2\phi^*(-x,t,q(-x,t),q^*(-x,t)) = 0. \tag{26}$$

The general solution of the above system is

$$\xi = -C_1 x + iC_2 t + C_4,\ \tau = -2C_1 t + C_5,\ \phi = \left(C_1 + iC_3 - \frac{1}{2}C_2 x\right)q, \tag{27}$$

where C_1, C_2, and C_3 are real-valued, while C_4 and C_5 are complex-valued. Therefore, the symmetries of the nonlocal NLS equation are generated by the following five infinitesimal generators

$$\partial_x,\ \partial_t,\ iq\partial_q,\ -x\partial_x - 2t\partial_t + q\partial_q,\ it\partial_x - \frac{1}{2}xq\partial_q. \tag{28}$$

They can, equivalently, be cast into evolutionary type (respectively), as follows

$$-q_x\partial_q,\ i\left(q_{xx} + 2q^2q^*(-x,t)\right)\partial_q,\ iq\partial_q,$$
$$\left(q + xq_x + 2it\left(q_{xx} + 2q^2q^*(-x,t)\right)\right)\partial_q,\ \left(-\frac{1}{2}xq - itq_x\right)\partial_q. \tag{29}$$

Alternatively, we can define $q(x,t) = u(x,t) - iv(x,t)$, where $u(x,t)$ and $v(x,t)$ are real-valued functions, and use the symmetry prolongation formula for real-valued differential equations to calculate the symmetries. Now, the infinitesimal generator is

$$\mathbf{v} = \xi(x,t,u,v)\partial_x + \tau(x,t,u,v)\partial_t + \phi(x,t,u,v)\partial_u + \eta(x,t,u,v)\partial_v, \tag{30}$$

and the equation becomes

$$\begin{cases} u_t - v_{xx} - 4uvu(-x,t) + 2\left(u^2 - v^2\right)v(-x,t) = 0, \\ v_t + u_{xx} + 4uvv(-x,t) + 2\left(u^2 - v^2\right)u(-x,t) = 0. \end{cases} \tag{31}$$

The following symmetries are obtained for the system above, using the linearized symmetry condition (18) again:

$$\partial_x,\ \partial_t,\ -v\partial_u + u\partial_v,\ -x\partial_x - 2t\partial_t + u\partial_u + v\partial_v. \tag{32}$$

They correspond to the first four generators of (28). The last one obtained above does not appear here, since it will transform the real-valued x to a complex-valued argument as $\xi = it$.

3.2. Symmetry Reductions

Next, we will use the symmetries to conduct possible reductions. We choose to use the symmetries (28) with complex variables. The simplest reduction one would expect is probably traveling-wave solutions, which are difficult to obtain here, as the invariant $x - at$ becomes $-x - at$ at $(-x, t)$.

Consider the most general infinitesimal generator

$$a\partial_x + b\partial_t + ciq\partial_q + d\left(-x\partial_x - 2t\partial_t + q\partial_q\right) + e\left(it\partial_x - \frac{1}{2}xq\partial_q\right), \tag{33}$$

where a, b, c, d, and e are arbitrary constants. The invariant variables can be found by solving the characteristic equations

$$\frac{dx}{a - dx + i et} = \frac{dt}{b - 2dt} = \frac{dq}{i cq + dq - \frac{1}{2} exq}, \quad (34)$$

and we summarize the results as follows. Note that the equation depends on $q(x,t)$ and $q^*(-x,t)$ simultaneously, and we must select the constants properly to make the invariants meaningful.

- If $d = b = 0$ (and $a^2 + e^2 \neq 0$), we have

$$y = t,$$
$$q(x,t) = \exp\left\{\frac{x(4ic - ex)}{4(i et + a)}\right\} p(t). \quad (35)$$

When $a = 0$ and $e \neq 0$, the reduced equation is

$$i e^2 p'(t) + \frac{ie^2}{2t} p(t) + \frac{c^2}{t^2} p(t) + 2|p(t)|^2 p(t) = 0. \quad (36)$$

- If $d = 0$ and $b \neq 0$, we have

$$y = bx - \frac{1}{2} i et^2 - at,$$
$$q(x,t) = \exp\left\{-\frac{ae}{4b^2} t^2 - \frac{e}{2b^2} ty + i\left(\frac{c}{b} t - \frac{e^2}{12b^2} t^3\right)\right\} p(y). \quad (37)$$

As $b \neq 0$, we must choose $a = e = 0$. Next, we consider the corresponding reductions to nonlocal and local ODEs separately (here and throughout).

- **Reduction to a nonlocal ODE.** If we choose $y = x$ and $q(x,t) = \exp(ict) p(y)$, we obtain the nonlocal Painlevé-type equation as shown in [7]:

$$p''(y) - cp(y) + 2p^2(y) p^*(-y) = 0. \quad (38)$$

Note that, since $p(y)$ is invariant, and so is $p(-y)$; namely, the nonlocal invariant is

$$p(-y) = \exp(-ict) q(-x,t). \quad (39)$$

- **Reduction to a local ODE.** Alternatively, we may choose the invariants as $y = x^2$ and $q(x,t) = \exp(ict) p(y)$. The reduced equation is a local ODE

$$4yp''(y) = cp(y) - 2p'(y) - 2|p(y)|^2 p(y). \quad (40)$$

If we assume that $p(y)$ is real, the solution of the above equation can be expressed, using the Jacobi elliptic function, as

$$p(y) = C_2 \sqrt{\frac{c}{C_2^2 + c - 1}} \, \text{sn}\left(\sqrt{\frac{c}{C_2^2 + c - 1}} \left(\sqrt{-(c-1)} y + C_1\right), \frac{C_2}{\sqrt{c-1}}\right), \quad (41)$$

where C_1 and C_2 are integration constants. The above equation can actually be written in a simpler form by introducing $y = z^2$ and $\hat{p}(z) = p(y)$; the resulting equation is

$$\hat{p}''(z) = c\hat{p}(z) - 2|\hat{p}(z)|^2 \hat{p}(z). \quad (42)$$

- If $d \neq 0$, we have

$$y = \frac{d^2 x - ad + ie(dt - b)}{d^2 \sqrt{|2dt - b|}},$$

$$q(x,t) = \exp\left\{\left(\frac{ae}{4d^2} - \frac{1}{2}\right) \ln|2dt - b| + \frac{e}{2d} y \sqrt{|2dt-b|}\right\} \times \qquad (43)$$

$$\exp\left\{-i\frac{e^2}{4d^2}t + i\left(\frac{be^3}{8d^3} - \frac{c}{2d}\right) \ln|2dt-b|\right\} p(y).$$

Now, we must set $a = e = 0$.

- **Reduction to a nonlocal ODE.** Let

$$y = \frac{x}{\sqrt{|2dt - b|}},$$

$$q(x,t) = \exp\left\{-\left(\frac{1}{2} + \frac{ic}{2d}\right) \ln|2dt - b|\right\} p(y). \qquad (44)$$

The reduced equation is a nonlocal ODE

$$p''(y) = (id - c)\, p(y) + i\, dy\, p'(y) - 2 p^2(y) p^*(-y). \qquad (45)$$

- **Reduction to a local ODE.** If we choose the invariant variables by

$$y = \frac{x^2}{2dt - b},$$

$$q(x,t) = \exp\left\{-\left(\frac{1}{2} + \frac{ic}{2d}\right) \ln|2dt - b|\right\} p(y), \qquad (46)$$

the reduced equation is local; that is,

$$4y p''(y) = (id - c) p(y) + (2i\, dy - 2) p'(y) - 2|p(y)|^2 p(y). \qquad (47)$$

Introducing $y = z^2$ and $\widehat{p}(z) = p(y)$ changes the equation to

$$\widehat{p}''(z) = (id - c)\widehat{p}(z) + i\, dz\, \widehat{p}(z) - 2|\widehat{p}(z)|^2 \widehat{p}(z). \qquad (48)$$

4. The Nonlocal mKdV Equation

The nonlocal mKdV equation we consider in this paper is (see, for example, [8])

$$u_t(x,t) + u(x,t)u(-x,-t)u_x(x,t) + u_{xxx}(x,t) = 0. \qquad (49)$$

Assuming that the infinitesimal generator reads

$$\xi(x,t,u)\partial_x + \tau(x,t,u)\partial_t + \phi(x,t,u)\partial_u + \phi(-x,-t,u(-x,-t))\partial_{u(-x,-t)}, \qquad (50)$$

its prolongation can be obtained using (17). From a similar procedure for applying the linearized symmetry condition (18) to the nonlocal NLS equation above, a straightforward calculation gives the following infinitesimal generators for symmetries of the nonlocal mKdV equation:

$$\partial_x, \quad \partial_t, \quad -x\partial_x - 3t\partial_t + u\partial_u. \qquad (51)$$

We follow the same approach as for the nonlocal NLS equation to search for symmetry reductions. The most general symmetry generator can be denoted by

$$a\partial_x + b\partial_t + c(-x\partial_x - 3t\partial_t + u\partial_u), \tag{52}$$

where a, b, and c are arbitrary constants. The characteristic equations read

$$\frac{dx}{a - cx} = \frac{dt}{b - 3ct} = \frac{du}{cu}. \tag{53}$$

- When $c = 0$, it corresponds to the traveling-wave case.

  – **Reduction to a nonlocal ODE.** The corresponding invariants are

  $$y = bx - at \text{ and } v(y) = u(x, t). \tag{54}$$

  The reduced equation is

  $$b^3 v'''(y) + bv(y)v(-y)v'(y) - av'(y) = 0. \tag{55}$$

  When $b = 0$, we obtain a constant solution; when $b \neq 0$, without loss of generality, it can be chosen as $b = 1$; namely

  $$v'''(y) + v(y)v(-y)v'(y) - av'(y) = 0. \tag{56}$$

  In principle, it can be integrated once, as it admits a symmetry generated by ∂_y, but will involve the inverse of nonlocal functions. We will show some of its special solutions with the assumption $a > 0$.

  * Exponential solutions:

    $$v(y) = C_1 \exp(C_2 y) \text{ subject to } C_1^2 + C_2^2 = a. \tag{57}$$

  * Soliton solutions:

    $$v(y) = \pm \frac{2\sqrt{6a}}{\exp(\sqrt{a}y) + \exp(-\sqrt{a}y)}. \tag{58}$$

  – **Reduction to a local ODE.** We may, alternatively, introduce the invariants in another way; namely, $y = (bx - at)^2$ and $v(y) = u(x, t)$. Now, the reduced equation reads

  $$4b^3 y v'''(y) + 6b^3 v''(y) + bv^2(y)v'(y) - av'(y) = 0, \tag{59}$$

  which can be integrated once:

  $$4b^3 y v''(y) + 2b^3 v'(y) + \frac{b}{3} v^3(y) - av(y) + C_1 = 0. \tag{60}$$

  This equation can be further simplified by introducing $y = z^2$ and $\hat{v}(z) = v(y)$, amounting to

  $$b^3 \hat{v}''(z) + \frac{b}{3} \hat{v}^3(z) - a\hat{v}(z) + C_1 = 0. \tag{61}$$

  The final equation is solvable by letting $\hat{v}(z) = w(\hat{v})$; the general solution is

  $$z + C_3 = \pm \int_0^{\hat{v}(z)} \frac{\sqrt{6} b^{3/2}}{\sqrt{-bs^4 + 6as^2 - 12 C_1 s + 6 C_2 b^3}} ds, \tag{62}$$

where C_1, C_2, and C_3 are integration constants.

- If $c \neq 0$, the invariants are

$$y = (cx - a)(3ct - b)^{-1/3} \text{ and } v(y) = (3ct - b)^{1/3} u(x,t). \tag{63}$$

Now, we must set $a = b = 0$; namely, reduction related to the generator $-x\partial_x - 3t\partial_t + u\partial_u$. The related invariants are $y = t^{-1/3}x$ and $v(y) = t^{1/3}u(x,t)$, and we obtain the reduced equation as a local ODE

$$v'''(y) - v^2(y)v'(y) - \frac{v(y) + yv'(y)}{3} = 0. \tag{64}$$

It can be integrated once to the second Painlevé equation

$$v''(y) = \frac{1}{3}v^3(y) + \frac{1}{3}yv(y) + C. \tag{65}$$

Now, we are able to conclude that all reduced local ODEs for the nonlocal mKdV equation are integrable, analogously to the local situation.

Remark 1. *In [7], the authors pointed out that similarity reduction of the nonlocal NLS equation may lead to nonlocal ODEs. However, as shown by the two illustrative examples, such an inconvenience can be overcome by choosing the invariant variables (or functions) in a proper manner and the reduced ODEs become local.*

5. A Remark on Transformations Between Nonlocal Differential Equations and Differential-Difference Equations

In [21], the authors introduced variable transformations to connect nonlocal and local integrable equations. For instance, the nonlocal NLS equation becomes a local NLS equation under the transformation

$$x = i\hat{x}, \quad t = -\hat{t}, \quad q(x,t) = \hat{q}(\hat{x}, \hat{t}). \tag{66}$$

The nonlocal complex mKdV equation becomes the local (classical) complex mKdV equation under the transformation

$$x = i\hat{x}, \quad t = -i\hat{t}, \quad u(x,t) = \hat{u}(\hat{x}, \hat{t}). \tag{67}$$

In this section, we will show the relations between nonlocal differential equations and DDEs through variable transformations.

For the nonlocal NLS equation, we consider the following transformations

$$x = \exp(\hat{x}), \quad t = \hat{t}, \quad q(x,t) = \hat{q}(\hat{x}, \hat{t}), \tag{68}$$

where the variable $\hat{x}$ is imaginary, making x imaginary too. Let us drop the hats (always) and the nonlocal NLS equation becomes a DDE

$$i q_t + \exp(-2x)(q_{xx} - q_x) + 2q^2 q^*(x + i\pi, t) = 0. \tag{69}$$

Let us introduce the following transformations

$$x = \exp(\hat{x}), \quad t = \exp(\hat{t}), \quad u(x,t) = \hat{u}(\hat{x}, \hat{t}), \tag{70}$$

where the variables $\hat{x}$ and $\hat{t}$ are both imaginary. The nonlocal mKdV equation becomes

$$\exp(-t)u_t + \exp(-x)uu(x + i\pi, t + i\pi)u_x + \exp(-3x)(u_{xxx} - 3u_{xx} + 2u_x) = 0. \tag{71}$$

Under the transformation $y = \exp(\hat{y})$, $v(y) = \hat{v}(\hat{y})$, the reduced Equation (55) becomes

$$b^3 \exp(-2y)\left(v'''(y) - 3v''(y) + 2v'(y)\right) + (bv(y)v(y+i\pi) - a)\,v'(y) = 0. \tag{72}$$

These DDEs can further be re-scaled and normalized. For example, taking $y = i\pi\hat{y}$ and $v(y) = \hat{v}(\hat{y})$, Equation (72) becomes

$$b^3 \exp(-i2\pi y)\left(-\frac{1}{\pi^2}v'''(y) + \frac{3i}{\pi}v''(y) + 2v'(y)\right) + (bv(y)v(y+1) - a)\,v'(y) = 0. \tag{73}$$

Similar DDEs were investigated in [22], but the variables were real-valued therein. In the same manner, the above DDEs transformed from the nonlocal NLS and mKdV equations can also be re-scaled, respectively, as follows:

$$i\,q_t + \exp(-i2\pi x)\left(-\frac{1}{\pi^2}q_{xx} + \frac{i}{\pi}q_x\right) + 2q^2 q^*(x+1,t) = 0, \tag{74}$$

and

$$\exp(-i\pi t)u_t + \exp(-i\pi x)uu(x+1,t+1)u_x + \exp(-i3\pi x)\left(-\frac{1}{\pi^2}u_{xxx} + \frac{3i}{\pi}u_{xx} + 2u_x\right) = 0. \tag{75}$$

Remark 2. *The above examples show that simple transformations allow us to transfer nonlocal equations to DDEs. Apparently, similar transformations can be immediately introduced for other nonlocal differential equations/systems using the same manner.*

6. Conclusions

In this paper, symmetry analysis was extended to study nonlocal differential equations. The general theory presented in Section 2 is applicable to any nonlocal differential equations involving space and/or time reflections. In particular, two integrable nonlocal equations—the nonlocal NLS equation and the nonlocal mKdV equation—served as illustrative examples. All Lie point symmetries of these two nonlocal PDEs were obtained and possible symmetry reductions to nonlocal and local ODEs were conducted. It was shown that, at least for the two illustrative examples, one can always carefully choose the invariant variables to ensure that all reduced differential equations are local.

Finally, we introduced some local transformations which transfer nonlocal differential equations to DDEs; there is potential, hence, to extend the existing theory for DDEs to nonlocal differential equations; for instance, the symmetries, conservation laws, and integrability of DDEs (see, for example, [22–27]). We will explore more in this direction in a separate project.

Funding: This work was partially supported by JSPS Grant-in-Aid for Scientific Research (No. 16KT0024), the MEXT 'Top Global University Project', and Waseda University Grant for Special Research Projects (Nos. 2019C-179, 2019E-036, 2019R-081).

Conflicts of Interest: The author declares no conflict of interest.

References

1. Ackerman, M.; Hermann, R. *Sophus Lie's 1880 Transformation Group Paper*; Math. Sci. Press: Brookline, MA, USA, 1975.
2. Bluman, G.W.; Cole, J.D. General similarity solution of the heat equation. *J. Math. Mech.* **1969**, *18*, 1025–1042.
3. Bluman, G.W.; Kumei, S. *Symmetries and Differential Equations*; Springer: New York, NY, USA, 1989.
4. Hydon, P.E. *Symmetry Methods for Differential Equations: A Beginner's Guide*; Cambridge University Press: Cambridge, UK, 2000.
5. Olver, P.J. *Applications of Lie Groups to Differential Equations*, 2nd ed.; Springer: New York, NY, USA, 1993.

6. Ablowitz, M.J.; Ramani, A.; Segur, H. Nonlinear evolution equations and ordinary differential equations of Painlevé type. *Lett. Nuovo Cimento* **1978**, *23*, 333–338. [CrossRef]
7. Ablowitz, M.J.; Musslimani, Z.H. Integrable nonlocal nonlinear Schrödinger equation. *Phys. Rev. Lett.* **2013**, *110*, 064105. [CrossRef] [PubMed]
8. Ji, J.L.; Zhu, Z.N. On a nonlocal modified Korteweg-de Vries equation: Integrability, Darboux transformation and soliton solutions. *Commun. Nonlinear Sci. Numer. Simul.* **2017**, *42*, 699–708. [CrossRef]
9. Ablowitz, M.J.; Musslimani, Z.H. Integrable nonlocal nonlinear equations. *Stud. Appl. Math.* **2017**, *139*, 7–59. [CrossRef]
10. Fokas, A.S. Integrable multidimensional versions of the nonlocal nonlinear Schrödinger equation. *Nonlinearity* **2016**, *29*, 319–324. [CrossRef]
11. Lou, S.Y.; Huang, F. Alice-Bob physics: Coherent solutions of nonlocal KdV systems. *Sci. Rep.* **2017**, *7*, 869. [CrossRef]
12. Song, C.Q.; Xiao, D.M.; Zhu, Z.N. Reverse space-time nonlocal Sasa–Satsuma equation and its solutions. *J. Phys. Soc. Jpn.* **2017**, *86*, 054001. [CrossRef]
13. Yan, Z. Integrable $\mathcal{PT}$-symmetric local and nonlocal vector nonlinear Schrödinger equations: A unified two-parameter model. *Appl. Math. Lett.* **2015**, *47*, 61–68. [CrossRef]
14. Ablowitz, M.J.; Musslimani, Z.H. Inverse scattering transform for the integrable nonlocal nonlinear Schrödinger equation. *Nonlinearity* **2016**, *29*, 915–946. [CrossRef]
15. Gurses, M.; Pekcan, A. Nonlocal nonlinear Schrödinger equations and their soliton solutions. *J. Math. Phys.* **2018**, *59*, 051501. [CrossRef]
16. Khare, A.; Saxena, A. Periodic and hyperbolic soliton solutions of a number of nonlocal nonlinear equations. *J. Math. Phys.* **2015**, *56*, 032104. [CrossRef]
17. Song, C.Q.; Xiao, D.M.; Zhu, Z.N. Solitons and dynamics for a general integrable nonlocal coupled nonlinear Schrödinger equation. *Commun. Nonlinear Sci. Numer. Simul.* **2017**, *45*, 13–28. [CrossRef]
18. Xu, Z.X.; Chow, K.W. Breathers and rogue waves for a third order nonlocal partial differential equation by a bilinear transformation. *Appl. Math. Lett.* **2016**, *56*, 72–77. [CrossRef]
19. Zhou, Z.X. Darboux transformations and global solutions for a nonlocal derivative nonlinear Schrödinger equation. *Commun. Nonlinear Sci. Numer. Simul.* **2018**, *62*, 480–488. [CrossRef]
20. Kumei, S. Group theoretic aspects of conservation laws of nonlinear dispersive waves: KdV type equations and nonlinear Schrödinger equations. *J. Math. Phys.* **1977**, *18*, 256–264. [CrossRef]
21. Yang, B.; Yang, J. Transformations between nonlocal and local integrable equations. *Stud. Appl. Math.* **2017**, *140*, 178–201. [CrossRef]
22. Quispel, G.R.W.; Capel, H.W.; Sahadevan, R. Continuous symmetries of differential-difference equations: The Kac–van Moerbeke equation and Painlevé reduction. *Phys. Lett. A* **1992**, *170*, 379–383. [CrossRef]
23. Kupershmidt, B.A. *Discrete Lax Equations and Differential-Difference Calculus*; Astérisque: Paris, France, 1985.
24. Levi, D.; Winternitz, P.; Yamilov, R.I. Lie point symmetries of differential-difference equations. *J. Phys. A Math. Theor.* **2010**, *43*, 292002. [CrossRef]
25. Mikhailov, A.V.; Wang, J.P.; Xenitidis, P. Cosymmetries and Nijenhuis recursion operators for difference equations. *Nonlinearity* **2011**, *24*, 2079–2097. [CrossRef]
26. Peng, L. Symmetries, conservation laws, and Noether's theorem for differential-difference equations. *Stud. Appl. Math.* **2017**, *139*, 457–502. [CrossRef]
27. Yamilov, R. Symmetries as integrability criteria for differential difference equations. *J. Phys. A Math. Gen.* **2006**, *39*, R541–R623. [CrossRef]

© 2019 by the authors. Licensee MDPI, Basel, Switzerland. This article is an open access article distributed under the terms and conditions of the Creative Commons Attribution (CC BY) license (http://creativecommons.org/licenses/by/4.0/).

Article

Quasi-Noether Systems and Quasi-Lagrangians

V. Rosenhaus [1,*] and Ravi Shankar [2]

[1] Department of Mathematics and Statistics, California State University, Chico, CA 95929, USA
[2] Department of Mathematics, University of Washington, Seattle, WA 98195, USA
* Correspondence: vrosenhaus@csuchico.edu

Received: 5 July 2019; Accepted: 2 August 2019; Published: 5 August 2019

Abstract: We study differential systems for which it is possible to establish a correspondence between symmetries and conservation laws based on Noether identity: quasi-Noether systems. We analyze Noether identity and show that it leads to the same conservation laws as Lagrange (Green–Lagrange) identity. We discuss quasi-Noether systems, and some of their properties, and generate classes of quasi-Noether differential equations of the second order. We next introduce a more general version of quasi-Lagrangians which allows us to extend Noether theorem. Here, variational symmetries are only sub-symmetries, not true symmetries. We finally introduce the critical point condition for evolution equations with a conserved integral, demonstrate examples of its compatibility, and compare the invariant submanifolds of quasi-Lagrangian systems with those of Hamiltonian systems.

Keywords: symmetries; conservation laws; Noether operator identity; quasi-Noether systems; quasi-Lagrangians

---

1. Introduction

For variational systems the relation between symmetries of the Lagrangian function and conservation laws was known from the classical Noether result [1]. It was shown that there is one-to-one correspondence between variational symmetries (symmetries of variational functional) and local conservation laws of a differential system, [2].

In this paper, we study differential systems that allow a Noether-type association between its conservation laws and symmetries (quasi-Noether systems). Our approach is based on the Noether operator identity [3] that relates the infinitesimal transformation operator to the Euler and divergence operators. The Noether operator identity has been shown to provide a Noether-type relation between symmetries and conservation laws not only for Lagrangian systems, but also for a large class of differential systems that may not have a well-defined variational functional, see [4,5].

Noether operator identity was also demonstrated to allow derivation of extension of Second Noether Theorem for non–Largangian systems possessing infinite symmetry algebras parametrized by arbitrary functions of all independent variables, [6]. These infinite symmetry algebras were shown to lead to differetial identities between the equations of the original differential system and their derivatives.

Recently Noether identity was used to generate relations between sub-symmetries [7] and corresponding local conservation laws [8] for quasi-Noether systems.

In this paper, we analyze quasi-Noether systems and some of their properties. In Section 2, we review known correspondence between symmetries and conservation laws for variational systems from the standpoint of the Noether identity. In Section 3, we discuss this correspondence for a more general class of differential systems (quasi-Noether systems) that includes non-variational problems. We review conservation laws obtained with the use of approach based on the Noether identity and compare them with the results obtained from the Lagrange (Green–Lagrange) identity. We also find the class of quasi-Noether evolution equations of the second order, and quasi-linear equations of the second order. In Section 4, we discuss the concept of quasi-Lagrangian, and use it to prove a new extension of the Noether theorem to non–Lagrangian systems. In this approach, variational symmetries of quasi-Lagrangians are only sub-symmetries, and need not be symmetries. We give an example where all lower conservation laws are generated in this way, but where the previous correspondence fails. As a geometric application, we compare the invariant submanifolds of quasi-Noether systems to those of Hamiltonian systems and show that they satisfy opposite containments. To address this fact we introduce the notion of a critical point of a conserved quantity, and demonstrate examples of the compatibility of the critical point condition with the time evolution of the PDE system.

2. Symmetries and Conservation Laws of Variational Systems

Let us briefly outline the approach we follow. A system of governing partial differential equations will be written as

$$\Delta^a(x, u, u_{(1)}, u_{(2)}, \dots) = 0, \qquad a = 1, \dots, q,$$

By a conservation law of the system Δ^a, we mean the p-tuple $(K^1, K^2, \dots K^p)$ such that

$$D_i K^i(x, u, u_{(1)}, u_{(2)}, \dots) \doteq 0, \tag{1}$$

on all solutions of the original system; we denote this type of equality by ($\doteq$). Here, $x = (x^1, x^2, \dots, x^p)$ and $u = (u^1, u^2, \dots, u^q)$ are the tuples of independent and dependent variables, respectively; $u_{(r)}$ is the tuple of rth-order derivatives of u, $r = 1, 2, \dots$; Δ^a and K^i are differential functions, i.e., smooth functions of x, u and a finite number of derivatives of u (see [2]); $i, j = 1, \dots, p, a = 1, \dots, q$. We assume summation over repeated indices.

We let

$$D_i = \partial_i + u_i^a \partial_{u^a} + u_{ij}^a \partial_{u_j^a} + \cdots = \partial_i + u_{iJ}^a \partial_{u_J^a}$$

be the i-th total derivative, $1 \leq i \leq p$, the sum extending over all (unordered) multi-indices $J = (j_1, j_2, \dots, j_k)$ for $k \geq 0$ and $1 \leq j_k \leq p$.

Two conservation laws K and $\tilde{K}$ are equivalent if they differ by a trivial conservation law [2]. A conservation law $D_i P^i \doteq 0$ is trivial if a linear combination of two kinds of triviality is taking place: 1. The p-tuple P vanishes on the solutions of the original system: $P^i \doteq 0$. 2. The divergence identity is satisfied for any point $[u] = (x, u_{(n)})$ in the jet space (e.g., div rot $u = 0$).

We consider smooth functions $u^a = u^a(x)$ defined on an open subset $D \subset \mathbb{R}^p$. Let

$$S = \int_D L(x, u, u_{(1)}, \dots) d^p x$$

be the action functional, where L is the Lagrangian density. The equations of motion are

$$E_a(L) \equiv \Delta^a(x, u, u_{(1)}, \dots) = 0, \qquad 1 \leq a \leq q, \tag{2}$$

where

$$E_a = \frac{\partial}{\partial u^a} - \sum_i D_i \frac{\partial}{\partial u_i^a} + \sum_{i \leq j} D_i D_j \frac{\partial}{\partial u_{ij}^a} + \cdots \tag{3}$$

is the a-th Euler (Euler–Lagrange) operator (variational derivative). We call the tuple $E = (E_1, \dots, E_n)$ the Euler operator. In the notation of [2], we could give it the following form:

$$E_a = (-D)_J \frac{\partial}{\partial u_J^a}, \qquad 1 \leq a \leq q, \tag{4}$$

The operator $(-D)_J$ is defined here as $(-D)_J = (-1)^k D_J = (-D_{j_1})(-D_{j_2})\cdots(-D_{j_k})$. The operator E_a annihilates total divergences.

Consider an infinitesimal (one-parameter) transformation with the canonical infinitesimal operator

$$X_\alpha = \alpha^a \frac{\partial}{\partial u^a} + \sum_i (D_i \alpha^a) \frac{\partial}{\partial u_i^a} + \sum_{i \leq j} (D_i D_j \alpha^a) \frac{\partial}{\partial u_{ij}^a} + \cdots = (D_J \alpha^a) \partial_{u_J^a}. \tag{5}$$

where $\alpha^a = \alpha^a(x, u, u_{(1)}, \ldots)$, and the sum is taken over all (unordered) multi-indices J. The variation of the functional S under the transformation with operator X_α is

$$\delta S = \int_D X_\alpha L \, d^p x. \tag{6}$$

X_α is a variational (Noether) symmetry if

$$X_\alpha L = D_i M^i, \tag{7}$$

where $M^i = M^i(x, u, u_{(1)}, \ldots)$ are smooth functions of their arguments. The Noether (operator) identity [3] (see also, e.g., [9] or [4]) relates the operator X_α to E_a,

$$X_\alpha = \alpha^a E_a + D_i R^i, \tag{8}$$

$$R^i = \alpha^a \frac{\partial}{\partial u_i^a} + \left\{ \sum_{k \geq i} (D_k \alpha^a) - \alpha^a \sum_{k \leq i} D_k \right\} \frac{\partial}{\partial u_{ik}^a} + \cdots. \tag{9}$$

The expression for R^i can be presented in a more general form [3,10]:

$$R^i = (D_K \alpha^a)(-D)_J \partial_{u_{iJK}^a}, \tag{10}$$

where J and K sum over multi-indices.

Applying the identity (8) with (9) to L and using (7), we obtain

$$D_i(M^i - R^i L) = \alpha^a \Delta^a, \tag{11}$$

which on the solution manifold ($\Delta = 0, D_i \Delta = 0, \ldots$)

$$D_i(M^i - R^i L) \doteq 0, \tag{12}$$

leads to the statement of the First Noether Theorem: any one-parameter variational symmetry transformation with infinitesimal operator X_α (5) gives rise to the conservation law (12).

Note that Noether [1] used the identity (11) and not the operator identity (8). The first mention of the Noether operator identity (8), to our knowledge, was made in [3].

We next consider differential systems that may not have well-defined Lagrangian functions.

3. Symmetries and Conservation Laws of Quasi-Noether Systems

For a general differential system, a relationship between symmetries and conservation laws is unknown. In [4,5], an approach based on the Noether operator identity (8) was suggested to relate symmetries to conservation laws for a large class of differential systems that may not have well-defined Lagrangian functions. In the current paper, we will follow this approach.

3.1. Approach Using the Noether Operator Identity

We consider q smooth functions $u = (u^1, u^2, ..., u^q)$ of p independent variables $x = (x^1, x^2, ..., x^p)$ defined on some nonempty open subset of $\mathbb{R}^p$. Consider a system of n ℓth order differential equations $\Delta = (\Delta^1, \Delta^2, ..., \Delta^n)$ for functions u:

$$\Delta^a(x, u, u_{(1)}, u_{(2)}, \ldots, u_{(l)}) = 0, \quad a = 1, 2, ..., n. \tag{13}$$

Here, each $\Delta^a(x, u, u_{(1)}, u_{(2)}, \ldots, u_{(l)})$, $a = 1, 2, ..., n$ is a smooth function of x, u, and all partial derivatives of each u^v, $(v = 1, \ldots, q)$ with respect to the x^i ($i = 1, \ldots, p$) up to the ℓth order (differential function [2]). We assume that system (13) is normal, and totally nondegenerate (locally solvable at every point, and of maximal rank) [2].

Let $\Delta(x, u, u_{(1)}, u_{(2)}, \ldots, u_{(l)}) \equiv \Delta[u] \equiv \Delta$, $x_J = (x_{j_1}, x_{j_2}, ..., x_{j_k})$ and u_J^v be partial derivatives, where $J = (j_1, j_2, ..., j_k)$. Applying the Noether operator identity (8) to a combination of original equations with some coefficients (differential operators) $\beta^a \Delta^a$, we obtain

$$X_\alpha(\beta^a \Delta^a) = \alpha^v E_v(\beta^a \Delta^a) + D_i R^i(\beta^a \Delta^a), \quad a = 1, \ldots n, \quad v = 1, \ldots, q, \quad i = 1, \ldots, p. \tag{14}$$

If our system (13) allows the existence of coefficients β^a (*cosymmetries*) such that

$$E_v(\beta^a \Delta^a) \doteq 0, \quad a = \ldots, n, \quad v = 1, \ldots, q \tag{15}$$

on the solution manifold ($\Delta = 0, D_i \Delta = 0, \ldots$), then according to (14), each symmetry of the system X_α will lead to a local conservation law (see [4,5])

$$D_i R^i(\beta^a \Delta^a) \doteq 0. \tag{16}$$

for any differential systems of class (15). In [4], the quantity $\beta^a \Delta^a$ was referred to as an alternative Lagrangian.

Let us note that the correspondence between symmetries and local conservation laws defined above for differential systems without well-defined Lagrangian functions may not be one-to-one or onto, as in the case of variational symmetries and local conservation laws [2], and non-trivial symmetries may lead to trivial conservation laws. If β generates a conservation law, i.e., $E(\beta \cdot \Delta) \equiv 0$, then it was shown in [11] that translation symmetries $\alpha = a^i u_i, a = const.$ lead to trivial conservation laws if $\beta_x = 0$.

In general, the nontriviality of a conservation law is determined by the characteristic. To compute it explicitly, let $X_\alpha(\beta \cdot \Delta) = A \cdot \Delta$ and $E(\beta \cdot \Delta) = B \cdot \Delta$ for A, B differential operators. Then the Noether identity and integration by parts yield

$$\begin{aligned} A \cdot \Delta = X_\alpha(\beta \cdot \Delta) &= \alpha \cdot E(\beta \cdot \Delta) + div \\ &= \alpha \cdot (B \cdot \Delta) + div \\ &= \Delta \cdot (B^* \cdot \alpha) + div, \end{aligned} \tag{17}$$

where $(B^*)_{va} = (-D)_I \circ B_{avI}$ is the adjoint operator. Integrating by parts the LHS and rearranging yields overall

$$[B^* \cdot \alpha - A^*(1)] \cdot \Delta = div, \tag{18}$$

where the divergence is equivalent to (16) on solutions, and $A^*(1)_v = (-D)_I A_{vI}$.

Thus, the conservation law obtained from Noether identity and corresponding to symmetry X_α is nontrivial if

$$B^* \cdot \alpha - A^*(1) \neq 0 \tag{19}$$

on solutions of the original system (13).

The condition (15) can be written in a somewhat more general form [5]

$$E_v(\beta^{ac}\Delta^a) \doteq 0, \quad a,c = 1,\ldots n, \quad v = 1,\ldots,q. \tag{20}$$

In terms of the Fréchet derivative operator D_Δ and its adjoint D_Δ^* [2]

$$(D_\Delta)_{av} = \sum_J \frac{\partial \Delta^a}{\partial u_J^v} D_J, \tag{21}$$

$$(D_\Delta^*)_{av}\beta^a = \sum_J (-1)^k D_J \left(\frac{\partial \Delta^a}{\partial u_J^v} \beta^a \right), \tag{22}$$

the condition (15) was shown in [12] to be related to the condition of self-adjointness of the operator D_Δ (generalized Helmholtz condition)

$$D_\Delta^* - D_\Delta = 0, \tag{23}$$

this relationship being:

$$E_v(\beta^a \Delta^a) \doteq (D_\Delta^* - D_\Delta)_{av} \beta^a, \quad a = 1,\ldots n, \quad v = 1,\ldots,n. \tag{24}$$

Expression (24) provides a relationship between the condition (15) for a system to be quasi-Noether and the existence of a variational functional for a transformed differential system.

The condition (15) can be considered as defining quasi-Noether systems, see also [6]. A system (13) is *quasi-Noether* if there exist functions (differential operators) β^a such that the condition (15) is satisfied. In [4], the quantity $\beta^v \Delta_v$ was referred to as an alternative Lagrangian. If coefficients β^a generate a conservation law, they are related to adjoint symmetries [13], and referred to as characteristics of a corresponding conservation law [2], generating functions [14] or multipliers [15]. In general, we call them cosymmetries [14] if they satisfy (15).

It should be noted that the condition (15) (for a system to be quasi-Noether and possess a correspondence between symmetries and conservation laws) was obtained earlier within an alternative approach based on the Lagrange identity. In [16], the classical Lagrange identity (Green's formula) was used for generating conservation laws for a linear differential system. A condition for the existence of a certain conservation law written in terms of an adjoint differential operator was presented in [14] within a general framework of algebraic geometry. In [17], a correspondence between symmetries and conservation laws based on Green's formula and a condition similar to (15) was obtained for evolution systems. For the case of point transformations, this correspondence was discussed in [18]. In [13], the geometric meaning of this condition was discussed for mechanical systems. In [10], a condition that can be reduced to (15) was presented for a general differential system.

Note that the condition (15) played a key role in later developed direct method [15] and the nonlinear self-adjointness approach [19].

3.2. Approach Using Lagrange Identity

Let us briefly describe the alternative approach based on the Lagrange identity and compare the results with those of the Noether identity approach.

The well-known Lagrange identity is as follows:

$$\beta^a (D_\Delta)_{av} \alpha^v - \alpha^v (D_\Delta^*)_{av} \beta^a = D_i Q^i[\alpha, \beta, \Delta], \tag{25}$$

where D_Δ is the Fréchet derivative of Δ, and D_Δ^* is its adjoint. An explicit expression for the trilinear fluxes $Q[\alpha, \beta, \Delta]$ was given in [10]:

$$Q^i[\alpha, \beta, \Delta] = (-1)^{|J|} D_K \alpha^v D_J \left(\beta^a \partial_{u^v_{ijK}} \Delta^a \right), \quad 1 \leq i \leq p.$$

Using the fact that

$$(D_\Delta)_{av} \alpha^v = X_\alpha \Delta^a, \tag{26}$$

(see (5), and (21)) we can express (25) as follows:

$$\beta^a X_\alpha \Delta^a - \alpha^v (D_\Delta^*)_{av} \beta^a = D_i Q^i[\alpha, \beta, \Delta]. \tag{27}$$

If, for a given system Δ, there exist functions β^a and operators $Y^{va} = v^{vaJ} D_J$ such that the following relationships hold:

$$(D_\Delta^*)_{av} \beta^a = Y^{va} \Delta^a, \quad 1 \leq v \leq q, \tag{28}$$

then $(D_\Delta^*)_{av} \beta^a \doteq 0$. Thus, for each symmetry X_α of the system $\Delta = 0$, ($X_\alpha \Delta^a = \Lambda^{ab} \Delta^b$ for some operators $\Lambda^{ab} = \lambda^{abJ} D_J$), Equation (27) provides a corresponding conservation law:

$$D_i Q^i[\alpha, \beta, \Delta] = (\beta^a \Lambda^{ab} - \alpha^v Y^{va}) \Delta^a \doteq 0. \tag{29}$$

We now show that the correspondence between symmetries and conservation laws in terms of Lagrange identity is equivalent to the one using the Noether identity and leads to the same conservation laws. Indeed, using the product rule [2]:

$$E_v(\beta^a \Delta^a) = (D_\Delta^*)_{av} \beta^a + (D_\beta^*)_{av} \Delta^a, \quad 1 \leq v \leq q, \tag{30}$$

in (27) gives:

$$\beta^a X_\alpha \Delta^a - \alpha^v E_v(\beta^a \Delta^a) + \alpha^v (D_\beta^*)_{av} \Delta^a = D_i Q^i.$$

Using $\beta^a X_\alpha \Delta^a = X_\alpha (\beta^a \Delta^a) - \Delta^a X_\alpha \beta^a$ we obtain

$$X_\alpha(\beta^a \Delta^a) = \alpha^v E_v(\beta^a \Delta^a) + D_i Q^i[\alpha, \beta, \Delta] + (X_\alpha \beta^a - \alpha^v (D_\beta^*)_{av}) \Delta^a. \tag{31}$$

The last term in (31) is a total divergence by (25):

$$\Delta^a X_\alpha \beta^a - \alpha^v (D_\beta^*)_{av} \Delta^a = \Delta^a (D_\beta)_{av} \alpha^v - \alpha^v (D_\beta^*)_{av} \Delta^a = D_i Q^i[\alpha, \Delta, \beta].$$

Thus, (31) provides the same result as the Noether identity:

$$X_\alpha(\beta^a \Delta^a) = \alpha^v E_v(\beta^a \Delta^a) + D_i(Q^i[\alpha, \Delta, \beta] + Q^i[\alpha, \beta, \Delta]). \tag{32}$$

Let us show that the condition used in direct method [15]

$$(D_\Delta^*)_{av} \beta^a \doteq 0 \tag{33}$$

is equivalent to the quasi-Noether condition (15). Indeed, using the identity [2]

$$E_v(\beta^a \Delta^a) = (D_\Delta^*)_{av} \beta^a + (D_\beta^*)_{av} \Delta^a, \tag{34}$$

and the fact that
$$(D_\beta^*)_{av}\Delta^a \doteq 0, \tag{35}$$

we obtain condition (33)
$$E_v(\beta^a \Delta^a) \doteq (D_\Delta^*)_{av}\beta^a \doteq 0. \tag{36}$$

An alternative key expression of the direct method
$$E_v(\beta^a \Delta^a) = 0 \tag{37}$$

is a special case of (15). Note also that the direct method aims at the generation of conservation laws for a differential system without regard to its symmetries while the goal of both approaches above taken earlier was to establish a correspondence between symmetries and conservation laws.

3.3. Quasi-Noether Systems

As noted above the condition (15) (or (20)) determines *quasi-Noether systems*.

It can be shown that the general case of differential systems (13) satisfying the condition (15), with β^a being differential operators

$$\beta^a = \beta^{aJ}[u]D_J, \quad D_J = D_{j_1}D_{j_2}\cdots D_{j_k}, \quad D_{j_r} = \frac{d}{dx_{j_r}}, \quad r = 1, 2, ..., k, \tag{38}$$

can be reduced to the case of the condition (15) with the β^{aJ} being differential functions. Indeed, (see [2])

$$\beta^a \Delta^a = \beta^{aJ} D_J \Delta^a = D_J\left(\beta^{aJ}\Delta^a\right) + \left((-D)_J \beta^{aJ}\right)\Delta^a. \tag{39}$$

Since the contribution of the first term in the RHS, being a total divergence, is zero after applying the Euler operator, we find that

$$E_v(\beta^a \Delta^a) \doteq E_v(\tilde{\beta}^a \Delta^a) \doteq 0, \quad a = 1, \ldots n, \quad v = 1, \ldots, q, \tag{40}$$

where each $\tilde{\beta}^a = ((-D)_J \beta^{aJ})$ is a differential function.

In the following theorem, we prove that local conservation laws exist only for quasi-Noether systems.

Theorem 1 (Quasi-Noether systems and conservation laws). *Any differential system (13) that possesses local conservation laws is quasi-Noether. A quasi-Noether system that admits continuous symmetries in general, possesses local conservation laws.*

Proof. Assume that the system (13) has some local conservation law
$$D_i P^i = \gamma^a \Delta^a, \tag{41}$$

where P^i and γ^a are differential functions. Then
$$E_v(\gamma^a \Delta^a) = 0, \quad a = 1, \ldots n, \quad v = 1, \ldots, q, \quad i = 1, \ldots, p, \tag{42}$$

which shows that the system is quasi-Noether.

Suppose now that the quasi-Noether system (15) possesses a continuous symmetry with the infinitesimal operator X_α (5). Rewriting condition (15) in the form

$$E_v(\beta^a \Delta^a) = \Gamma^{va}\Delta^a \doteq 0, \quad v = 1, \ldots q, \tag{43}$$

where the sum is taken over $1 \leq a \leq n$, β^a are differential functions, and $\Gamma^{va} = \Gamma^{vaJ}D_J$ are differential operators. Multiplying the Equation (43) by α^v and taking a sum over v, we obtain

$$\alpha^v E_v(\beta^a \Delta^a) = \alpha^v \Gamma^{va} \Delta^a. \tag{44}$$

Applying the Noether identity (8) in the LHS, we obtain

$$\left(X_\alpha - D_i R^i\right)(\beta^a \Delta^a) = \alpha^v \Gamma^{va} \Delta^a, \quad a = 1, \ldots n, \quad v = 1, \ldots, q, \quad i = 1, \ldots, p. \tag{45}$$

Since X_α is a symmetry operator for the system Δ^a, we obtain

$$D_i R^i (\beta^a \Delta^a) \doteq 0, \quad a = 1, \ldots n, \quad v = 1, \ldots, q, \quad i = 1, \ldots, p, \tag{46}$$

and therefore, a quasi-Noether system with continuous symmetries in general, possesses local conservation law (46). □

Note that in some cases the conservation law (46) may be trivial.

Note also that condition (15) allows one to find and classify quasi-Noether equations of a certain type. For example, it can be shown that equations of the following class:

$$u_t = u_{xx} + u_x^n, \quad n = 0, 1, 2, 3, \ldots \tag{47}$$

possess conservation law(s) only for $n = 0, 1, 2$. Correspondingly, for $n \geq 3$, such equations are not quasi-Noether and, hence, do not admit any conservation laws.

3.4. Classes of Quasi-Noether Equations

The quasi-Noether class of equations for which it is possible to establish a Noether type correspondence between symmetries and conservation laws is quite large, and covers practically all interesting differential systems of mathematical physics, and all systems possessing conservation laws. Many examples of equations of the class were given in [4,5,12]. Quasi-Noether systems include differential systems in the form of conservation laws, e.g., KdV, mKdV, Boussinesq, Kadomtsev-Petviashvili equations, nonlinear wave and heat equations, Euler equations, and Navier-Stokes equations; as well as the homogeneous Monge-Ampere equation, and its multi-dimensional analogue, see [4]. In [6], the approach based on the Noether identity was applied to quasi-Noether systems possessing infinite symmetries involving arbitrary functions of all independent variables, in order to generate an extension of the Second Noether theorem for systems that may not have well-defined Lagrangian functions.

3.4.1. Evolution Equations

Consider evolution equations of the form

$$u_t = A(u, u_x)u_{xx} + B(u, u_x). \tag{48}$$

Assuming $\beta = \beta(t, x, u, u_x, u_{xx}, \ldots, u_{nx})$, and requiring our equation to be quasi-Noether (15) we obtain the following condition

$$E_u[\beta u_t - \beta(Au_{xx} + B)] \doteq 0. \tag{49}$$

It can be shown that $\beta = \beta(t, x, u)$. Moreover, it can be shown that for this system, the above equality holds for all u (=) rather than strictly solutions ($\doteq$).

We obtain the following classes of equations:

1.
$$A(u, u_x) = \alpha H_{u_x}/S'(u),$$
$$B(u, u_x) = \alpha(u_x H_u - S(u))/S'(u), \qquad (50)$$
$$\beta(t, u) = e^{\alpha t} S'(u),$$

where $H(u, u_x)$ and $S(u)$ are arbitrary functions, and α is a constant.

In this case the equation $\Delta = u_t - Au_{xx} - B$ is obviously, quasi-Noether since the left hand side turns into total divergence upon multiplication by β:

$$\beta \Delta = D_t[e^{\alpha t} S(u)] - \alpha D_x[e^{\alpha t} H(u, u_x)].$$

2.
$$A(u, u_x) = (u_x G_{u_x} - G)/u_x^2,$$
$$B(u, u_x) = G_u + (b/u_x + c'(u))G + h + u_x(a + h_u + hc_u)/b, \qquad (51)$$
$$\beta(t, x, u) = \exp(at + bx + c(u)),$$

where $G(u, u_x)$, $c(u)$, and $h(u)$ are arbitrary functions, and a and b are constants.

a. The special case $G(u, u_x) = u_x^2$, $c(u) = u$, $h(u) = 0$, and $a = -b^2$, leads to the following equation:

$$u_t - u_{xx} - u_x^2 = 0. \qquad (52)$$

Multiplication by $\beta = \exp(bx - b^2 t + u)$ turns the LFS of Equation (52) into a total divergence:

$$D_t(\beta u) - D_x[\beta(u_x - b)] = 0.$$

b. Choosing $a = -b^2$, $c'(u) = 0$, $G(u, u_x) = u_x^2 + pu_x$, $p = const$, $h(u) = 1 - bp$ we obtain heat equation, $A = 1$, $B = 1$

$$u_t = u_{xx} \qquad (53)$$

c. Choosing $a = -b^2$, $c'(u) = 0$, $G(u, u_x) = u_x^2 + u^2 u_x/2$, $h(u) = -bu^2/2$ we obtain Burgers equation, $A = 1$, $B = -uu_x$

$$u_t = u_{xx} - uu_x. \qquad (54)$$

d. Choosing $a = -b^2 = -1$, $c'(u) = 0$, $G(u, u_x) = u_x^2 + (u - u^2)u_x$, $h(u) = 0$ we obtain Fisher equation, $A = 1$, $B = u - u^2$

$$u_t = u_{xx} + u(1 - u). \qquad (55)$$

3.
$$A(u, u_x) = a,$$
$$B(u, u_x) = au_x^2(\Phi_{uu} + b\Phi_u^2)/\Phi_u + 1/\Phi_u + \epsilon u_x, \qquad (56)$$
$$\beta(t, x, u) = v(t, x) \exp(-bt + b\Phi),$$

where $\Phi(u)$ is an arbitrary function, a and b are constants, and $v(x,t)$ is a solution of the following equation:
$$v_t - bv_x + av_{xx} = 0.$$

4.
$$\begin{aligned} A(u, u_x) &= a, \\ B(u, u_x) &= au_x^2 \ell_u + \epsilon u_x, \\ \beta(t, x, u) &= v(x, t) e^{\ell}, \end{aligned} \quad (57)$$

where $\ell(u)$ is an arbitrary function, and v satisfies the same linear Equation (57).

5.
$$\begin{aligned} A(u, u_x) &= G_u/\Phi_u, \\ B(u, u_x) &= u_x^2 (G_{uu} + \tau G_u^2 + \epsilon G_u \Phi_u)/\Phi_u + u_x(\delta + 2\sigma G_u/\Phi_u) + 1/\Phi_u, \\ \beta(t, x, u) &= \exp[-\epsilon t + \sigma(x + \delta t) + \tau G + \epsilon \Phi] \cosh[(x + \delta t - \mu)\sqrt{\sigma^2 - \tau}]\Phi_u, \end{aligned} \quad (58)$$

where $\Phi(u)$ and $G(u)$ are arbitrary functions, and $\delta, \epsilon, \mu, \sigma$, and τ are arbitrary constants.

We do not pursue the remaining cases, since these necessarily involve dependence on $1/u_x$. However, the above analysis gives insight into, for instance, equations of the following class:

$$u_t = u_{xx} + u_x^n, \quad n = 0, 1, 2, 3, \ldots \quad (59)$$

We have shown that such an equation possesses a conservation law only for $n = 0, 1, 2$. For $n \geq 3$, such equations are not quasi-Noether and, hence, do not admit conservation laws.

3.4.2. Quasi-Linear Equations

Consider now quasi-linear equations of the form

$$\Delta = u_{tt} - A(u, u_x)u_{xx} - C(u, u_x) = 0. \quad (60)$$

We require our equation to be quasi-Noether (15)

$$E_u[\beta(u_{tt} - Au_{xx} - C)] \doteq 0. \quad (61)$$

Using the identity (34)
$$E_u(\beta \Delta) = (D_\Delta^*)\beta + (D_\beta^*)\Delta, \quad (62)$$

we obtain
$$E_u(\beta \Delta) \doteq (D_\Delta^*)\beta \doteq 0. \quad (63)$$

For Equation (60) condition (63) takes a form

$$\begin{aligned} D_t^2 \beta - D_x^2(\beta A) + D_x(\beta u_{xx} A_{u_x}) + D_t(\beta u_{xx} A_{u_t}) + \\ D_x(\beta C_{u_x}) + D_t(\beta C_{u_t}) - \beta C_u - \beta A_u u_{xx} = 0. \end{aligned} \quad (64)$$

Solving (64) for $\beta = \beta(x, t, u)$, we obtain

$$\begin{aligned} A &= u_t L + M, \\ C &= u_t R + P + f(u, u_t), \end{aligned} \quad (65)$$

where

$$R(u, u_x) = \int \left[\frac{\beta_x}{\beta} L + u_x L_u\right] du_x \qquad (66)$$

$$P(u, u_x) = \int K(u, u_x) du_x,$$

and $L = L(u, u_x)$, $K = K(u, u_x)$, $f = f(u, u_t)$, $\beta = \beta(x, t, u)$ are functions to be determined. We get the following solutions:

1.

$$L = \frac{1}{m}[2M_u + lM_{u_x} + u_x M_{uu_x} - K_{u_x}]$$
$$R = \int [lL + u_x L_u] \, du_x = l^2 M + 2lu_x M_u + u_x{}^2 M_{uu} - lK - u_x K_u + \int K_u du_x, \qquad (67)$$

where l, m are abrbitrary constants, $M(u, u_x)$, $K(u, u_x)$ are arbitrary functions, $\beta = e^{lx+mt}$, and $f(u, u_t)$ satisfies

$$-f_u + mf_{u_t} + u_t f_{uu_t} = -m^2. \qquad (68)$$

2.

$$R = L = 0$$
$$K = (l + qu_x)M + u_x M_u + \int [qM + M_u] du_x + r(u), \qquad (69)$$

where l, m, q are abrbitrary constants, function $M(u, u_x)$ is arbitrary, $\beta = e^{lx+mt+qu}$, and

$$f(u, u_t) = -qu_t^2 + s(u)u_t + v(u), \quad s(u) = -m + \frac{qv(u) + v'(u)}{m}. \qquad (70)$$

Examples:

a. If $M(u, u_x) = c$, we obtain

$$u_{tt} = Mu_{xx} + (lM + r(u))u_x + qMu_x{}^2 + R(u) + (-qu_t{}^2 + su_t + v), \qquad (71)$$

where $s(u)$ satisfies (70) and functions $r(u)$ and $R(u)$ are arbitrary. Choosing $q = s = v = 0$, $r = -lM$ we obtain

$$u_{tt} = Mu_{xx} + R(u). \qquad (72)$$

The class (72) includes Liouville equation $(R(u) = ke^{\lambda u})$

$$u_{tt} = Mu_{xx} + ke^{\lambda u}, \qquad (73)$$

and Sine-Gordon equation $((R(u) = k\sin\lambda u)$

$$u_{tt} = Mu_{xx} + k\sin\lambda u. \qquad (74)$$

b. Choosing $M(u, u_x) = g(u)$, $q = s = v = 0$, $r(u) = -lg(u)$, we obtain

$$u_{tt} = g(u)u_{xx} + g'(u)u_x{}^2. \qquad (75)$$

The Equation (75) is a nonlinear wave equation.

4. Quasi-Lagrangians

For β a cosymmetry, the quantity $L = \beta \cdot \Delta$ serves an analogous role to a Lagrangian. In this section, we examine the applications of such a *quasi-Lagrangian L* in more detail. We show that equations with a quasi-Lagrangian have a Noether correspondence on a subspace. We then demonstrate the incompleteness of the correspondence from Section 3.1 with an example. We conclude with a comparison of the invariant submanifolds of quasi-Noether systems with those of Hamiltonian systems.

4.1. A Noether Correspondence

Let us introduce the operator T depending on β such that $E(\beta \cdot \Delta) = T \cdot \Delta$. If $\mathbb{D}_\Delta$ is an endomorphism (same number of equations as dependent variables), then $T = \mathbb{D}_\beta^* - \mathbb{D}_\beta$ by (24).

We say an operator T is nondegenerate if its restriction to $\Delta = 0$ is not the zero operator. If β is a characteristic for a conservation law (generator), then $E(\beta \cdot \Delta) = 0$, so $T = 0$ (degenerate). Thus β, in this section, will be a cosymmetry which is *not* a characteristic.

Recall that a functional $\mathscr{L}[u] = \int L[u] dx$ is an equivalence class modulo total divergences. We say a smooth functional $\mathscr{L}[u] = \int L[u] dx$ is nondegenerate if it is nonzero and some representative $L[u]$ (hence all of them) does not vanish quadratically with Δ. We identify nondegenerate functionals with their affine terms: $L \sim L|_{\Delta=0} + \beta^I \cdot D_I \Delta$. Mod a divergence, we then have $L \sim L_0 + \beta \cdot \Delta$, where $\beta = (-D)_I \beta^I$ comes from integration by parts. We will also identify $\mathscr{L}$ with L. As a non-example, consider, for example, a second order scalar nonlinear PDE $\Delta = \Delta^1(u_{ij})$. Then for $L = \Delta^2$, we have $E(L) = D_{ij}(A^{ij}\Delta)$, where $A_{ij} = 2\partial\Delta/\partial u_{ij}$.

The concept of *sub-symmetry* was introduced in [7]. We say X_α generates a *sub-symmetry* of Δ if $X_\alpha(T \cdot \Delta)|_{\Delta=0} = 0$ for some nondegenerate T. In the special case that $X_\alpha(T \cdot \Delta)|_{T \cdot \Delta=0} = 0$, clearly X_α is an "ordinary" symmetry of a sub-system $T \cdot \Delta$; the more general definition will not be needed here. Roughly speaking, in the case of sub-symmetry transformations, only some (not all) combinations of original equations are required to be invariant. We note that in [8], sub-symmetries of sub-systems generated by cosymmetries yield conservation laws through the mechanism described in Section 3.1. In the present work, we find a new appearance of sub-symmetries.

We now present an extension of Noether theorem. Every variational symmetry X_α of a quasi-Lagrangian L corresponds to a conservation law characteristic $T^*\alpha$ in the image of T^*. The variational symmetry is only a sub-symmetry of the PDE Δ. If $L = L_0 + \beta \cdot \Delta$ and cosymmetry β is also in the image of T^*, then a family of equations $\Delta + ker[T]$ shares variational symmetries, hence conservation laws generated by the quasi-Lagrangian.

Theorem 2. *Let Δ be a normal, totally nondegenerate PDE system, and suppose there exist smooth and nondegenerate operator T and functional L such that $E(L) = T \cdot \Delta$.*

1. *$X_\alpha L = div$, if and only if $(T^*\alpha) \cdot \Delta = div$.*
2. *If $X_\alpha L = div$, then α is a sub-symmetry of Δ.*

    Let $L = L_0 + \beta \cdot \Delta$. Define $\Delta_f = \Delta + f$ and $L_f = L_0 + \beta \cdot \Delta_f$ for $f \in ker[T]$.
3. *If $\beta \in im[T^*]$, $X_\alpha L_f = div$ if and only if $X_\alpha L = div$.*

Proof. 1. The Noether theorem states $X_\alpha L = div$ if and only if $\alpha \cdot E(L) = div$; thus, $\alpha \cdot (T \cdot \Delta) = div$. If we integrate by parts, we obtain $(T^*\alpha) \cdot \Delta = div$, as desired.

2. It is well known $X_\alpha L = div$ implies X_α is a symmetry of $E(L)$. Since $E(L) = T \cdot \Delta$ is a sub-system of a prolongation of Δ, this means X_α is only a sub-symmetry of Δ.

3. Since $\beta = T^*\gamma$ for some smooth γ, we have $\beta \cdot f = (T^*\gamma) \cdot f = \gamma \cdot (Tf) + div = div$, so we conclude $X_\alpha L_f = X_\alpha L + div$. □

Remark 1. *Part 3 illustrates that a variational symmetry α of L need not correspond to a symmetry of Δ, since if $X_\alpha \Delta_f = A \Delta_f$, then $X_\alpha f = A f$ need not be true for all $f \in \ker[T]$. We will demonstrate this failure in Section 4.2.*

Therefore, the Green–Lagrange-Noether approach from Section 3 is incomplete: it requires vector field X_α to be a symmetry of Δ, while Part 2 indicates we must instead consider sub-symmetries. The incompleteness was partially demonstrated in [11] in the case β is a characteristic of a conservation law, but the case where cosymmetry β is not a characteristic was not indicated.

Note that Ibragimov's approach [19] is distinct; it uses variational symmetries of an extended Lagrangian system obtained by, essentially, treating the β's as dependent variables. Such variational symmetries must be symmetries of Δ, which is one reason why our result is different. Another reason follows from a direct computation of the characteristic; there is an additional term due to the symmetric action of X_α on L, which in our approach is simply a divergence not contributing to the characteristic.

Remark 2. *The main difference of our extension is that the vector field X_α is only a sub-symmetry of the PDE Δ, not a true symmetry, since we are considering variational symmetries of L. This highlights the incompleteness of the previous approach from Section 3, which requires using symmetries of the PDE. We will present an example in Section 4.2 where all lower conservation laws arise according to the correspondence in Theorem 2, many only from sub-symmetries.*

Remark 3. *Upon restricting said symmetries of Δ to variational symmetries of L, the results of the two approaches are equivalent, of course. Let $A\Delta = X_\alpha \Delta$. The Green–Lagrange-Noether approach yields the following conservation law:*

$$(T^*\alpha - X_\alpha \beta - A^* \beta)\Delta = \text{div}. \tag{76}$$

But in fact, $(X_\alpha \beta + A^* \beta) = X_\alpha L + \text{div} = \text{div}$ since α is a variational symmetry. Thus,

$$(T^*\alpha - \mu)\Delta = \text{div} \tag{77}$$

for some generating function μ. This implies $T^*\alpha$ is also a generating function, as claimed.

Remark 4. *We note that $L_0 = L|_{\Delta=0}$ is nonzero (mod divergence) in general. For example, if $E(L) = \Delta$ for $L = uv_t - \lambda[u,v]$ and the two-component evolution system*

$$\begin{aligned} E_u(L) &= v_t - E_u \lambda = 0, \\ E_v(L) &= -u_t - E_v \lambda = 0, \end{aligned} \tag{78}$$

then unless λ is of a special form, L need not vanish on $\Delta = 0$.

4.2. An Example

It is well known [2] that the second order Burgers equation

$$u_t = u u_x + u_{xx} = D_x\left(\frac{1}{2}u^2 + u_x\right) \tag{79}$$

has exactly one conservation law and no cosymmetries other than $\beta = 1$. It thus has no quasi-Lagrangians of the form $\beta \cdot \Delta$ and its conservation law does not arise from such. We next consider the opposite situation.

Consider a third order evolution equation $u_t = F[u]$ of the form

$$u_t = F[u] := \frac{3 u_{xx} u_{xxx}}{u_x} - \frac{u_{xx}^3}{u_x^2} + f(t) = \frac{1}{u_{xx}} D_x\left(\frac{u_{xx}^3}{u_x} + f(t) u_x\right). \tag{80}$$

Even if $f(t) = 0$, this equation cannot be written as a Hamiltonian system, since although $u_t = \mathscr{D} \cdot E(H)$ for $H = -\frac{1}{2}u_x^2$ and the skew-adjoint operator

$$\mathscr{D} = 2\frac{u_{xx}}{u_x}D_x + \frac{u_{xxx}}{u_x} - \frac{u_{xx}^2}{u_x^2},$$

it can be shown that $\mathscr{D}$ does not verify the Jacobi identity. We do, however, see a quasi-Lagrangian structure:

$$E(L) := E(-\frac{1}{2}u_x(u_t - F)) = D_x(u_t - F) =: T\Delta. \tag{81}$$

We present the cosymmetries of order $\beta(t, x, u, \ldots, u_5)$:

$$u_x, \quad -u_{xx}, \quad -D_x(xu_x + tF), \quad -D_xF. \tag{82}$$

The even-order cosymmetries in (82) are conservation law generators which arise from variational symmetries of the quasi-Lagrangian

$$L := -\frac{1}{2}u_x(u_t - F[u]).$$

Indeed, if we rewrite them using (80),

$$-D_xu_x, \quad -D_x(xu_x + tu_t), \quad -D_xu_t, \tag{83}$$

then we see they generate a translation $x \to x + \varepsilon$, a scaling $(x,t) \to e^\varepsilon(x,t)$, and a translation $t \to t + \varepsilon$, respectively. These (three) cosymmetries comprise the (order 4) conservation law generators for Equation (80).

There are two differences from the classical Noether theorem:
1. The variational symmetry $u \to u + \varepsilon$ does not lead to a conservation law since for $\alpha = 1$, we have $T^*(\alpha) = -D_x(1) = 0$.
2. Time translation is *not* a symmetry of $\Delta = 0$ unless $f'(t) = 0$. Nevertheless, since

$$\frac{1}{2}u_x f(t)$$

is a divergence expression, we see $\alpha = u_t$ generates a variational symmetry of L, and then a conservation law of Δ.

Remark 5. *We present a comparison with the quasi-Noether/nonlinear self-adjointness/Green–Lagrange approach. Suppose the symmetry condition $X_\alpha \Delta = A\Delta$, and integrate by parts:*

$$X_\alpha L = \beta X_\alpha \Delta + (X_\alpha \beta)\Delta = \beta A\Delta + (X_\alpha \beta)\Delta = (A^*\beta + X_\alpha \beta)\Delta + \text{div}. \tag{84}$$

Also recall the Noether identity and quasi-Lagrangian structure (81)

$$X_\alpha L = \alpha E(L) + \text{div} = \alpha D_x \Delta + \text{div} = (-D_x\alpha)\Delta + \text{div}. \tag{85}$$

Combining these two formulas gives the characteristic for the conservation law generated using this approach:

$$B(\alpha) := -D_x\alpha - X_\alpha \beta - A^*\beta. \tag{86}$$

The only symmetry in (83) is $\alpha = u_x$. In this case, it can be verified that

$$-X_\alpha \beta = \frac{1}{2} D_x \alpha, \quad X_\alpha \Delta = D_x \Delta, \quad -A^*\beta = -\frac{1}{2} D_x \alpha,$$

so that $B(\alpha) = -D_x \alpha = T^*(\alpha)$. Recall the characteristic generated by a variational symmetry is also $-D_x \alpha$. Thus, the two approaches agree when α is a symmetry of (80).

Remark 6. Observe that $\alpha = u_t$ and $\alpha = tu_t + xu_x$ do not generate symmetries of (80) if $f'(t) \neq 0$. However, a simple modification yields the desired conservation law.

Combining (84) with (85) without the symmetry condition $X\Delta \doteq 0$ yields

$$(-D_x \alpha - X_\alpha \beta)\Delta - \beta X_\alpha \Delta = \text{div}. \tag{87}$$

For $\alpha = u_t$, it can be shown that

$$X_\alpha \Delta = D_t \Delta - f'(t) =: R\Delta - f'(t). \tag{88}$$

Therefore, after an integration by parts, we recover the conservation law plus an error term:

$$(-D_x \alpha - X_\alpha \beta - A^*\beta)\Delta + \beta f'(t) = \text{div}. \tag{89}$$

In fact, the error term is a divergence: $f'(t)\beta = D_x(-\frac{1}{2} u f'(t))$, and we derive an analogous conservation law. Again, the characteristic is $-D_x \alpha$.

Let us apply the same analysis to $\alpha = tu_t + xu_x$. We have

$$X_\alpha \Delta = -\Delta + tD_t\Delta + xD_x\Delta + (1 - tD_t)f(t) =: A\Delta + f(t) - tf'(t).$$

As before, we obtain

$$(-D_x \alpha - X_\alpha \beta - A^*\beta)\Delta + \beta(f(t) - tf'(t)) = \text{div}.$$

The second expression on the left hand side is a divergence, so we obtain a conservation law. Using $-A^*\beta = -\frac{1}{2} D_x \alpha$, we conclude that the characteristic is again $-D_x \alpha$.

4.3. Critical Points and Symmetries

We first recall the notion of invariant submanifolds of an evolutionary system, introduce the notion of critical points of conservation laws, then show that the critical points and symmetry invariant submanifolds satisfy opposite containments in the quasi-Lagrangian and Hamiltonian cases, see Theorem 3.

Let $\Delta = u_t - P[u]$ be an evolution system with evolutionary operator

$$D_t = \partial_t + u_{It}^a \partial_{u_I^a} \doteq \partial_t + D_I P^a \partial_{u_I^a} = \partial_t + X_P. \tag{90}$$

Each of these objects satisfies special determining equations: α symmetry, β cosymmetry. Symmetry α satisfies the equation

$$D_t \alpha = \mathbb{D}_P \alpha. \tag{91}$$

Cosymmetry β satisfies the equation

$$D_t \beta = -\mathbb{D}_P^* \beta. \tag{92}$$

These equations imply that *symmetry invariance* $\alpha = 0$ and *cosymmetry invariance* $\beta = 0$ are compatible with $\Delta = 0$. By *compatible*, we mean if $u_0(x)$ solves $\gamma(t_0, x, u_0, \ldots) = 0$, then for time evolution $u(t) = U(t; t_0)u_0$, we have $\gamma(t, x, u(t), \ldots) = 0$. This follows for P, u_0 analytic by a (local) power series expansion, since the invariance conditions imply $D_t^k \gamma|_{\gamma=0} = 0$ for all $k \geq 0$. The same holds for systems in Cauchy-Kovalevskaya form, which can be rewritten as evolution systems.

Remark 7. *It was observed in [11] that a co-symmetry β induces an invariant one-form $\beta[u] \cdot du$ with respect to time evolution, in an analogous way that a symmetry α induces an invariant vector field $\alpha[u] \cdot \partial/\partial u$. The determining equations are simply vanishing Lie derivatives of these tensors. The critical sets of these tensors are therefore time invariant, which gives a geometric meaning to the compatibility of these invariance conditions.*

Let us note an interesting phenomenon. If $E(L) = T\Delta$ is quasi-Lagrangian and X_α is a variational symmetry, then X_α is a symmetry of $T \cdot \Delta$, but $T^*\alpha$ generates a conservation law of Δ. We thus find a "duality" relation between the existence of compatible systems.

Proposition 1. *If $E(L) = T\Delta$ and $X_\alpha L = \text{div}$, then both $\{T^*\alpha = 0, \Delta = 0\}$ and $\{\alpha = 0, T\Delta = 0\}$ are compatible systems.*

Next, if cosymmetry β also generates a conservation law, we call the equation $\beta = 0$ the *critical point condition* for the conserved integral generated by β. To justify this terminology, we recall the well known fact that cosymmetry β is a characteristic if and only if $\mathbb{D}_\beta^* - \mathbb{D}_\beta = 0$ (self-adjointness), i.e., $\beta = E(M^t)$ for the conserved density M^t of the conservation law

$$D_t M^t + D_i M^i = \beta \cdot \Delta = 0. \tag{93}$$

It follows that the condition $\beta = 0$ is the Euler–Lagrange equation for the (possibly time-dependent) functional $\int M^t[u]dx$, i.e., the equation for critical points of this conserved integral.

By the discussion for cosymmetries, the critical point condition is compatible with time evolution $u_t = P$. This is sensible for two reasons: (1) if $u(0)$ is a minimizer of $\int M^t[u]dx$ for suitable boundary conditions, then the conservation of $\int M^t[u]dx$ implies $u(t)$ is also a minimizer at later times. (2) If $\partial_t M^t = 0$, then

$$X_P M^t = D_t M^t = -D_i M^i = \text{div}, \tag{94}$$

so P actually generates a variational symmetry of the functional $\int M^t[u]dx$. It follows that X_P is a symmetry of $E(M^t[u]) = 0$, hence it preserves the solution space.

Remark 8. *Although many current papers are devoted to the construction of Lie-type invariant solutions and conservation laws for non-Hamiltonian systems, we are not aware of any which constructed the critical points of these conservation laws. The possibility for such was raised in [20].*

Example 1 (Time-dependent conservation law). *The KdV equation*

$$u_t + uu_x + u_{xxx} = 0, \tag{95}$$

has conservation law

$$D_t M^t + D_x M^x \doteq 0 \tag{96}$$

of the form

$$\frac{\partial}{\partial t}\left(xu - \frac{tu^2}{2}\right) + \frac{\partial}{\partial x}\left(t\left(\frac{u_x^2}{2} - uu_{xx} - \frac{u^3}{3}\right) + \frac{xu^2}{2} + xu_{xx} - u_x\right) = 0, \tag{97}$$

such that $M^t[u] = xu - \frac{tu^2}{2}$, and $E_1 M^t[u] = x - tu$. If $\sigma \neq 0$, then $u_*(x) = x/\sigma$ is a solution of $E_1 M^t(\sigma, x, u, \ldots) = 0$, and $u(t, x) = x/t$ is the solution of $\{u_t = P, E_1 M^t = 0\}$ which satisfies $u(\sigma, x) = u_*(x)$.

Example 2 (Time-dependent evolution). *The generalized KdV equation* [21]

$$u_t + f(t,u)u_x + u_{xxx} = 0, \tag{98}$$

where $f(t,u) = at^{-1/3}u + bu + cu^2$ *for* a,b,c *constant, has explicit time dependence if* $a \neq 0$. *It has a conservation law* $D_t M^t + D_x M^x \doteq 0$ *with conserved density*

$$M^t = \frac{1}{2}ctu_x^2 - \frac{1}{12}t(cu^2+bu)^2 + \frac{1}{6}x(cu^2+bu) - \frac{1}{2}at^{2/3}(\frac{1}{3}cu^3+\frac{1}{4}bu^2), \tag{99}$$

and characteristic (note that [21] *has a typographical error in the* u_{xx} *term)*

$$E_u(M^t) = -ctu_{xx} - \frac{1}{6}t(2c^2u^3 + 3bcu^2 + b^2u) - \frac{1}{4}at^{2/3}(2cu^2+bu) + \frac{1}{6}x(2cu+b). \tag{100}$$

Let us show that the system $\{u_t + fu_x + u_{xxx} = 0, E_u(M^t) = 0\}$ is compatible. Suppose first that $c = 0$. Then the solution of $E_u(M^t) = 0$ is

$$u(t,x) = \frac{2x}{(3a + 2bt^{1/3})t^{2/3}}, \tag{101}$$

which solves (98) for $c = 0$. Suppose now that $c \neq 0$. If we solve $E_u(M^t) = 0$ for u_{xx}, then it is easy to show that $(u_{xx})_t = (u_t)_{xx}$, or that (98) is consistent with $E_u(M^t) = 0$. Alternatively, substituting $E_u(M^t)$ into (98) gives a first order PDE, which has solution

$$u(t,x) = -\frac{b}{2c} + t^{-1/3}g(\xi), \quad \xi = t^{-1/3}x + (4c)^{-1}(3abt^{1/3} + b^2 t^{2/3}), \tag{102}$$

where g is an arbitrary function. Substitution into $E_u(M^t) = 0$ gives an ODE for g.

Example 3 (A non-Hamiltonian example). *The nonlinear telegraph system* [22]

$$\begin{aligned} v_t + ke^u u_x - e^u &= 0, \\ u_t - v_x &= 0, \end{aligned} \tag{103}$$

where $k \neq 0$ *is a constant, has a conservation law with density (note the small error in* [22])

$$M^t = e^{-x/k}[(tv/2 + 2x)v - k(uv + te^u)] \tag{104}$$

and characteristic

$$\begin{aligned} E_u(M^t) &= -ke^{-x/k}(v + te^u), \\ E_v(M^t) &= e^{-x/k}(2x + tv - ku). \end{aligned} \tag{105}$$

A critical point (u_*, v_*) satisfies the system $E_u(M^t) = E_v(M^t) = 0$, whose solution is

$$\begin{aligned} u_*(x,t) &= 2x/k - W(\frac{t^2}{k}e^{2x/k}), \\ v_*(x,t) &= -\frac{k}{t}W(\frac{t^2}{k}e^{2x/k}), \end{aligned} \tag{106}$$

where W is the Lambert W function. It is straightforward to show that (u_*, v_*) solves (103).

Before turning to our next result, we recall some properties of Hamiltonian systems, see e.g., [2]. The two types of invariant submanifolds are coupled in this case. Suppose

$$P = \mathcal{D} \cdot E(H) \tag{107}$$

for some Hamiltonian $H[u]$ and skew-symmetric operator $\mathcal{D}$ which verifies the Jacobi identity, in the sense that if $\{\mathcal{P}, \mathcal{Q}\} = \int E(P) \cdot \mathcal{D} \cdot E(Q) dx$ is the Poisson bracket induced by $\mathcal{D}$, then $\{,\}$ verifies the Jacobi identity. Then the Noether relation (Theorem 7.15 in [2]) states that every conserved integral $\int M^t[u] dx$ yields a symmetry of Hamiltonian form: $\alpha = \mathcal{D} \cdot E(M^t)$. In other words,

$$\alpha = \mathcal{D} \cdot \beta, \tag{108}$$

where $\beta = E(M^t)$ is the associated characteristic. Therefore,

$$\{\beta = 0\} \subset \{\alpha = 0\}, \tag{109}$$

where $\{\beta = 0\}$ is an invariant submanifold of critical points, and $\{\alpha = 0\}$ is an invariant submanifold of symmetry-invariant functions.

The converse is not true: if the Hamiltonian vector field generated by $\alpha = \mathcal{D} \cdot E(M^t)$ is a symmetry, then $\beta = E(M^t)$ need not be a characteristic. In general, there exists a time-dependent $C[t; u]$ with $C[t_0; .] \in \ker \mathcal{D}$ for each t_0, such that $\beta - C$ generates a conservation law.

Let us show that the containment in the quasi-Lagrangian case is opposite to (109). We summarize the comparison below.

Theorem 3. *Let L be a quasi-Lagrangian for an evolutionary system with $E(L) = T \cdot \Delta$, and let α be a variational symmetry α with conservation law characteristic $\beta = T^* \alpha$. Then*

$$\{\alpha = 0\} \subset \{\beta = 0\}, \tag{110}$$

so that vector field invariant solutions are critical points of a conserved integral $\int M^t[u] dx$.

Let H be the Hamiltonian for the system $u_t = \mathcal{D} \cdot E(H)$, and let $\alpha = \mathcal{D} \cdot \beta$ be a Hamiltonian symmetry with $\beta = E(M^t)$. Then

$$\{\beta = 0\} \subset \{\alpha = 0\} \tag{111}$$

such that critical points of the integral $\int M^t[u] dx$ are symmetry invariant.

Remark 9. *The two situations overlap in the Lagrangian case. If $T, \mathcal{D}$ commute with ∂_t, then their invertibility implies that the modified systems $T^{-1} \Delta, \mathcal{D}^{-1} \Delta$ are Lagrangian. In this case, the symmetry invariant submanifolds coincide with the critical points of conserved integrals.*

Remark 10. *In the quasi-Lagrangian case, vector field X_α need not be a symmetry, so $\alpha = 0$ may not be compatible with time evolution. In Section 4.2, compatibility of $\alpha = 0$ fails for all symmetries induced by (83) (e.g., $u_t = 0$) if $f'(t) \neq 0$ in (80).*

Conversely, critical point condition $\beta = 0$ is compatible, but $T^* \alpha = \beta = 0$ need not imply symmetry invariance, as the example in Section 4.2 shows for $\beta = D_x F$. However, symmetry invariance occurs after dividing by the kernel of T^*, which is $\{g(t)\}$ in Section 4.2. This is analogous to the Hamiltonian case, for which if $\mathcal{D} \cdot \beta = \alpha = 0$, then $\beta = 0$ holds after dividing by the kernel of $\mathcal{D}$.

Remark 11. *After this paper was written we learned about the paper [23] where "symplectic operator" $\mathcal{E}$ was introduced. The operator $\mathcal{E}$ is similar to our operator T determining quasi-Lagrangians.*

5. Conclusions

In this paper, we discussed the problem of correspondence between symmetries and conservation laws for a general class of differential systems (quasi-Noether systems). Our approach is based on the Noether operator identity. We discussed some properties of Noether identity, and showed that it leads to the same conservation laws as Green–Lagrange identity. We generated classes of quasi-Noether equations of the second order of evolutionary and quasilinear form.

We introduced and analyzed the notion of a *quasi-Lagrangian*, which generalizes the quantity $L := \beta \cdot \Delta$ used in previous works to the case where $L|_{\Delta=0}$ may not be zero, and we studied variational symmetries with respect to the quasi-Lagrangian, which generalizes the previous work's restriction to symmetries of differential system Δ. Because this recovers the Lagrangian case $\Delta = E(L)$, we extended Noether's theorem (Theorem 2).

The previous work was not able to achieve both of the following: (1) extend Noether's theorem while still involving invariant vector fields (see [11] for an extension using only cosymmetries), and (2) recover the case of a Lagrangian system $\Delta = E(L)$, unlike, in general, the quasi-Noether approach of Section 3. Moreover, in Section 4.2, we presented a system for which variational symmetries of a quasi-Lagrangian do not correspond to symmetries, such that the previous work is unable to associate the conservation laws to vector fields using the given cosymmetry.

Based on the notion of invariant submanifolds we introduced critical points of conservation laws, and gave examples of the compatibility of cosymmetry invariance and the critical point condition, including for a non-Hamiltonian system.

We concluded with a comparison of the invariant submanifolds of quasi-Noether and Hamiltonian evolution equations, and showed these systems are "opposite" in some sense. For Lagrangian systems, the cosymmetry and symmetry invariant submanifolds coincide, while in general there is a containment in one direction or the other.

Author Contributions: Conceptualization, V.R. and R.S.; methodology, V.R. and R.S.; software, V.R. and R.S.; validation, V.R. and R.S.; formal analysis, V.R. and R.S.; investigation, V.R. and R.S.; resources, V.R. and R.S.; data curation, V.R. and R.S.; writing—original draft preparation, V.R. and R.S.; writing—review and editing, V.R. and R.S.; visualization, V.R. and R.S.; supervision, V.R. and R.S.; project administration, V.R. and R.S.; funding acquisition, V.R. and R.S.

Funding: This research received no external funding.

Acknowledgments: R. Shankar was partially supported by the National Science Foundation Graduate Research Fellowship Program under grant No. DGE-1762114.

Conflicts of Interest: The authors declare no conflicts of interest.

References

1. Noether, E. Invariantevariationsprobleme, Nachr. König. Gessell. Wissen. Göttingen. *Math. Phys.* **1918**, *K1*, 235–257.
2. Olver, P.J. *Applications of Lie Groups to Differential Equations*; Springer Science & Business Media: Berlin, Germany, 2000; Volume 107.
3. Rosen J. *Some Properties of the Euler–Lagrange Operators*; Preprint TAUP-269-72; Tel-Aviv University: Tel-Aviv, Israel, 1972.
4. Rosenhaus, V.; Katzin, G.H. On symmetries, conservation laws, and variational problems for partial differential equations. *J. Math. Phys.* **1994**, *35*, 1998–2012. [CrossRef]
5. Rosenhaus, V.; Katzin, G.H. Noether operator relation and conservation laws for partial differential equations. In *Proceedings of the First Workshop: Nonlinear Physics: Theory and Experiment, Gallipoli, Italy, 29 June–7 July 1995*; World Scientific: Singapore, 1996; pp. 286–293.
6. Rosenhaus, V.; Shankar, R. Second Noether theorem for quasi-Noether systems. *J. Phys. A Math. Theor.* **2016**, *49*, 175205. [CrossRef]
7. Rosenhaus, V.; Shankar, R. Sub-symmetries, and their properties. *Theor. Math. Phys.* **2018**, *197*, 1514–1526, [CrossRef]
8. Rosenhaus, V.; Shankar, R. Sub-Symmetries and Conservation Laws. *Rep. Math. Phys.* **2019**, *83*, 21–48. [CrossRef]
9. Ibragimov, N.H. *Transformation Groups Applied to Mathematical Physics*; Translated from the Russian; Mathematics and its Applications (Soviet Series); D. Reidel Publishing Co.: Dordrecht, The Netherlands, 1985.
10. Lunev, F.A. An analogue of the Noether theorem for non-Noether and nonlocal symmetries. *Theor. Math. Phys.* **1991**, *84*, 816–820. [CrossRef]

11. Anco, S. On the incompleteness of Ibragimov's conservation law theorem and its equivalence to a standard formula using symmetries and adjoint-symmetries. *Symmetry* **2017**, *9*, 33. [CrossRef]
12. Rosenhaus, V. Differential Identities, Symmetries, Conservation Laws and Inverse Variational Problems. In *Proceedings of the XXI Intern. Colloquium on Group Theoretical Methods in Physics, Goslar, Germany, 15–20 July 1996*; World Scientific: Singapore, 1997; pp. 1037–1044.
13. Sarlet, W.; Cantrijn, F.; Crampin, M. Pseudo-symmetries, Noether's theorem and the adjoint equation. *J. Phys. A* **1987**, *20*, 1365–1376. [CrossRef]
14. Vinogradov, A.M. Local Symmetries and Conservation Laws. *Acta Appl. Math.* **1984**, *2*, 21–78. [CrossRef]
15. Anco, S.C.; Bluman, G. Direct construction of conservation laws from field equations. *Phys. Rev. Lett.* **1997**, *78*, 2869–2873. [CrossRef]
16. Vladimirov, V.S.; Zharinov, V.V. Closed forms associated with linear differential operators. *Differ. Equ.* **1980**, *16*, 534–552.
17. Zharinov, V.V. Conservation laws of evolution systems. *Theor. Math. Phys.* **1986**, *68*, 745–751. [CrossRef]
18. Caviglia, G. Symmetry transformations, isovectors, and conservation laws. *J. Math. Phys.* **1986**, *27*, 973–978. [CrossRef]
19. Ibragimov, N.H. Nonlinear self-adjointness and conservation laws. *J. Phys. A Math. Theor.* **2011**, *44*, 432002. [CrossRef]
20. Ma, W.X.; Zhou, R. Adjoint symmetry constraints leading to binary nonlinearization. *J. Nonlinear Math. Phys.* **2002**, *9* (Suppl. 1), 106–126. [CrossRef]
21. Anco, S.; Rosa, M.; Gandarias, M. Conservation laws and symmetries of time-dependent generalized KdV equations. *arXiv* **2017**, arXiv:1705.04999.
22. Bluman, G.W. Conservation laws for nonlinear telegraph equations. *J. Math. Anal. Appl.* **2005**, *310*, 459–476. [CrossRef]
23. Fels, M.E.; Yasar, E. Variational Operators, Symplectic Operators, and the Cohomology of Scalar Evolution Equations. *J. Nonlinear Math. Phys.* **2019**, *26*, 604–649. [CrossRef]

© 2019 by the authors. Licensee MDPI, Basel, Switzerland. This article is an open access article distributed under the terms and conditions of the Creative Commons Attribution (CC BY) license (http://creativecommons.org/licenses/by/4.0/).

Article

Comparison of Noether Symmetries and First Integrals of Two-Dimensional Systems of Second Order Ordinary Differential Equations by Real and Complex Methods

Muhammad Safdar [1,*], Asghar Qadir [2] and Muhammad Umar Farooq [3]

[1] School of Mechanical and Manufacturing Engineering (SMME), National University of Sciences and Technology, Campus H-12, Islamabad 44000, Pakistan
[2] School of Natural Sciences (SNS), National University of Sciences and Technology, Campus H-12, Islamabad 44000, Pakistan; asgharqadir46@gmail.com
[3] Department of Basic Sciences & Humanities, College of E & ME, National University of Sciences and Technology, H-12, Islamabad 44000, Pakistan; m_ufarooq@yahoo.com
* Correspondence: msafdar@smme.nust.edu.pk

Received: 27 June 2019; Accepted: 19 August 2019; Published: 17 September 2019

Abstract: Noether symmetries and first integrals of a class of two-dimensional systems of second order ordinary differential equations (ODEs) are investigated using real and complex methods. We show that first integrals of systems of two second order ODEs derived by the complex Noether approach cannot be obtained by the real methods. Furthermore, it is proved that a complex method can be extended to larger systems and higher order.

Keywords: systems of ODEs; Noether operators; Noether symmetries; first integrals

1. Introduction

Lie developed a symmetry method for solving differential equations (DEs) [1–4]. Noether [5] used these methods to prove that, for DEs obtained from a variational principle, for each symmetry generator there is a corresponding invariant, first integral. These symmetries are called Noether and, if they exist, then Noether's theorem readily provides the associated first integrals. Since they provide a double reduction of the order of the equation, and a sufficient number can actually be used to solve the equation, it is worthwhile to obtain them. Furthermore, they are useful for studying the physical aspects of the dynamical systems, like time translational symmetry gives energy conservation, spatial translation provides momentum conservation and rotational symmetry implies conservation of angular momentum. For a scalar ODE, the corresponding Lagrangian has a five-dimensional maximal Noether symmetry algebra, as guaranteed by a theorem [6], and all the lower dimensions (obviously except 4).

Though Lie methods involved complex functions of complex variables, they did not make explicit use of the Cauchy–Riemann (CR) equations. These conditions provide an auxiliary system of DEs satisfied by the corresponding system of DEs obtained by splitting the complex functions of the scalar or systems of DEs into the two real ones. One obtains either a system of partial differential equations (PDEs), if the independent variable is complex or a system of ODEs if it is real. The explicit use of complex functions of complex or real variables is demonstrated in [7–10] where solvability of systems of DEs is achieved through Noether symmetries and corresponding first integrals. Furthermore, by employing complex symmetry procedures: the energy stored in the field of a coupled harmonic oscillator was studied in [11] and linearizability of systems of two second order ODEs was addressed in [12,13]. The complex procedure, indeed, has been extended to higher dimensional systems of second order ODEs [14] and two-dimensional, systems of third order ODEs [15].

In this paper, we extend the use of complex symmetry methods further to obtain invariants of systems of ODEs and demonstrate that we can obtain new invariants not obtainable by the usual, non-complex, methods. The new invariants for systems arise due to complex Lagrangians and first integrals of the base ODEs involving complex dependent functions of the real independent variables. Complex symmetries have already been used to construct first integrals through Noether symmetries and derive invariants for two-dimensional, systems of second order ODEs [8–10]. We first compare the usual (real) and complex Noether approaches developed to derive first integrals for systems of two second order ODEs. We find that the latter yields more first integrals than the former for these systems. The first integrals derived using a complex procedure also satisfy the conditions of the real Noether's theorem that exists for systems of ODEs. Next, we prove that Lagrangians and corresponding first integrals of the complex scalar ODEs will always split into two real Lagrangians and first integrals for the corresponding system of two equations. For this purpose, we use the CR-equations, which are satisfied by the Lagrangians and first integrals provided by the complex procedure. Furthermore, we show that the complex Noether symmetries do not, in general, split into two Noether symmetries of the corresponding systems. The thrust is not to find directly applicable invariants, which could turn up but to demonstrate how a complex method can provide new invariants and insights into Noether symmetries and first integrals. This work also suggests that the class of systems presented here should, indeed, be singled out when classifying systems of ODEs on the basis of their Noether symmetries and first integrals as it may not follow the classifications presented by employing real symmetry methods. Theorems and their proofs in the later part of this paper show that the method adopted here can trivially be extended to higher dimensions and order of ODEs.

The plan of the paper is as follows: the next section gives the procedures to derive Noether symmetries, operators and corresponding first integrals for systems of two second order ODEs. In the third section, we obtain Noether symmetries and first integrals for two-dimensional, systems of second order ODEs using a real symmetry method. In the subsequent section, first integrals for these systems are derived by employing complex procedures. We end with a concluding section, which also gives the proofs of the claims given in the previous section.

2. Preliminaries

For a system of two coupled (in general) nonlinear ODEs

$$y'' = S_1(x, y, z, y', z'), \quad z'' = S_2(x, y, z, y', z'), \qquad (1)$$

where prime denotes derivative with respect to x, and the point symmetry generator is

$$\mathbf{X} = \xi(x, y, z)\partial_x + \eta_1(x, y, z)\partial_y + \eta_2(x, y, z)\partial_z, \qquad (2)$$

where ξ, η_1, and η_2, are the functions that appear in the infinitesimal coordinate transformations of the dependent and independent variables, $\partial_x = \partial/\partial x$, etc. The first extension of $\mathbf{X}$ is

$$\mathbf{X}^{[1]} = \mathbf{X} + \left(\frac{d}{dx}\eta_1 - y'\frac{d}{dx}\xi\right)\partial_{y'} + \left(\frac{d}{dx}\eta_2 - z'\frac{d}{dx}\xi\right)\partial_{z'}, \qquad (3)$$

where $d/dx = \partial_x + y'\partial_y + z'\partial_z + \cdots$. If system (1) admits a Lagrangian $L(x, y, z, y', z')$, then it is equivalent to the Euler–Lagrange equations

$$\frac{d}{dx}\left(\frac{\partial L}{\partial y'}\right) - \frac{\partial L}{\partial y} = 0, \quad \frac{d}{dx}\left(\frac{\partial L}{\partial z'}\right) - \frac{\partial L}{\partial z} = 0. \qquad (4)$$

The vector field (2) is called a *Noether symmetry generator* corresponding to the Lagrangian $L(x, y, z, y', z')$ for system (1) if there exists a *gauge function* $B(x, y, z)$, such that

$$\mathbf{X}^{[1]}(L) + D(\xi)L = D(B), \tag{5}$$

where D is the total differentiation operator defined by

$$D = \partial_x + y'\partial_y + z'\partial_z + y''\partial_{y'} + z''\partial_{z'} + \cdots. \tag{6}$$

Theorem 1. *If $\mathbf{X}$ is a Noether point symmetry generator corresponding to a Lagrangian $L(x, y, z, y', z')$ of (1), then the corresponding first integral is:*

$$I = \xi L + (\eta_1 - \xi y')\frac{\partial L}{\partial y'} + (\eta_2 - \xi z')\frac{\partial L}{\partial z'} - B. \tag{7}$$

For a first integral I, of system (1), the following equations

$$\mathbf{X}^{[1]} I = 0, \tag{8}$$

$$DI = 0, \tag{9}$$

where $\mathbf{X}^{[1]}$, and D, given in (3) and (6), are satisfied identically. Though the construction of the variational form of (1) along with Noether symmetries to determine the conserved quantities is nontrivial, the complex method converts a class of systems (1) into variational form trivially [8,10,11], which is obtainable from a single (base) scalar complex equation $u'' = S(x, u, u')$. This class is derived by considering $u(x) = y(x) + \iota z(x)$, and $S(x, u, u') = S_1(x, y, z, y', z') + \iota S_2(x, y, z, y', z')$. Such system admits a pair of Lagrangians $L_1(x, y, z, y', z')$ and $L_2(x, y, z, y', z')$ as the Lagrangian $L(x, u, u')$, of the complex base equation also involves the complex function $u(x)$ and its derivative, hence $L = L_1 + \iota L_2$. With these assumptions, (1) can be obtained from

$$\frac{\partial L_1}{\partial y} + \frac{\partial L_2}{\partial z} - \frac{d}{dx}\left(\frac{\partial L_1}{\partial y'} + \frac{\partial L_2}{\partial z'}\right) = 0, \quad \frac{\partial L_2}{\partial y} - \frac{\partial L_1}{\partial z} - \frac{d}{dx}\left(\frac{\partial L_2}{\partial y'} - \frac{\partial L_1}{\partial z'}\right) = 0. \tag{10}$$

These are obtained by splitting the complex Euler–Lagrange equation of the scalar complex second order ODEs. They are different from (4); however, in the later part of this work, their reduction to (4) is done. The operators

$$\mathbf{X}^{[1]} = \xi_1 \partial_x + \tfrac{1}{2}(\eta_1 \partial_y + \eta_2 \partial_z + \eta_1' \partial_{y'} + \eta_2' \partial_{z'}),$$
$$\mathbf{Y}^{[1]} = \xi_2 \partial_x + \tfrac{1}{2}(\eta_2 \partial_y - \eta_1 \partial_z + \eta_2' \partial_{y'} - \eta_1' \partial_{z'}) \tag{11}$$

are said to be *Noether operators* corresponding to $L_1(x, y, z, y', z')$ and $L_2(x, y, z, y', z')$ of (1), if there exist gauge functions $B_1(x, y, z)$, and $B_2(x, y, z)$, such that

$$\mathbf{X}^{[1]} L_1 - \mathbf{Y}^{[1]} L_2 + (D\xi_1)L_1 - (D\xi_2)L_2 = DB_1,$$
$$\mathbf{X}^{[1]} L_2 + \mathbf{Y}^{[1]} L_1 + (D\xi_1)L_2 + (D\xi_2)L_1 = DB_2. \tag{12}$$

Theorem 2. *If $\mathbf{X}^{[1]}$ and $\mathbf{Y}^{[1]}$ are Noether operators corresponding to the Lagrangians $L_1(x, y, z, y', z')$ and $L_2(x, y, z, y', z')$ of (1), then the first integrals for (1) are*

$$I_1 = \xi_1 L_1 - \xi_2 L_2 + \tfrac{1}{2}(\eta_1 - y'\xi_1 + z'\xi_2)\left(\frac{\partial L_1}{\partial y'} + \frac{\partial L_2}{\partial z'}\right)$$
$$- \tfrac{1}{2}(\eta_2 - y'\xi_2 - z'\xi_1)\left(\frac{\partial L_2}{\partial y'} - \frac{\partial L_1}{\partial z'}\right) - B_1, \tag{13}$$

$$I_2 = \xi_1 L_2 + \xi_2 L_1 + \tfrac{1}{2}(\eta_1 - y'\xi_1 + z'\xi_2)\left(\tfrac{\partial L_2}{\partial y'} - \tfrac{\partial L_1}{\partial z'}\right) \\ + \tfrac{1}{2}(\eta_2 - y'\xi_2 - z'\xi_1)\left(\tfrac{\partial L_1}{\partial y'} + \tfrac{\partial L_2}{\partial z'}\right) - B_2. \tag{14}$$

Theorem 3. *The first integrals I_1 and I_2, associated with the Noether operators $X^{[1]}$ and $Y^{[1]}$, satisfy*

$$X^{[1]} I_1 - Y^{[1]} I_2 = 0, \quad X^{[1]} I_2 + Y^{[1]} I_1 = 0, \tag{15}$$

and

$$D_1 I_1 - D_2 I_2 = 0, \quad D_1 I_2 + D_2 I_1 = 0, \tag{16}$$

where $D_1 = \partial_x + \tfrac{1}{2}(y'\partial_y + z'\partial_z + y''\partial_{y'} + z''\partial_{z'} + \cdots)$, $D_2 = \partial_x + \tfrac{1}{2}(z'\partial_y - y'\partial_z + z''\partial_{y'} - y''\partial_{z'} + \cdots)$.

3. Noether Symmetries and Corresponding First Integrals

In this section, we reconsider a class of two-dimensional, systems of second order ODEs that is solved using complex methods [13]. There it was shown that, for this class of systems, dimensions of the Lie point symmetry algebra remain less than 5, while the base complex equations in most of the cases possess an eight-dimensional Lie and a five-dimensional Noether algebra. This will help us here in showing that the number of first integrals for such systems using the real Noether approach remains less than that generated through the complex procedures. This class of systems of two second order cubically semi-linear ODEs reads as:

$$\begin{aligned} y'' &= A_{10} y'^3 - 3A_{20} y'^2 z' - 3A_{10} y' z'^2 + A_{20} z'^3 + B_{10} y'^2 - 2B_{20} y' z' - B_{10} z'^2 + C_{10} y' - C_{20} z' + D_{10}, \\ z'' &= A_{20} y'^3 + 3A_{10} y'^2 z' - 3A_{20} y' z'^2 - A_{10} z'^3 + B_{20} y'^2 + 2B_{10} y' z' - B_{20} z'^2 + C_{20} y' + C_{10} z' + D_{20}, \end{aligned} \tag{17}$$

where $A_{j0}, B_{j0}, C_{j0}, D_{j0}$, ($j = 1, 2$), are analytic functions of x, y, and z. In order to apply the complex Noether approach, we establish correspondence of the above system with the complex scalar second order ODE

$$u'' = A_0(x, u) u'^3 + B_0(x, u) u'^2 + C_0(x, u) u' + D_0(x, u), \tag{18}$$

by considering $y(x) + \iota z(x) = u(x)$, $A_{10} + \iota A_{20} = A_0$, $B_{10} + \iota B_{20} = B_0$, $C_{10} + \iota C_{20} = C_0$, and $D_{10} + \iota D_{20} = D_0$.

Example 1. *The system of two second order quadratically semi-linear ODEs*

$$y'' = x(y'^2 - z'^2), \quad z'' = 2xy'z' \tag{19}$$

admits a three-dimensional Lie algebra spanned by the following symmetry generators

$$X_1 = \partial_y, \quad X_2 = \partial_z, \quad X_3 = -x\partial_x + y\partial_y + z\partial_z, \tag{20}$$

and the Lagrangians

$$L_1 = x + \frac{x^2 y'}{2} + \frac{1}{2} \ln(y'^2 - z'^2), \quad L_2 = \frac{x^2 z'}{2} + \arctan(z', y'). \tag{21}$$

These Lagrangians yield the first integrals

$$I_1^r = \frac{1}{2} x^2 + \frac{y'}{y'^2 + z'^2}, \quad I_2^r = -\frac{z'}{y'^2 + z'^2}, \tag{22}$$

corresponding to X_1, and X_2.

Example 2. For the system of two second order semi-linear ODEs,

$$y'' = -3yy' + 3zz' - y^3 + 3yz^2, \quad z'' = -3yz' - 3zy' - 3y^2z + z^3, \tag{23}$$

there are two Lagrangians

$$L_1 = \frac{3y' + 3y^2 - 3z^2}{(3y' + 3y^2 - 3z^2)^2 + (3z' + 6yz)^2}, \quad L_2 = \frac{3z' + 6yz}{(3y' + 3y^2 - 3z^2)^2 + (3z' + 6yz)^2}, \tag{24}$$

coming from the base complex equation

$$u'' = -3uu' - u^3, \tag{25}$$

and its Lagrangian $L = \frac{1}{3(u'+u^2)}$. System (23) has three Lie point symmetries

$$X_1 = \partial_x, \quad X_2 = x\partial_x - y\partial_y - z\partial_z, \quad X_3 = \frac{x^2}{2}\partial_x + (1 - xy)\partial_y - xz\partial_z. \tag{26}$$

Only one first integral for (23), corresponding to X_1, exists

$$I_1^r = \frac{1}{3\delta_1}\{y^6 + (z^2 + 4y')y^4 + 8y^3zz' + (5y'^2 - 8y'z^2 - z^4 + 3z'^2)y^2 - 8z(z^2 - \frac{1}{2}y')yz' \\ -z^6 + 4z^4y' - (5y'^2 + 3z'^2)z^2 + 2y'z'^2 + 2y'^3\}, \tag{27}$$

where

$$\delta_1 = (y^4 + (2z^2 + 2y')y^2 + 4yzz' + y'^2 + z^4 - 2y'z^2 + z'^2)^2. \tag{28}$$

Example 3. For the system of two second order cubically semi-linear ODEs

$$y'' = y'^3 - 3y'z'^2, \quad z'' = 3y'^2z' - z'^3, \tag{29}$$

the Lagrangians are

$$L_1 = 2y + \frac{y'}{y'^2 + z'^2}, \quad L_2 = 2z - \frac{z'}{y'^2 + z'^2}. \tag{30}$$

The above system possesses four Lie point symmetries

$$X_1 = \partial_x, \quad X_2 = \partial_y, \quad X_3 = \partial_z, \quad X_4 = 2x\partial_x + y\partial_y + z\partial_z. \tag{31}$$

There are three gauge functions, $B_1 = C_1$, $B_2 = 2x$, $B_3 = C_2$, for X_1, X_2, and X_3, respectively, for L_1. Similarly, L_2 generates $B_1 = C_3$, $B_2 = C_4$, $B_3 = 2x$, with the same point symmetries as mentioned above. Thus, there is a three-dimensional Noether algebra for (29) associated with L_1,

$$I_1^r = 2y + \frac{2y'}{y'^2+z'^2} - C_1, \quad I_2^r = \frac{1}{y'^2+z'^2} - \frac{2y'^2}{(y'^2+z'^2)^2} - 2x, \\ I_3^r = \frac{-2y'z'}{(y'^2+z'^2)^2} - C_2. \tag{32}$$

Similarly, for L_2, the first integrals are

$$I_1^r = 2z - \frac{2z'}{y'^2+z'^2} - C_3, \quad I_2^r = \frac{2y'z'}{(y'^2+z'^2)^2} - C_4, \\ I_3^r = \frac{-1}{y'^2+z'^2} + \frac{2z'^2}{(y'^2+z'^2)^2} - 2x. \tag{33}$$

Example 4. A nonlinear system of two second order cubically semi-linear ODEs

$$y'' = \alpha^2 xy'^3 - 3\alpha^2 xy'z'^2, \quad z'' = 3\alpha^2 xy'^2z' - \alpha^2 xz'^3, \tag{34}$$

where α is a constant and possesses the Lagrangians

$$L_1 = 2\alpha^2 xy + \frac{y'}{y'^2 + z'^2}, \quad L_2 = 2\alpha^2 xy - \frac{z'}{y'^2 + z'^2}. \tag{35}$$

The above system (34) has three Lie point symmetries

$$X_1 = \partial_y, \quad X_2 = \partial_z, \quad X_3 = x\partial_x. \tag{36}$$

For X_1, and X_2, we obtain gauge functions with L_1, and L_2 that are $B_1 = C_1 x^2$, $B_2 = C_2$, and $B_1 = C_2$, $B_2 = C_1 x^2$, respectively. Thus, two-dimensional Noether algebra is found to exist for (34) and the first integrals corresponding to L_1, and L_2, are

$$I_1^r = \frac{1}{y'^2 + z'^2} - \frac{2y'^2}{(y'^2 + z'^2)^2} - C_1 x^2, \quad I_2^r = \frac{-2y'z'}{(y'^2 + z'^2)^2} - C_2, \tag{37}$$

and

$$I_1^r = \frac{2y'z'}{(y'^2 + z'^2)^2} - C_2, \quad I_2^r = \frac{-1}{y'^2 + z'^2} + \frac{2z'^2}{(y'^2 + z'^2)^2} - C_1 x^2, \tag{38}$$

respectively.

Example 5. The system of two second order cubically semi-linear ODEs

$$y'' = \alpha yy'^3 - 3\alpha z y'^2 z' - 3\alpha yy' z'^2 + \alpha z z'^3,$$
$$z'' = \alpha z y'^3 + 3\alpha yy'^2 z' - 3\alpha z y' z'^2 - \alpha y z'^3 \tag{39}$$

has the Lagrangians

$$L_1 = \alpha y^2 - \alpha z^2 + \frac{y'}{y'^2 + z'^2}, \quad L_2 = 2\alpha yz - \frac{z'}{y'^2 + z'^2}, \tag{40}$$

where α is a constant. This system admits a two-dimensional Lie point symmetry algebra

$$X_1 = \partial_x, \quad X_2 = 3x\partial_x + y\partial_y + z\partial_z. \tag{41}$$

The gauge term for X_1, with both L_1, L_2, is $B = C_1$. Thus, there is a one-dimensional Noether algebra that provides the following first integrals

$$I_1^r = \alpha(y^2 - z^2) + \frac{2y'}{y'^2+z'^2} - C_1,$$
$$I_2^r = 2\alpha yz - \frac{2z'}{y'^2+z'^2} - C_2. \tag{42}$$

Example 6. The system of two second order cubically semi-linear ODEs

$$y'' = xyy'^3 - 3xzy'^2z' - 3xyy'z'^2 + xzz'^3,$$
$$z'' = xzy'^3 + 3xyy'^2z' - 3xzy'z'^2 - xyz'^3 \tag{43}$$

has two Lagrangians

$$L_1 = xy^2 - xz^2 + \frac{y'}{y'^2 + z'^2}, \quad L_2 = 2xyz - \frac{z'}{y'^2 + z'^2}, \tag{44}$$

and one Lie point symmetry

$$X_1 = x\partial_x. \tag{45}$$

There is no gauge function corresponding to X_1, for L_1, or L_2, which implies that there is a 0-dimensional Noether algebra. Hence, no first integral exists by a real method.

4. Noether Operators and Corresponding First Integrals

In this section, we obtain first integrals for all systems considered in the previous section, by employing complex Noether procedure.

Example 7. *By considering $y(x) + \iota z(x) = u(x)$, system (19) corresponds to a scalar ODE*

$$u'' = xu'^2, \tag{46}$$

which has only one symmetry $Z_1 = \partial_u$. A complex Lagrangian, $L = x + \frac{x^2 u'}{2} + \ln u'$, is admitted by (46), yielding the first integral

$$I_1 = \frac{x^2}{2} + \frac{1}{u'}. \tag{47}$$

This complex first integral splits into the following real first integrals

$$I_1^c = \frac{1}{2}x^2 + \frac{y'}{y'^2 + z'^2}, \quad I_2^c = -\frac{z'}{y'^2 + z'^2}, \tag{48}$$

of system (19). Notice that both these first integrals are the same as those obtained earlier in (22) by the real method. It shows an agreement between the complex and real Noether approaches.

Example 8. *The base scalar ODE (25) has a five-dimensional Noether symmetry algebra spanned by*

$$\begin{aligned}
&Z_1 = \partial_x, \quad Z_2 = u\partial_x - u^3\partial_u, \quad Z_3 = xu\partial_x + (u^2 - xu^3)\partial_u,\\
&Z_4 = (x - \frac{3x^2 u}{2})\partial_x + (2u - 3xu^2 + \frac{3x^2 u^3}{2})\partial_u,\\
&Z_5 = (\frac{x^3 u}{2} - \frac{x^2}{2})\partial_x + (1 - 2xu + \frac{3x^2 u^2}{2} - \frac{x^3 u^3}{2})\partial_u.
\end{aligned} \tag{49}$$

The first integrals corresponding to the above Noether symmetries are

$$\begin{aligned}
&I_1 = \frac{2u' + u^2}{3(u^2 + u')^2}, \quad I_2 = x - \frac{u}{u^2 + u'}, \quad I_3 = \frac{(-u + xu^2 + xu')^2}{(u^2 + u')^2},\\
&I_4 = \frac{1}{3}\frac{((u' + u^2)x - u)(2 + (u' + u^2)x^2 - 2xu)}{(u^2 + u')^2},\\
&I_5 = \frac{3(2 - 2xu + x^2 u^2 + x^2 u')}{u^2 + u'}.
\end{aligned} \tag{50}$$

Putting $u(x) = y(x) + \iota z(x)$, these complex first integrals split to provide the first integrals for the system (23)

$$\begin{aligned}
&I_1^c = \frac{1}{3}\frac{(2y' + y^2 - z^2)J_1 + (2z' + 2yz)J_2}{J_1^2 + J_2^2}, \quad I_2^c = \frac{1}{3}\frac{(2z' + 2yz)J_1 - (2y' + y^2 - z^2)J_2}{J_1^2 + J_2^2},\\
&I_3^c = x - \frac{yJ_3 - zJ_4}{J_3^2 + J_4^2}, \quad I_4^c = \frac{yJ_4 - zJ_3}{J_3^2 + J_4^2}, \quad I_5^c = \frac{J_5 J_6 + J_7 J_8}{J_6^2 + J_8^2}, \quad I_6^c = \frac{J_6 J_7 - J_5 J_8}{J_6^2 + J_8^2},\\
&I_7^c = \frac{1}{3}\frac{(J_9 J_{10} - J_{11} J_{12}) J_{13} + (J_{10} J_{11} + J_9 J_{12}) J_{14}}{J_{13}^2 + J_{14}^2},\\
&I_8^c = \frac{1}{3}\frac{(J_{10} J_{11} + J_9 J_{12}) J_{13} - (J_9 J_{10} - J_{11} J_{12}) J_{14}}{J_{13}^2 + J_{14}^2},\\
&I_9^c = 3\frac{J_3 J_{15} + J_4 J_{16}}{J_3^2 + J_4^2}, \quad I_{10}^c = 3\frac{J_3 J_{16} - J_4 J_{15}}{J_3^2 + J_4^2},
\end{aligned} \tag{51}$$

where

$$\begin{aligned}
J_1 &= -4yzz' - 6y^2z^2 + z^4 + y^4 + 2y^2y' - 2z^2y' + y'^2 - z'^2, \\
J_2 &= 4yzy' - 4yz^3 + 2y'z' + 4y^3z + 2y^2z' - 2z^2z', \\
J_3 &= y' + y^2 - z^2, \\
J_4 &= z' + 2yz, \\
J_5 &= -4x^2yzz' + 2xzz' + y^2 - z^2 + 6xyz^2 - 2xyy' - 6x^2y^2z^2 + 2x^2y^2y' \\
    &\quad - 2x^2z^2y' - 2xy^3 + x^2y^4 + x^2z^4 + x^2y'^2 - x^2z'^2, \\
J_6 &= y^4 + z^4 + y'^2 - 6y^2z^2 - 2z^2y' + 2y^2y' - 4yzz' - z'^2, \\
J_7 &= 2x^2y'z' + 4x^2y^3z + 2x^2y^2z' - 2x^2z^2z' - 4x^2yz^3 - 2xzy' - 6xy^2z \\
    &\quad - 2xyz' + 2yz + 2xz^3 + 4x^2yzy', \\
J_8 &= 4y^3z + 2y^2z' - 4yz^3 - 2z^2z' + 2y'z' + 4yzy', \\
J_9 &= (y^2 - z^2 + y')x - y, \\
J_{10} &= 2 + (y^2 - z^2 + y')x^2 - 2xy, \\
J_{11} &= (2yz + z')x - z, \\
J_{12} &= (2yz + z')x^2 - 2xz, \\
J_{13} &= -6y^2z^2 - 4yzz' + 2y^2y' - 2z^2y' + y'^2 - z'^2 + y^4 + z^4, \\
J_{14} &= -2z^2z' - 4yz^3 + 2y^2z' + 4y^3z + 4yzy' + 2y'z', \\
J_{15} &= 2 - 2xy + x^2(y^2 - z^2) + x^2y', \\
J_{16} &= -2xz + 2x^2yz + x^2z'.
\end{aligned} \qquad (52)$$

In the following examples, we show that a complex symmetry approach provides 10 first integrals for systems of two second order ODEs. In particular, there is a system (43) that has a 0-dimensional Noether symmetry algebra, but 10 first integrals are generated by Noether operators obtained by complex methods.

Example 9. *The base complex scalar ODE*

$$u'' = u'^3, \qquad (53)$$

for system (29) *has the Lagrangian* $L = 2u + \frac{1}{u'}$. *For the five Lie point symmetries*

$$\begin{aligned}
Z_1 &= \partial_x, \quad Z_2 = \partial_u, \quad Z_3 = u\partial_x, \quad Z_4 = (u^3 - 2xu)\partial_x - 2u^2\partial_u, \\
Z_5 &= (3u^2 - 2x)\partial_x - 4u\partial_u
\end{aligned} \qquad (54)$$

of the ODE (53), *the gauge terms found are* $B_1 = C$, $B_2 = 2x$, $B_3 = 2x + u^2$, $B_4 = \frac{3u^4}{2} - 2xu^2 - 2x^2$, $B_5 = 4u^3$, *respectively. The corresponding first integrals are*

$$\begin{aligned}
I_1 &= 2u + \frac{2}{u'} - C, \quad I_2 = \frac{-1}{u'^2} - 2x, \quad I_3 = u^2 - 2x + \frac{2u}{u'}, \\
I_4 &= \frac{1}{2u'^2}((u^2 - 2x)u' + 2u)^2, \quad I_5 = 2u^3 - 4xu + \frac{(6u^2 - 4x)}{u'} + \frac{4u}{u'^2}.
\end{aligned} \qquad (55)$$

Splitting (54) *into real and imaginary parts yields the 10 Noether operators that provide the following first integrals:*

$$\begin{aligned}
I_1^c &= 2y + \frac{2y'}{y'^2+z'^2} - C_1, \quad I_2^c = 2z - \frac{2z'}{y'^2+z'^2} - C_2, \\
I_3^c &= -2x - \frac{y'^2-z'^2}{(y'^2+z'^2)^2}, \quad I_4^c = \frac{2y'z'}{(y'^2+z'^2)^2}, \\
I_5^c &= y^2 - z^2 - 2x + 2\frac{yy'+zz'}{y'^2+z'^2}, \quad I_6^c = 2\left(yz + \frac{y'z-yz'}{y'^2+z'^2}\right), \\
I_7^c &= y^4 + z^4 - 6y^2z^2 + 4x^2 - 4x(y^2 - z^2) + 4\frac{(y^3-3yz^2-2xy)y'+(3y^2z-z^3-2xz)z'}{y'^2+z'^2} \\
    &\quad + 4\frac{(y^2-z^2)(y'^2-z'^2)+4yzy'z'}{(y'^2+z'^2)^2},
\end{aligned} \qquad (56)$$

$$
\begin{aligned}
I_8^c &= 4yz(y^2-z^2) - 8xyz + 4\tfrac{(3y^2z-z^3-2xz)y'-(y^3-3yz^2-2xy)z'}{y'^2+z'^2} \\
&\quad + 4\tfrac{2yz(y'^2-z'^2)-2(y^2-z^2)y'z'}{(y'^2+z'^2)^2}, \\
I_9^c &= y^3 - 3yz^2 - 2xy + \tfrac{(3y^2-3z^2-2x)y'+6yzz'}{y'^2+z'^2} + 2\tfrac{y(y'^2-z'^2)+2zy'z'}{(y'^2+z'^2)^2}, \\
I_{10}^c &= 3y^2z - z^3 - 2xz + \tfrac{6yzy'+(3z^2-3y^2+2x)z'}{y'^2+z'^2} + 2\tfrac{z(y'^2-z'^2)-2yy'z'}{(y'^2+z'^2)^2}.
\end{aligned}
\tag{57}
$$

Example 10. *System (34) is obtainable from a scalar second order complex ODE*

$$u'' = \alpha^2 x u'^3. \tag{58}$$

The Lagrangian associated with this equation is $L = 2\alpha^2 xu + \frac{1}{u'}$. The following gauge functions

$$
\begin{aligned}
B_1 &= \alpha^2 x^2, \quad B_2 = 2\alpha^2 xu \sin(\alpha u) + 2\alpha x \cos(\alpha u), \quad B_3 = 2\alpha^2 xu \cos(\alpha u) - 2\alpha x \sin(\alpha u), \\
B_4 &= \alpha^2 x^2 (2\alpha u \cos(2\alpha u) - \sin(2\alpha u)), \quad B_5 = \alpha^2 x^2 (2\alpha u \sin(2\alpha u) + \cos(2\alpha u)),
\end{aligned}
\tag{59}
$$

correspond to the respective Lie point symmetries

$$
\begin{aligned}
Z_1 &= \partial_u, \quad Z_2 = \sin(\alpha u)\partial_x, \quad Z_3 = \cos(\alpha u)\partial_x, \\
Z_4 &= \alpha x \cos(2\alpha u)\partial_x + \sin(2\alpha u)\partial_u, \quad Z_5 = \alpha x \sin(2\alpha u)\partial_x - \cos(2\alpha u)\partial_u.
\end{aligned}
\tag{60}
$$

Hence, there are five Noether symmetries and corresponding first integrals

$$
\begin{aligned}
I_1 &= \tfrac{-1}{u'^2} - \alpha^2 x^2, \\
I_2 &= 2\left(\tfrac{\sin(\alpha u)}{u'} - \alpha x \cos(\alpha u)\right), \\
I_3 &= 2\left(\tfrac{\cos(\alpha u)}{u'} + \alpha x \sin(\alpha u)\right), \\
I_4 &= \alpha^2 x^2 \sin(2\alpha u) - \tfrac{\sin(2\alpha u)}{u'^2} + 2\tfrac{\alpha x \cos(2\alpha u)}{u'}, \\
I_5 &= -\alpha^2 x^2 \cos(2\alpha u) + \tfrac{\cos(2\alpha u)}{u'^2} + 2\tfrac{\alpha x \sin(2\alpha u)}{u'}.
\end{aligned}
\tag{61}
$$

The real and imaginary parts of (61) yield 10, first integrals for (34).

Example 11. *The system (39) and Lagrangian (40) correspond to the complex scalar linearizable ODE*

$$u'' = \alpha u u'^3, \tag{62}$$

and Lagrangian $L = \alpha u^2 + \frac{1}{u'}$, which have the following gauge functions $B_1 = C$, $B_2 = 2x + \frac{\alpha u^3}{3}$, $B_3 = \frac{\alpha^2 u^4}{2}$, $B_4 = \frac{2}{3}\alpha^2 u^6 - \alpha x u^3 - 3x^2$, $B_5 = -3\alpha^2 u^5$, for the following Lie point symmetries

$$
\begin{aligned}
Z_1 &= \partial_x, \quad Z_2 = u\partial_x, \quad Z_3 = \alpha u^2 \partial_x - 2\partial_u, \\
Z_4 &= (\alpha u^4 - 3xu)\partial_x - 3u^2 \partial_u, \quad Z_5 = (-5\alpha u^3 + 6x)\partial_x + 12u\partial_u.
\end{aligned}
\tag{63}
$$

Thus, there are five complex first integrals

$$
\begin{aligned}
I_1 &= \alpha u^2 + 2/u' - C, \quad I_2 = \tfrac{1}{3}(2\alpha u^3 - 6x) + \tfrac{2u}{u'}, \\
I_3 &= \tfrac{(2+\alpha u^2 u')^2}{2u'^2}, \quad I_4 = \tfrac{1}{3}\left(\tfrac{(\alpha u^3 - 3x)u' + 3u}{u'}\right)^2, \\
I_5 &= -2\alpha^2 u^5 + 6\alpha x u^2 - 2\tfrac{(5\alpha u^3 - 6x)}{u'} - 12\tfrac{u}{u'^2},
\end{aligned}
\tag{64}
$$

which split into 10 real first integrals for system (39).

Example 12. *System (43) is obtainable from the complex linearizable ODE*

$$u'' = xuu'^3, \tag{65}$$

with the Lagrangian $L = xu'^2 + \frac{1}{u'}$. It admits a five-dimensional Noether symmetry algebra, which yields 10 first integrals for system (43). In the previous section, we showed that there is no Noether symmetry for system (43) by real methods, but the complex method yields 10 first integrals.

5. Conclusions

In this paper, we demonstrated, by considering explicit examples that the complex methods provide Noether invariants that do not appear by real methods. While the examples, in themselves, do prove the point, one would like to understand why this should be the case. For this purpose, we state and prove the following theorems that summarize our results and provide insight into how the complex methods work and go beyond the real methods.

Theorem 4. *The Lagrangian and associated first integrals of complex nth ($n \geq 2$) order ODEs with complex dependent and real independent variable provide Lagrangians and first integrals, respectively, for corresponding two-dimensional systems of nth order ODEs.*

Proof. We prove the result for $n = 2$, as its extension to higher orders is trivial. In this case, i.e., for a scalar second order ODE, the Euler–Lagrange equation reads as $\frac{d}{dx}\left(\frac{\partial L}{\partial u'}\right) - \frac{\partial L}{\partial u} = 0$, which expands to

$$L_{xu'} + u'L_{uu'} + u''L_{u'u'} - L_u = 0.$$

By considering $u(x) = y(x) + \imath z(x)$, $L(x, u, u') = L_1(x, y, z, y', z') + \imath L_2(x, y, z, y', z')$, in the above equation and splitting it into the real and imaginary parts, one obtains

$$\frac{1}{2}(L_{1,xy'} + L_{2,xz'}) + \frac{y'}{4}(L_{1,yy'} + L_{2,y'z} + L_{2,yz'} - L_{1,zz'}) - \frac{z'}{4}(L_{2,yy'} - L_{1,y'z} - L_{1,yz'} - L_{2,zz'})$$
$$+ \frac{y''}{4}(L_{1,y'y'} + L_{2,y'z'} + L_{2,y'z'} - L_{1,z'z'}) - \frac{z''}{4}(L_{2,y'y'} - L_{1,y'z'} - L_{1,y'z'} - L_{2,z'z'}) - \frac{1}{2}(L_{1,y} + L_{2,z}) = 0,$$
$$\frac{1}{2}(L_{2,xy'} - L_{1,xz'}) + \frac{y'}{4}(L_{2,yy'} - L_{1,y'z} - L_{1,yz'} - L_{2,zz'}) + \frac{z'}{4}(L_{1,yy'} + L_{2,y'z} + L_{2,yz'} - L_{1,zz'})$$
$$+ \frac{y''}{4}(L_{2,y'y'} - L_{1,y'z'} - L_{1,y'z'} - L_{2,z'z'}) + \frac{z''}{4}(L_{1,y'y'} + L_{2,y'z'} + L_{2,y'z'} - L_{1,z'z'}) - \frac{1}{2}(L_{2,y} - L_{1,z}) = 0.$$

Both of the above equations reduce to

$$L_{i,xy'} + y'L_{i,yy'} + z'L_{i,y'z} + y''L_{i,y'y'} + z''L_{i,y'z'} - L_{i,y} = 0,$$
$$L_{i,xz'} + y'L_{i,yz'} + z'L_{i,zz'} + y''L_{i,y'z'} + z''L_{i,z'z'} - L_{i,z} = 0,$$

for $i = 1, 2$, by employing the CR-equations $L_{1,y} = L_{2,z}$, $L_{1,z} = -L_{2,y}$, $L_{1,y'} = L_{2,z'}$, and $L_{1,z'} = -L_{2,y'}$. Notice that these are Euler–Lagrange Equations (4) for two-dimensional systems of second order ODEs. Hence, the real and imaginary parts L_i, for $i = 1, 2$, of a complex Lagrangian $L(x, u, u')$ satisfy the Euler–Lagrange equations for systems obtainable from complex scalar equations. In other words, the Euler–Lagrange Equations (10) become the Euler–Lagrange Equations (4). A similar argument applies to first integrals I_i, for $i = 1, 2$, obtained for a system of two second order ODEs from complex first integral $I(x, u, u')$, of a scalar second order complex equation which satisfy $DI = 0$, i.e.,

$$I_x + u'I_u + u''I_{u'} = 0.$$

Considering $u(x) = y(x) + \iota v(x)$, $I(x, u, u') = I_1(x, y, z, y', z') + \iota I_2(x, y, z, y', z')$, $D = D_1 + \iota D_2$, and splitting into the real and imaginary parts and employing CR-equations $I_{1,y} = I_{2,z}$, $I_{1,z} = -I_{2,y}$, $I_{1,y'} = I_{2,z'}$, $I_{1,z'} = -I_{2,y'}$, yields

$$I_{i,x} + y' I_{i,y} + z' I_{i,z} + y'' I_{i,y'} + z'' I_{i,z'} = 0, \quad i = 1, 2.$$

This is exactly the criterion whose first integrals of a system of two second order ODEs satisfy $DI_1 = DI_2 = 0$, where D is the derivative operator for such systems given in (6). □

The above result is extendable to higher dimensional systems of ODEs of order more than two as the CR-equations and their derivatives establish a connection between Lagrangians and first integrals of the complex base equations and the corresponding systems. Therefore, for those systems (of nth order ODEs) that correspond to complex DEs (of the same order), their Lagrangians and first integrals are obtainable from the complex Lagrangian and first integrals of the base equations.

The base complex equations in Examples (2)–(6) and (8)–(12) admit an eight-dimensional Lie and five-dimensional Noether symmetry algebras. It implies that there exist five first integrals for these scalar equations, which, when considered complex, convert into ten first integrals (as guaranteed by above theorem) of the corresponding two-dimensional systems of second order ODEs. Based on these observations, we can state the following result.

Corollary 1. *For two-dimensional systems of second order ODEs with symmetry algebras of dimension d, (d < 5) that are obtainable from complex linearizable scalar ODEs, a complex Noether approach provides more first integrals than the real symmetry method.*

Theorem 5. *The real and imaginary parts of the complex Noether symmetries of the complex scalar second order ODEs are not necessarily the Noether symmetries of the corresponding two-dimensional systems of second order ODEs.*

Proof. A complex first integral $I(x, u, u')$ satisfies the invariance criterion $\mathbf{Z}^{[1]} I = 0$, where

$$\mathbf{Z}^{[1]} = \xi \partial_x + \eta' \partial_{u'} + \eta'' \partial_{u''}$$

is the first extension of the Noether symmetry of a second order complex ODE. Splitting it into the real and imaginary parts leads to two invariance conditions (15) that expand to

$$\xi_1 I_{1,x} - \xi_2 I_{2,x} + \frac{1}{2}\{\eta_1(I_{1,y} + I_{2,z}) - \eta_2(I_{2,y} - I_{1,z}) + \eta_1'(I_{1,y'} + I_{2,z'}) - \eta_2'(I_{2,y'} - I_{1,z'})\} = 0,$$

$$\xi_1 I_{2,x} + \xi_2 I_{1,x} + \frac{1}{2}\{\eta_1(I_{2,y} - I_{1,z}) + \eta_2(I_{1,y} + I_{2,z}) + \eta_1'(I_{2,y'} - I_{1,z'}) + \eta_2'(I_{1,y'} + I_{2,z'})\} = 0,$$

respectively, where $\mathbf{X}^{[1]}$, and $\mathbf{Y}^{[1]}$, are the operators given in (11). Applying the CR-equations on I_1, and I_2, the above equations become

$$\xi_1 I_{1,x} - \xi_2 I_{2,x} + \eta_1 I_{1,y} + \eta_2 I_{1,z} + \eta_1' I_{1,y'} + \eta_2' I_{1,z'} = 0,$$

$$\xi_1 I_{2,x} + \xi_2 I_{1,x} + \eta_1 I_{2,y} + \eta_2 I_{2,z} + \eta_1' I_{2,y'} + \eta_2' I_{2,z'} = 0,$$

while the real invariance criterion for systems reads as $\mathbf{X}^{[1]} I_i = 0$, for $i = 1, 2$, which yields two equations

$$\xi I_{1,x} + \eta_1 I_{1,y} + \eta_2 I_{1,z} + \eta_1' I_{1,y'} + \eta_2' I_{1,z'} = 0,$$

$$\xi I_{2,x} + \eta_1 I_{2,y} + \eta_2 I_{2,z} + \eta_1' I_{2,y'} + \eta_2' I_{2,z'} = 0.$$

A comparison of these equations with the previous two implies that the real and imaginary parts of a complex Noether symmetry of the base scalar equation split into two Noether symmetries for the corresponding system of

ODEs only if $\xi_2 = 0$, which implies that, if the infinitesimal coordinate ξ, of a complex Noether symmetry is a function of both the real independent variable x, and the complex dependent variable $u(x)$, then it does not split into Noether symmetries for the corresponding system. □

Author Contributions: Conceptualization, M.S. and A.Q.; methodology, M.S. and A.Q.; software, M.S.; validation, M.U.F.; formal analysis, M.U.F. and M.S.; investigation, M.S. and M.U.F.; resources, M.S.; writing—original draft preparation, M.S. and A.Q.; review and editing, A.Q.; supervision, A.Q.

Funding: This research received no external funding.

Acknowledgments: We thank the anonymous referees for their useful suggestions, which helped with refining the results and presentation of the article.

Conflicts of Interest: The authors declare no conflict of interest.

References

1. Lie, S. Klassifikation und integration von gewohnlichen differentialgleichungen zwischen x, y, die eine gruppe von transformationen gestatten I, II, II and IV. *Archiv. Math.* **1883**, *8*, 187–224, 249–288, 371–458.
2. Lie, S. Uber differentialinvarianten. *Math. Ann.* **1884**, *24*, 537. [CrossRef]
3. Lie, S. *Theorie der Transformationsgruppen I, II and III, Teubner, Leipzig, 1888*; Chelsea Publishing Company: New York, NY, USA, 1970
4. Ibragimov, N.H. *Elementary Lie Group Analysis and Ordinary Differential Equations*; Wiley: New York, NY, USA, 1999.
5. Noether, E. Invariante Variations probleme. *Nachrichten Der Akadmie Der Wiss. Gottingen, -Mathemtisch-Phys. Klasse* **1918**, *2*, 235–237.
6. Ibragimov, N.H. *CRC Handbook of Lie Group Analysis of Differential Equations, 1, 2, 3*; CRC Press: Boca Raton, FL, USA, 1994–1996.
7. Ali, S. Complex Lie Symmetries for Differential Equations. Ph.D. Thesis, National University of Sciences and Technology, Islamabad, Pakistan, 2009.
8. Ali, S.; Mahomed, F.M.; Qadir, A. Complex Lie symmetries for variational problems. *J. Nonlinear Math. Phys.* **2008**, *15*, 25–35. [CrossRef]
9. Ali, S.; Mahomed, F.M.; Qadir, A. Complex Lie symmetries for Scalar Second order ordinary differential equations. *Nonlinear Anal. Real World Appl.* **2009**, *10*, 3335–3344. [CrossRef]
10. Farooq, M.U.; Ali, S.; Mahomed, F.M. Two dimensional systems that arise from the Noether classification of Lagrangian on the line. *Appl. Math. Comput.* **2011**, *217*, 6959–6973. [CrossRef]
11. Farooq, M.U.; Ali, S.; Qadir, A. Invariants of two-dimensional systems via complex Lagrangians with applications. *Commun. Nonlinear Sci. Numer. Simul.* **2011**, *16*, 1804–1810. [CrossRef]
12. Safdar, M.; Qadir, A.; Ali, S. Linearizability of systems of ordinary differential equations obtained by complex symmetry analysis. *Math. Probl. Eng.* **2011**, *2011*, 171834. [CrossRef]
13. Ali, S.; Safdar, M.; Qadir, A. Linearization from complex Lie point transformations. *J. Appl. Math.* **2014**, *2014*, 793247. [CrossRef]
14. Safdar, M.; Ali, S.; Mahomed, F.M. Linearization of systems of four second order ordinary differential equations. *PRAMANA J. Phys.* **2011**, *77*, 581–594. [CrossRef]
15. Dutt, H.M.; Safdar, M.; Qadir, A. Linearization criteria for two-dimensional, systems of third order ordinary differential equations by complex approach. *Arab. J. Math.* **2019**, *8*, 163–170. [CrossRef]

© 2019 by the authors. Licensee MDPI, Basel, Switzerland. This article is an open access article distributed under the terms and conditions of the Creative Commons Attribution (CC BY) license (http://creativecommons.org/licenses/by/4.0/).

Article

Optimal System and New Approximate Solutions of a Generalized Ames's Equation

Marianna Ruggieri [1] and Maria Paola Speciale [2],*

[1] Faculty of Engineering and Architecture, University of Enna "Kore", 94100 Enna, Italy; marianna.ruggieri@unikore.it
[2] Department of Mathematics and Computer Science, Physical Sciences and Earth Science, University of Messina, 98166 Messina, Italy
* Correspondence: mpspeciale@unime.it

Received: 30 July 2019; Accepted: 20 September 2019; Published: 1 October 2019

Abstract: In this paper, by applying Valenti's theory for the approximate symmetry, we introduce and define the concept of a one-dimensional optimal system of approximate subalgebras for a generalized Ames's equation; furthermore, the algebraic structure of the approximate Lie algebra is discussed. New approximately invariant solutions to the equation are found.

Keywords: partial differential equations; approximate symmetry and solutions

1. Introduction

In [1], Ames et al. performed the symmetries classification of the model

$$u_{tt} = [f(u)u_x]_x \tag{1}$$

that can describes the flow of one-dimensional gas, longitudinal wave propagation on a moving threadline, dynamics of a finite nonlinear string. The study of Equation (1) gave impetus to later investigations. In 2007, Bluman et al. [2] made an interesting nonlocal analysis of Equation (1). In this paper, our aim is to investigate the generalized model of Equation (1)

$$u_{tt} = [f(u)u_x]_x + \varepsilon[\lambda(u)\,u_t]_{xx}, \tag{2}$$

where f and λ are smooth functions. $\varepsilon \ll 1$ is the small parameter for the perturbative analysis, while, when $\varepsilon = 0$, we recover the unperturbed Equation (1).

As in the exact symmetries, even in the approximate one, an important task in determing approximately invariant solutions is to employ the concept of an optimal system of approximate subalgebras in order to obtain all the essentially different approximate invariant solutions.

In this manuscript, in the context of Valenti's theory [3], we define the definition of one-dimensional optimal system of approximate subalgebras for Equation (2). By its application, we get new approximate solutions for the generalized Ames's equation.

The plan of the manuscript is the following: in Section 2, after a brief introduction of the main concepts of Lie theory, we introduce the definition of Approximate Subalgebra and, finally, recall the main results of the approximate symmetry analysis of Equation (2). The Optimal Systems of one-dimensional approximate subalgebras are introduced and defined in Section 3 and the application of the method to the model is provided. Section 4 presents the reductions of Equation (2) to ordinary differential equations (ODEs) through the approximate optimal operators and new approximate solutions are obtained. Finally, in Section 5, the optimal system is used in order to construct new approximate non-invariant solutions for two other models linked to Equation (2) by a nonlocal transformation.

2. On the Approximate Symmetry Classifications

Various and different theories on approximate symmetries have been developed over the years; the first contribution on this argument is due to Baikov et al. [4], but the method does not consider an approximation in the perturbation meaning well, since, if we utilise the first order operator, the corresponding approximate solution could include higher order terms. Later, another interesting method was suggested by Fushchich et al. [5], where the strategy is consistent in the perturbation sense and produces correct terms for the approximate solutions, but it is impossible to work in hierarchy; then, the algebra could really grow. In 2004, Pakdemirli et al. [6] have compared these two methods. Afterwards, Valenti in [3] introduced his method, where, following the technique proposed in [5], he removes the "obstacle" of the impossibility of working in hierarchy, in accordance with the perturbation theory and, for this reason, we will apply this method.

Before entering in the details of approximate theory, we recall briefly the main concepts of Lie theory for a system of PDEs.

For a given system of differential equation

$$\Delta(t, x, u, v, u^{(k)}, v^{(h)}) = 0, \tag{3}$$

where t, x represents the independent variables. u, v are the dependent variables and $u^{(k)}, v^{(h)}$ stand for all partial derivatives of u and v up to order k and h, respectively.

The invertible transformations of t, x, u, v

$$T = T(t, x, u, v, a), \quad X = X(t, x, u, v, a), \quad U = U(t, x, u, v, a), \quad V = V(t, x, u, v, a), \tag{4}$$

depending on a continuous parameter a, are defined as *one-parameter (a) (exact) Lie point symmetry transformations* of Equation (3) if Equation (3) has the same form in the new variables T, X, U, V.

By expanding Equation (4) in Taylor's series around $a = 0$, we obtain the infinitesimal transformations, according to the Lie theory:

$$T = t + a\,\xi^1(t, x, u, v) + o(a^2), \quad X = x + a\,\xi^2(t, x, u, v) + o(a^2), \tag{5}$$

$$U = u + a\,\eta^1(t, x, u, v) + o(a^2), \quad V = v + a\,\eta^2(t, x, u, v) + o(a^2), \tag{6}$$

where their infinitesimals ξ^1, ξ^2, η^1 and η^2 are given by

$$\xi^1(t, x, u, v) = \left.\frac{\partial T}{\partial a}\right|_{a=0}, \quad \xi^2(t, x, u, v) = \left.\frac{\partial X}{\partial a}\right|_{a=0}, \quad \eta^1(t, x, u, v) = \left.\frac{\partial U}{\partial a}\right|_{a=0}, \quad \eta^2(t, x, u, v) = \left.\frac{\partial V}{\partial a}\right|_{a=0}.$$

The corresponding operator

$$\Xi = \xi^1(t, x, u, v)\partial_t + \xi^2(t, x, u, v)\partial_x + \eta^1(t, x, u, v)\partial_u + \eta^2(t, x, u, v)\partial_v \tag{7}$$

is known in the literature as the infinitesimal operator or *generator* of the Lie group.

The Lie group of point transformations, which leave a differential Equation (3) invariant, is obtained by means of the Lie's algorithm, with the requirement that the $\bar{k}$-order prolongation of Equation (7), which acts on Equation (3), is zero along the solutions, i.e.:

$$\Xi^{\bar{k}}\Delta = 0|_{\Delta=0}, \tag{8}$$

where $\bar{k} = max(k, h)$. The invariance condition Equation (8) produces an overdetermined system of linear differential equations (called determining equations) for the infinitesimals whose integration provides the generators of Lie Algebra admitted by Equation (3).

Now, we are able to define the concepts of *Approximate Subalgebra* of Equation (2).

Definition 1. We call Approximate Subalgebra of Equation (2) the (exact) subalgebra of the following system of PDEs:

$$L_0 := u_{0tt} - f(u_0) u_{0xx} - f'(u_0) u_{0x}^2 = 0, \tag{9}$$

$$\begin{aligned}L_1 := {} & u_{1tt} - f(u_0) u_{1xx} - f'(u_0) u_{0xx} u_{1x} \\ & - 2 f'(u_0) u_{0x} u_{1x} - f''(u_0) u_{0x}^2 u_1 \\ & - \lambda''(u_0) u_{0x}^2 u_{0t} - \lambda'(u_0) u_{0xx} u_{0t} \\ & - 2 \lambda'(u_0) u_{0x} u_{0tx} - \lambda(u_0) u_{0txx} = 0.\end{aligned} \tag{10}$$

Systems (9)–(10) have been obtained considering $u(t, x, \varepsilon)$ analytics in ε and expanded it in power series of ε, i.e.,

$$u(t, x, \varepsilon) = u_0(t, x) + \varepsilon u_1(t, x) + \mathcal{O}(\varepsilon^2), \tag{11}$$

where u_0 is the solution of the "unperturbed equation" (9) while u_1 can be obtained from the linear Equation (10).

For the sake of clarity, we briefly recall the main results of the symmetry classifications of Equation (2), which are obtained in [7].

The approximate generator of Equation (2) is written in the form

$$\begin{aligned}X = {} & \xi_0^1(t, x, u_0) \frac{\partial}{\partial t} + \xi_0^2(t, x, u_0) \frac{\partial}{\partial x} + \eta_0^0(t, x, u_0) \frac{\partial}{\partial u_0} \\ & + [\eta_0^1(t, x, u_0) + \eta_1^1(t, x, u_0) u_1] \frac{\partial}{\partial u_1}\end{aligned} \tag{12}$$

and the associate *Approximate Principal Lie Algebra* of Equation (2), obtained when $f(u_0)$ and $\lambda(u_0)$ are arbitrary functions of u_0, is

$$X_1 = \frac{\partial}{\partial t}, \quad X_2 = \frac{\partial}{\partial x}, \quad X_3 = t\frac{\partial}{\partial t} + x\frac{\partial}{\partial x} - u_1 \frac{\partial}{\partial u_1}, \tag{13}$$

denoted with $L_\mathcal{P}^A$. For suitable forms of functions $f(u_0)$ and $\lambda(u_0)$, we also get the symmetries summarized in the following Table 1:

Table 1. Classification of $f(u_0)$ and $\lambda(u_0)$ with the corresponding extensions of $L_\mathcal{P}^A$ of Equation (2). f_0, λ_0, p, q and s are constitutive constants with $p \neq 0$.

Case	Forms of $f(u_0)$ and $\lambda(u_0)$	Extensions of $\mathcal{A}pprox\tilde{L}_\mathcal{P}$
I	$f(u_0) = f_0 e^{\frac{1}{p} u_0}$ $\lambda(u_0) = \lambda_0 e^{\frac{1+s}{p} u_0}$	$X_4 = x \frac{\partial}{\partial x} + 2p \frac{\partial}{\partial u_0} + 2s u_1 \frac{\partial}{\partial u_1}$
II	$f(u_0) = f_0 (u_0 + q)^{\frac{1}{p}}$ $\lambda(u_0) = \lambda_0 (u_0 + q)^{\frac{1+s}{p} - 1}$	$X_4 = x \frac{\partial}{\partial x} + 2p (u_0 + q) \frac{\partial}{\partial u_0} + 2s u_1 \frac{\partial}{\partial u_1}$
III	$f(u_0) = f_0 (u_0 + q)^{-\frac{4}{3}}$ $\lambda(u_0) = \lambda_0 (u_0 + q)^{-\frac{4}{3}}$	$X_4 = x \frac{\partial}{\partial x} - \frac{3}{2} (u_0 + q) \frac{\partial}{\partial u_0} - \frac{3}{2} u_1 \frac{\partial}{\partial u_1}$ $X_5 = x^2 \frac{\partial}{\partial x} - 3x (u_0 + q) \frac{\partial}{\partial u_0} - 3x u_1 \frac{\partial}{\partial u_1}$

3. Optimal System of Approximate Subalgebras

In this section, we introduce the definition of Optimal Systems of one-dimensional approximate subalgebras for Equation (2), in the context of Valenti's theory of approximate symmetries; we begin by introducing the following definition:

Definition 2. *We call the One-Dimensional Optimal system of Approximate Subalgebras of Equation (2) the one-dimensional, Optimal system of (exact) Subalgebras of systems (9)–(10).*

As in the classical symmetries, even in the approximate one, an important instrument to get approximate solutions is to find the optimal system of approximate subalgebras in order to obtain all the essentially different approximate invariant solutions. In order to obtain reductions and to construct classes of group-invariant approximate solutions for Equation (2) in a systematic way, we will get an optimal system of one-dimensional approximate subalgebras for Equation (2).

Following [8] by using both the adjoint table [9] and the global matrix [10], we get one dimensional optimal system of subalgebras. We present only the details for the *approximate principal Lie Algebra L_P* of Equation (2). Thus, as a basis of L_P^A, we take adjoint operators, namely

$$\mathcal{A}_i = -[X_i, X_j]\frac{\partial}{\partial X_j}, \quad (i,j = 1,2,3), \tag{14}$$

where, if X_i and X_j are vector fields, then their Lie bracket $[X_i, X_j]$ is the unique vector field satisfying

$$[X_i, X_j] = X_i(X_j) - X_j(X_i).$$

The commutator table of Approximate Principal Lie Algebra of case (13) is shown in Table 2.

Table 2. Commutator Table of the Approximate Lie Algebra of Equation (2). The (i,j)-th entry indicates $[X_i, X_j] = X_i(X_j) - X_j(X_i)$.

	X_1	X_2	X_3
X_1	0	0	X_1
X_2	0	0	X_2
X_3	$-X_1$	$-X_2$	0

The adjoint representation can be denoted as $Ad(exp(\varepsilon_i X_i))X_j$ and is written by summing the Lie series

$$Ad(exp(\varepsilon_i X_i))X_j = \sum_{n=0}^{+\infty} \frac{\varepsilon^n}{n!}(ad\, X_i)^n(X_j) = X_j - \varepsilon[X_i, X_j] + \frac{\varepsilon^2}{2}[X_i, [X_i, X_j]] - \ldots$$

The Adjoint Table of Approximate Principal Lie Algebra spanned by operators (13) is written in Table 3:

Table 3. Adjoint Table of the Approximate Lie Algebra L_P of Equation (2). The (i,j)-th entry indicates $Ad(exp(\varepsilon_i X_i))X_j = X_j - \varepsilon[X_i, X_j] + \frac{\varepsilon^2}{2}[X_i, [X_i, X_j]] - \ldots$

Ad	X_1	X_2	X_3
X_1	X_1	X_2	$X_3 - \varepsilon_1 X_1$
X_2	X_1	X_2	$X_3 - \varepsilon_2 X_2$
X_3	$X_1(1+\varepsilon_3)$	$X_2(1+\varepsilon_3)$	X_3

According the method given in [8], keeping in consideration the commutators, in the first line of Table 2, we get for instance

$$\mathcal{A}_1 = -X_1 \frac{\partial}{\partial X_3}, \quad (15)$$

which generates the linear transformations

$$X'_1 = X_1, \qquad X'_2 = X_2, \qquad X'_3 = -\varepsilon_1 X_1 + X_3. \quad (16)$$

Moreover, the linear transformations generated by each $\mathcal{A}_i$ can be obtained simply by checking the first, second, etc ... row of the Adjoint table (Table 3) of approximate principal Lie algebra $\mathcal{L}_\mathcal{P}$ of Equation (2); for example, in the case of (16), the linear transformation is represented by the matrix

$$M_1(\varepsilon_1) = \begin{pmatrix} 1 & 0 & 0 \\ 0 & 1 & 0 \\ -\varepsilon_1 & 0 & 1 \end{pmatrix}.$$

Following [9], the global matrix M of the adjoint transformations is the product of matrices $M_i(\varepsilon_i)$ associated with each $\mathcal{A}_i$. For $\mathcal{L}_\mathcal{P}$ of Equation (2), we have

$$M = \Pi_{i=1}^3 M_i(\varepsilon_i) = \begin{pmatrix} 1+\varepsilon_3 & 0 & 0 \\ 0 & 1+\varepsilon_3 & 0 \\ -\varepsilon_1(1+\varepsilon_3) & -\varepsilon_2(1+\varepsilon_3) & 1 \end{pmatrix}.$$

In order to obtain the global action of operators $\mathcal{A}_i$, $(i = 1, ..., 3)$, we apply the matrix M^T, transposed matrix of M, to an element of $L_\mathcal{P}$, i.e., $X_o = \sum_{i=1}^3 a_i X_i$. Actually, it is preferable to work with the vector $\mathbf{a} \equiv (a_1, a_2, a_3)$; the coordinates of the transformed vector of $\mathbf{a}$ are

$$\tilde{a}_1 = (1+\varepsilon_3)a_1 - \varepsilon_1(1+\varepsilon_3)a_3,$$

$$\tilde{a}_2 = (1+\varepsilon_3)a_2 - \varepsilon_2(1+\varepsilon_3)a_3,$$

$$\tilde{a}_3 = a_3,$$

and firstly we underline that these transformations leave invariant the component a_3 and provide the adjoint group $G^{\mathcal{A}}$ of $L_\mathcal{P}^{\mathcal{A}}$.

We can determine the optimal system of $\mathcal{L}_\mathcal{P}$ by using a simple approach. We simplify any given vector $\mathbf{a} \equiv (a_1, a_2, a_3)$ through the above transformations. Then, distinguish the obtained vectors into nonequivalent classes, where we choose the one with the simplest form by which we obtain the following non-trivial operator of the optimal system of $\mathcal{L}_\mathcal{P}$:

$$X_{o1} = c_1 X_1 + c_2 X_2 + X_3 = (c_1 + t)\frac{\partial}{\partial t} + (c_2 + x)\frac{\partial}{\partial x} - u_1\frac{\partial}{\partial u_1}, \quad (17)$$

where c_1, c_2 are real parameters.

Finally, starting from Case III of Table 1, we are able to construct the corresponding extensions of the optimal system of approximate subalgebras $\mathcal{L}_p$ of Case III for Equation (2), by using the Adjoint Table in the Appendix A (Table A1), which in this case reads as:

$$X_{o2} = c_1 X_1 + X_3 + X_4 = (c_1 + t)\frac{\partial}{\partial t} + 2x\frac{\partial}{\partial x} - \frac{3}{2}(u_0 + q)\frac{\partial}{\partial u_0} - \frac{5}{2}u_1\frac{\partial}{\partial u_1}, \quad (18)$$

$$X_{o3} = c_1 X_1 + c_2 X_2 + X_5 = c_1\frac{\partial}{\partial t} + (c_2 + x^2)\frac{\partial}{\partial x} - 3x(u_0 + q)\frac{\partial}{\partial u_0} - 3xu_1\frac{\partial}{\partial u_1}. \quad (19)$$

4. Reduction to ODEs and New Approximate Invariant Solution for the Generalized Ames's Equation

Based on the results obtained in the previous section, thanks to the approximate generators, we can construct the corresponding reduced ODEs of Equation (2); indeed, as we know from literature, group classification problems are interesting not only from a purely mathematical point of view but above all in the applications that can be achieved [9–12].

We obtain an approximate solution for Equation (2) considering the approximate operator (17) of $\mathcal{L}_p$, i.e.,

$$X_{o1} = c_1 X_1 + c_2 X_2 + X_3 = (c_1 + t)\frac{\partial}{\partial t} + (c_2 + x)\frac{\partial}{\partial x} - u_1 \frac{\partial}{\partial u_1},$$

we get the following transformation

$$z = \frac{c_2 + x}{c_1 + t}, \quad u_0 = \phi(z), \quad u_1 = \frac{\psi(z)}{c_1 + t}, \tag{20}$$

which maps Equation (2) into the following ODEs:

$$2z\phi' - f'\phi'^2 + (z^2 - f)\phi'' = 0,$$
$$\lambda z \phi''' - (f'\psi - 2\lambda - 3z\lambda'\phi')\phi'' + (z^2 - f)\psi''$$
$$+ (z\lambda''\phi' + 2(\lambda' - f'')\psi)\phi'' - 2(f'\phi' - 2z)\psi' + 2\psi = 0,$$

$f = f(u)$ and $\lambda = \lambda(u)$ being arbitrary functions of its argument. When $f = f_0 =$ constant and $\lambda = \lambda_0 =$ constant, by integration of the reduced equations, we get

$$\phi = k_1 + \frac{k_2}{f_0} \text{arctanh}\left(\frac{z}{\sqrt{f_0}}\right), \quad \psi = \frac{(k_3 + k_4 z)(f_0 - z^2) + k_2 \lambda_0 z}{(f_0 - z^2)^2},$$

where k_i, $(i = 1, \ldots, 4)$ are arbitrary constants, and, as a consequence, we obtain the following solution of Equation (2)

$$u = k_1 + \frac{k_2}{f_0} \text{arctanh}\left(\frac{c_2 + x}{\sqrt{f_0}(c_1 + t)}\right)$$
$$+ \varepsilon \frac{(k_3(c_1 + t) + k_4(c_2 + x))(f_0(c_1 + t)^2 - (c_2 + x)^2) + k_2 \lambda_0 (c_1 + t)^2 (c_2 + x)}{(f_0(c_1 + t)^2 - (c_2 + x)^2)^2}. \tag{21}$$

Another solution can be obtained by the operator

$$X_{o2} = (c_1 + t)\frac{\partial}{\partial t} + 2x\frac{\partial}{\partial x} - \frac{3}{2}(u_0 + q)\frac{\partial}{\partial u_0} - \frac{5}{2}u_1 \frac{\partial}{\partial u_1}$$

when $f = f_0(u_0 + q)^{-4/3}$ and $\lambda = \lambda_0(u_0 + q)^{-4/3}$ (Case III of Table 1), which leads to get the following transformation of variables:

$$z = \frac{x}{(c_1 + t)^2}, \quad u_0 = \frac{\phi(\xi)}{(c_1 + t)^{3/2}} - q, \quad u_1 = \frac{\psi(\xi)}{(c_1 + t)^{5/2}}, \tag{22}$$

and to the following reduced equations:

$$12(4z^2 \phi^{\frac{4}{3}} - f_0)\phi\phi'' + 16(f_0\phi' + 9z\phi^{\frac{7}{3}})\phi' + 45\phi^{\frac{10}{3}} = 0,$$
$$72\lambda_0 z \phi^2 \phi''' - 6\phi(48\lambda_0 z \phi' - 21\lambda_0 \phi - 8f_0 \psi)\phi'' + 36(4z^2 \phi^{\frac{4}{3}} - f_0)\phi^2 \psi''$$
$$+ 96(f_0 \phi' + 6z\phi^{\frac{7}{3}})\phi\psi' + 56(4\lambda_0 z \phi' - 2f_0 \psi - 21\lambda_0 \phi)\phi'^2 + 315\phi^{\frac{10}{3}}\psi = 0$$

that admit as a solution

$$\phi = (\frac{\sqrt{f_0}}{z})^{3/2}, \qquad \psi = \frac{1}{z^{3/2}}\left(k_1 + \frac{3}{2}k_2 \log z - \lambda_0 \left(\frac{\log z}{4\sqrt{f_0}}\right)^2\right),$$

k_1 and k_2 being arbitrary constants. Thus, a solution of (2) is

$$u = \sqrt{\frac{c_1+t}{x^3}}\left(f_0^{3/4}(c_1+t) - q + \varepsilon\left(k_1 + \frac{3}{2}k_2 \log \frac{x}{(c_1+t)^2} - \lambda_0 \left(\frac{\log \frac{x}{(c_1+t)^2}}{4\sqrt{f_0}}\right)^2\right)\right). \qquad (23)$$

Finally, if we consider the approximate operator:

$$X_{o3} = c_1 X_1 + c_2 X_2 + X_5 = c_1 \frac{\partial}{\partial t} + (c_2 + x^2)\frac{\partial}{\partial x} - 3x(u_0 + q)\frac{\partial}{\partial u_0} - 3x u_1 \frac{\partial}{\partial u_1}, \qquad (24)$$

we get the following transformation:

$$z = \frac{1}{\sqrt{c_2}}\arctan\left(\frac{c_1}{c_2}x\right) - \frac{t}{c_1}, \qquad u_0 = \left(\frac{c_2}{c_2+c_1 x^2}\right)^{\frac{3}{2}}\phi(z) - q, \qquad u_1 = \left(\frac{c_2}{c_2+c_1 x^2}\right)^{\frac{3}{2}}\psi(z),$$

which maps Equation (2) into the following ODEs:

$$3(c_2^2 \phi^{\frac{4}{3}} - c_1^4 f_0)\phi\phi'' + c_1^4 f_0(4\phi'^2 + 9c_2\phi^2) = 0, \qquad (25)$$

$$9c_1^3 \lambda_0 \phi^2 \phi''' + 12\phi(c_1 f_0 \psi - 3\lambda_0 \phi')\phi'' + 9(c_2^2 \phi^{\frac{4}{3}} - c_1^4 f_0)\phi^2 \psi'' \qquad (26)$$
$$+24 c_1^4 f_0 \phi \phi' \psi' - c_1^3 (c_1 f_0 \psi - \lambda_0 \phi')(9 c_2 \phi^2 + 28\phi'^2) = 0. \qquad (27)$$

A particular solution is given by $\phi = 0$, so that the solution of Equation (2) is

$$u = -q + \varepsilon \left(\frac{c_2}{c_2+c_1 x^2}\right)^{\frac{3}{2}} \psi(z) \qquad (28)$$

with $\psi(z)$ arbitrary function of

$$z = \frac{1}{\sqrt{c_2}}\arctan\left(\frac{c_1}{c_2}x\right) - \frac{t}{c_1}.$$

A simple solution that can be obtained from (24) by setting $c_1 = c_2 = 0$; the similarity solution and the variables become, respectively,

$$z = t, \qquad u_0 = x^{-3}\phi(t) - q, \qquad u_1 = x^{-3}\psi(t), \qquad (29)$$

while the corresponding reduced ODEs of (2) assume the simple form:

$$\phi'' \phi^{\frac{1}{3}} = 0, \qquad \psi'' \phi^{\frac{4}{3}} = 0,$$

which admit the solution

$$\phi = h_1 t + h_0, \qquad \psi = k_1 t + k_0 \qquad (30)$$

with h_1, h_0, k_1 and k_0 arbitrary constants of integration. Then, we can write the invariant approximate solution for Equation (2)

$$u(t, x) = \frac{h_1 t + h_0}{x^3} - q + \varepsilon \frac{k_1 t + k_0}{x^3} + \mathcal{O}(\varepsilon^2). \qquad (31)$$

5. The Potential System Associated with the Generalized Ames's Equation

We recall that, for any given system of PDEs, it is possible to construct nonlocally related potential systems that have the same solutions of the given system [2]. The transformation that allows for mapping the given system of PDEs into nonlocally potential systems is defined "nonlocal transformation".

Thus, we have that Equation (2) leads to getting the following system:

$$u_t - v_x = 0, \tag{32}$$

$$v_t - \left(\int^u f(s)\, ds + \varepsilon \lambda(u)\, v_x \right)_x = 0, \tag{33}$$

studied in [13], which can be treated as the *potential system* of Equation (2).

In addition, if we consider the nonlocal transformation $u = w_x$, $v = w_t$, systems (32)–(33) are equivalent to

$$w_{tt} = f(w_x)w_{xx} + \varepsilon[\lambda(w_x)w_{tx}]_x \tag{34}$$

studied in [8,14]. Special cases of models belonging to the class of Equation (34) can be found in [15–19]. Furthermore, some questions related to global existence, uniqueness and stability of solutions have been addressed in [20,21].

In Ref. [7], the authors have proved a Theorem affirming that For any f and λ, an approximate symmetry admitted by the systems (32)–(33) and Equation (34) defines an approximate symmetry admitted by Equation (2); conversely, some approximate symmetries of Equation (2) do not induce approximate symmetries of Equation (34) and the systems (32)–(33).

In a few words, the classification of Equation (2) is richer than those of systems (32)–(33) and of Equation (34). In fact, while there is a correspondence between the approximate symmetry operators $X_1, \ldots, X_4$ admitted by Equation (2) with the ones admitted by (34), the operator X_5 reported in case III of Table 2 is a new approximate operator, admitted only by Equation (2).

According to this theorem, in light of the determination of the one-dimensional optimal system of approximate subalgebras, we are able to get approximate *invariant* and *non-invariant* solutions for the systems (32)–(33) and Equation (34).

For better clarification: the correspondence between the approximate symmetry operators $X_1, \ldots, X_4$ admitted by Equation (2) with the ones admitted by (34), allows us to get approximate *invariant* solutions for the systems (32)–(33) and Equation (34); for instance, the general solution (21) which we have obtained in this paper, when $k_2 = 0$, includes the solution obtained considering an approximate operator admitted by Equation (34) and found in [8]. Instead, starting from the approximate solution (31), obtained by means of the operator X_{03} that involves the generator X_5, we are able to obtain approximate *non-invariant* solutions for the systems (32)–(33) and Equation (34) that could not have been obtained from the symmetry analysis and reductions performed in [13,14] because the operator X_5 is new and it is admitted only by Equation (2).

New Approximate Non-Invariant Solutions for Equation (35) and the Potential System

Starting from the approximate solution (31), keeping in consideration the nonlocal transformation $u = w_x$, by integrating (31) with respect to x, we obtain:

$$w(t, x) = -\frac{h_1 t + h_0}{2x^2} - qx + \chi_0(t) + \varepsilon \left(\frac{k_1 t + k_0}{2x^2} + \chi_1(t) \right), \tag{35}$$

where $\chi_0(t)$ and $\chi_1(t)$, at this stage, are arbitrary functions of t. When we substitute them into (34), we get that it must be linear in their arguments, so we obtain the following new non-invariant approximate solution for Equation (34):

$$w(t,x) = -\frac{h_1 t + h_0}{2x^2} - qx - \frac{27 f_0}{10 h_1^2}(h_1 t + h_0)^{\frac{5}{3}} + K_1 t$$
$$- \varepsilon \left[\frac{k_1 t + k_0}{2x^2} - \frac{9(h_0 + h_1 t)^{\frac{2}{3}}}{10 h_1^3} \left[45 \lambda_0 h_1^2 + f_0(h_1 k_1 t - 5 h_1 k_0 + 6 h_0 k_1) \right] - K_2 t \right] + \mathcal{O}(\varepsilon^2), \tag{36}$$

where K_1 and K_2 are arbitrary constants.

Finally, the non-invariant approximate solution for the system (32)–(33) reads as

$$u = \frac{h_1 t + h_0}{x^3} - q + \varepsilon \frac{k_1 t + k_0}{x^3},$$
$$v = -\frac{h_1}{2x^2} - \frac{9 f_0}{2 h_1}(h_1 t + h_0)^{\frac{2}{3}} + K_1 \tag{37}$$
$$- \varepsilon \left[\frac{k_1 t + k_0}{2x^2} - \frac{9(h_0 + h_1 t)^{\frac{2}{3}}}{10 h_1^3} \left[45 \lambda_0 h_1^2 + f_0(h_1 k_1 t - 5 h_1 k_0 + 6 h_0 k_1) \right] - K_2 t \right].$$

6. Conclusions

In this manuscript, we worked following the method of Valenti's approximate symmetries; moreover, thanks to the link between the approximate symmetries of the three related models (2), (32)–(33), (34) and the definition of one-dimensional optimal system of approximate subalgebras, we are also capable in this manuscript to get new approximate, invariant and non-invariant, solutions not only for the generalized Ames's Equation (2), but also for Equation (34) and the systems (32)–(33) which could not be obtained from the approximate symmetry analysis performed in [13,14].

Author Contributions: M.R. and M.P.S. worked together in the derivation of the mathematical results. Both authors provided critical feedback and helped shape the research, analysis and manuscript.

Funding: This research received no external funding.

Acknowledgments: The authors acknowledge the support by G.N.F.M. of INdAM.

Conflicts of Interest: The authors declare no conflict of interest.

Appendix A

In this Appendix, the Adjoint Table (Table A1) of the case III is reported.

Table A1. Adjoint Table of the Approximate Lie Algebra of Equation (2).

Ad	X_1	X_2	X_3	X_4	X_5
X_1	X_1	X_2	$X_3 - \varepsilon_1 X_1$	X_4	X_5
X_2	X_1	X_2	$X_3 - \varepsilon_2 X_2$	$X_4 - \varepsilon_2 X_2$	$X_5 - 2\varepsilon_2 X_4$
X_3	$X_1(1+\varepsilon_3)$	$X_2(1+\varepsilon_3)$	X_3	X_4	$X_5(1-\varepsilon_3)$
X_4	X_1	$X_2(1+\varepsilon_4)$	X_3	X_4	$X_5(1-\varepsilon_4)$
X_5	X_1	$X_2 + 2\varepsilon_5 X_4$	$X_3 + \varepsilon_5 X_5$	$X_4 + \varepsilon_5 X_5$	X_5

References

1. Ames, W.F.; Lohner, R.J.; Adams, E. Group properties of $u_{tt} = [f(u)u_x]_x$. *Int. J. Non-Linear Mech.* **1981**, *16*, 439–447. [CrossRef]
2. Bluman, G.; Cheviakov, A.F. Nonlocally related systems, linearization and nonlocal symmetries for the nonlinear wave equation. *J. Math. Anal. Appl.* **2007**, *333*, 93–111. [CrossRef]
3. Valenti, A. Approximate symmetries for a model describing dissipative media. In Proceedings of the MOGRAN X, Larnaca, Cyprus, 24–31 October 2004; pp. 236–242.
4. Baĭkov, V.A.; Gazizov, R.K.; Ibragimov, N.H. Approximate Symmetries. *Math. USSR Sb.* **1989**, *64*, 427–441. [CrossRef]
5. Fushchich, W.I.; Shtelen, W.M. On approximate symmetry and approximate solutions of the nonlinear wave equation with a small parameter. *J. Phys. A Math. Gen.* **1989**, *22*, 887–890. [CrossRef]
6. Pakdemirli, M.; Yurusoy, M.; Dolapc, I. Comparison of approximate symmetry methods for differential equation. *Acta Appl. Math.* **2004**, *80*, 243–271. [CrossRef]
7. Ruggieri, M.; Speciale, M.P. Lie group analysis of a wave equation with a small nonlinear dissipation. *Ric. Di Mat.* **2017**, *66*, 27–34. [CrossRef]
8. Ruggieri, M.; Speciale, M.P. Optimal System and Approximate Solutions of Nonlinear Dissipative Media. *Math. Methods Appl. Sci.* **2019**. [CrossRef]
9. Olver, P.J. *Applications of Lie Groups to Differential Equations*; Springer: New York, NY, USA, 1986.
10. Ovsiannikov, L.V. *Group Analysis of Differential Equations*; Academic Press: New York, NY, USA, 1982.
11. Ibragimov, N.H. *CRC Hanbook of Lie Group Analysis of Differential Equations*; CRC Press: Boca Raton, FL, USA, 1994.
12. Ruggieri, M.; Speciale, M.P. Quasi Self-adjoint Coupled KdV-like Equations. *AIP Conf. Proc.* **2013**, *1558*, 1220–1223.
13. Ruggieri, M.; Speciale, M.P. Approximate symmetries in viscoelasticity. *Theor. Math. Phys.* **2016**, *189*, 1500–1508. [CrossRef]
14. Ruggieri, M.; Valenti A. Approximate symmetries in nonlinear viscoelastic media. *Bound. Value Probl.* **2013**, *2013*, 143. [CrossRef]
15. Ruggieri, M.; Speciale, M.P. Approximate Analysis of a Nonlinear Dissipative Model. *Acta Appl. Math.* **2014**, *132*, 549–559. [CrossRef]
16. Valenti, A. Approximate symmetries of a viscoelastic model. In Proceedings of the WASCOM 2007 Baia Samuele, Sicily, Italy, 30 June–7 July 2008; Manganaro, N., Monaco, R., Rionero, S., Eds.; World Science Publishing: Singapore, 2008; pp. 582–588.
17. Ruggieri, M.; Valenti, A. Symmetries and reduction techniques for dissipative models. *J. Math. Phys.* **2009**, *50*, 063506. [CrossRef]
18. Ruggieri, M.; Valenti, A. Exact solutions for a nonlinear model of dissipative media. *J. Math. Phys.* **2011**, *52*, 043520. [CrossRef]
19. Ruggieri, M. Kink solutions for a class of generalized dissipative equation. *Abstr. Appl. Anal.* **2012**. [CrossRef]

20. Dafermos, C.M. The mixed initial-boundary value problem for the equations of nonlinear one-dimensional viscoelasticity. *J. Differ. Equ.* **1969**, *6*, 71–86. [CrossRef]
21. MacCamy, R.C. Existence, uniqueness and stability of $u_{tt} = \frac{\partial}{\partial x}[\sigma(u_x) + \lambda(u_x)u_{xt}]$. *Indiana Univ. Math. J.* **1970**, *20*, 231–238. [CrossRef]

© 2019 by the authors. Licensee MDPI, Basel, Switzerland. This article is an open access article distributed under the terms and conditions of the Creative Commons Attribution (CC BY) license (http://creativecommons.org/licenses/by/4.0/).

Article

Hamiltonian Structure, Symmetries and Conservation Laws for a Generalized (2 + 1)-Dimensional Double Dispersion Equation

Elena Recio *, Tamara M. Garrido, Rafael de la Rosa and María S. Bruzón

Department of Mathematics, Universidad de Cádiz, Puerto Real, 11510 Cádiz, Spain
* Correspondence: elena.recio@uca.es

Received: 15 July 2019; Accepted: 6 August 2019; Published: 9 August 2019

Abstract: This paper considers a generalized double dispersion equation depending on a nonlinear function $f(u)$ and four arbitrary parameters. This equation describes nonlinear dispersive waves in 2 + 1 dimensions and admits a Lagrangian formulation when it is expressed in terms of a potential variable. In this case, the associated Hamiltonian structure is obtained. We classify all of the Lie symmetries (point and contact) and present the corresponding symmetry transformation groups. Finally, we derive the conservation laws from those symmetries that are variational, and we discuss the physical meaning of the corresponding conserved quantities.

Keywords: Lie symmetry; conservation law; double dispersion equation; Boussinesq equation

1. Introduction

A well-known equation that models the motion of long dispersive shallow water waves, which are propagated in both directions, is the Boussinesq equation [1,2], given by:

$$u_{tt} = u_{xx} + bu_{xxxx} + \alpha(u^2)_{xx}. \tag{1}$$

This equation is an integrable system, which is well-posed for $b = -1$ and ill-posed for $b = 1$.

Many modified and generalized Boussinesq equations have been considered in the literature. A generalization of the Boussinesq equation depending on a nonlinearity power $p \neq 0$ consists of replacing in (1) the nonlinear term $(u^2)_{xx}$ by $(u^p)_{xx}$. The resulting equation is given by:

$$u_{tt} = u_{xx} + au_{xxxx} + \alpha(u^p)_{xx}. \tag{2}$$

A modified Boussinesq equation is obtained by substituting the fourth-order term u_{xxxx} in (1) by u_{ttxx}, yielding:

$$u_{tt} = u_{xx} + au_{ttxx} + \alpha(u^2)_{xx}. \tag{3}$$

This equation is well-posed. In [3–8], the Cauchy problem and initial boundary value problem were considered for these equations and also for a double dispersion equation that unifies the previous equations and depends on a nonlinear function $f(u)$. This generalized double dispersion equation is given by:

$$u_{tt} = u_{xx} + au_{ttxx} + bu_{xxxx} + du_{txx} + (f(u))_{xx}. \tag{4}$$

Another interesting variant of the Boussinesq equation is given by the sixth-order equation:

$$u_{tt} = u_{xx} + au_{xxxxxx} + bu_{xxxx} + (u^3)_{xx}, \tag{5}$$

and its generalization:
$$u_{tt} = cu_{xx} + au_{xxxxxx} + bu_{xxxx} + (f(u))_{xx}, \qquad (6)$$

where $f(u)$ is a nonlinear function. These equations model long gravity-capillary surface waves with a short amplitude, propagating in both directions in shallow water. The Cauchy problem was considered for Equation (5) in [9], and the Hamiltonian formulation and a complete classification of Lie point symmetries and conservation laws were obtained for Equation (6) in [10].

Recently, a generalization of Equation (2) to two spatial dimensions, given by:
$$u_{tt} = u_{xx} + bu_{xxxx} + \alpha(u^{p+1})_{xx} + \beta u_{yy}, \quad p \neq 0 \qquad (7)$$

was considered in [11], where the point symmetries, conservation laws, and line soliton solutions were derived.

In this work, we consider a $(2+1)$-dimensional generalization of the double dispersion Equation (4), given by:
$$u_{tt} = u_{xx} + au_{ttxx} + bu_{xxxx} + du_{txx} + (f(u))_{xx} + \beta u_{yy}, \qquad (8)$$

where $f(u)$ is a nonlinear arbitrary function and a, b, d, and β are arbitrary parameters. The 2D generalized double dispersion (2D gDD) family of Equations (8) unifies the previous Equations (1)–(4) and (7). Some $(2+1)$-dimensional equations in this family were considered in recent literature and were shown to admit interesting exact solutions such as line solitons and lump solutions [11–13].

For any given partial differential equation (PDE), symmetries are transformations that leave invariant the whole space of solutions of the equation. Symmetries can be used to obtain reductions and exact group-invariant solutions. These invariant solutions play a key role in the investigation of certain analytical properties, e.g., asymptotic and blow-up behavior. In addition, explicit solutions can be used to assess the accuracy and reliability of numerical solution methods. For a given PDE, all admitted Lie symmetries can be determined by applying the Lie method.

A conservation law of a given evolution equation is a continuity equation that yields basic conserved quantities for all solutions. Some important uses, among others, are that they allow detecting and constructing mappings of nonlinear evolution equations to linear equations. Moreover, they can be used for studying integrability.

In Section 2, we write the 2D gDD Equation (8) as a potential equation, and we find a Lagrangian formulation for $d = 0$. The associated Hamiltonian formulation is also included in Section 2. By using the Lie method, in Section 3, we classify all point and contact symmetries of the potential equation, and we include the corresponding symmetry groups. We provide an Appendix A with a summary of the computations. In Section 4, from the previous classification of Lie symmetries of the potential equation, the variational symmetries are found. Next, in Section 5, we derive the conservation laws of the potential equation from the variational symmetries by using Noether's theorem. Furthermore, we discuss the physical meaning of the associated conserved quantities. Finally, in Section 6, we give some conclusions.

2. Potential Form and Hamiltonian Formulation

The generalized double dispersion Equation (8) can be expressed in potential form by using a potential:
$$u = v_x. \qquad (9)$$

The resulting equation is then given by:
$$G = v_{tt} - v_{xx} - av_{ttxx} - bv_{xxxx} - dv_{txx} - (f(v_x))_x - \beta v_{yy} = 0. \qquad (10)$$

This potential Equation (10) admits a local Lagrangian structure:

$$\frac{\delta L}{\delta v} = 0 \qquad (11)$$

if and only if $d = 0$, where $\delta/\delta v$ is the variational derivative with respect to v:

$$\frac{\delta}{\delta v} = \partial_v - D_t \partial_{v_t} - D_x \partial_{v_x} - D_y \partial_{v_y} + D_t^2 \partial_{v_{tt}} + D_x^2 \partial_{v_{xx}} + D_y^2 \partial_{v_{yy}} + D_t D_x \partial_{v_{tx}} + D_t D_y \partial_{v_{ty}} \\ + D_x D_y \partial_{v_{xy}} + \cdots . \qquad (12)$$

Indeed, the double dispersion equation in potential form (10) can be written as the Euler-Lagrangian equation of a local Lagrangian if and only if the Helmholtz conditions are satisfied [14,15], i.e., the Fréchet derivative of Equation (10) is self-adjoint. We recall [14,15] that the Fréchet derivative of a differential function $G(t, x, y, v, v_t, v_x, v_y, \ldots)$ acting on $P(t, x, y)$ is given by:

$$\delta_P G = P\frac{\partial G}{\partial v} + D_t P \frac{\partial G}{\partial v_t} + D_x P \frac{\partial G}{\partial v_x} + D_y P \frac{\partial G}{\partial v_y} + D_t^2 P \frac{\partial G}{\partial v_{tt}} + D_x^2 P \frac{\partial G}{\partial v_{xx}} + D_y^2 P \frac{\partial G}{\partial v_{yy}} \\ + D_t D_x P \frac{\partial G}{\partial v_{tx}} + D_t D_y P \frac{\partial G}{\partial v_{ty}} + D_x D_y P \frac{\partial G}{\partial v_{xy}} + \cdots , \qquad (13)$$

and the adjoint Fréchet derivative is given by:

$$\delta_P^* G = P\frac{\partial G}{\partial v} - D_t\left(P\frac{\partial G}{\partial v_t}\right) - D_x\left(P\frac{\partial G}{\partial v_x}\right) - D_y\left(P\frac{\partial G}{\partial v_y}\right) + D_t^2\left(P\frac{\partial G}{\partial v_{tt}}\right) + D_x^2\left(P\frac{\partial G}{\partial v_{xx}}\right) \\ + D_y^2\left(P\frac{\partial G}{\partial v_{yy}}\right) + D_t D_x\left(P\frac{\partial G}{\partial v_{tx}}\right) + D_t D_y\left(P\frac{\partial G}{\partial v_{ty}}\right) + D_x D_y\left(P\frac{\partial G}{\partial v_{xy}}\right) - \cdots . \qquad (14)$$

To show that the Fréchet derivative of the potential 2D gDD Equation (10) is self-adjoint, we compute the Fréchet derivative of Equation (10):

$$\delta_P G = -f''(v_x) v_{xx} D_x P + D_t^2 P - (1 + f'(v_x)) D_x^2 P - \beta D_y^2 P - d D_t D_x^2 P - a D_t^2 D_x^2 P - b D_x^4 P, \qquad (15)$$

and the adjoint Fréchet derivative of Equation (10):

$$\delta_P^* G = D_x(f''(v_x) v_{xx} P) + D_t^2 P - D_x^2((1 + f'(v_x)) P) - \beta D_y^2 P + d D_t D_x^2 P - a D_t^2 D_x^2 P - b D_x^4 P, \qquad (16)$$

where $P = P(t, x, y)$, and we verify that Expression (15) coincides with its adjoint (16) for all $P(t, x, y)$. It is straightforward to show that this only occurs iff $d = 0$.

There are two main implications when a PDE admits a Lagrangian structure: the existence of conserved energy, momentum, etc., from Noether's theorem and the existence of a Hamiltonian formulation. From the physical point of view, the restriction $d = 0$ on Equation (10) means that the dynamics is being restricted such that dissipative processes are being removed, leading to conservation of energy, momentum, etc.

The Lagrangian corresponding to the Lagrangian formulation (11) of Equation (10) is then given by:

$$L = -\tfrac{1}{2} v_t^2 + \tfrac{1}{2} v_x^2 - \tfrac{1}{2} a v_{tt} v_{xx} - \tfrac{1}{2} b v_{xx}^2 + F(v_x) + \tfrac{1}{2} \beta v_y^2, \qquad (17)$$

where $F'(v_x) = f(v_x)$.

The potential Equation (10) can be expressed as an equivalent evolution system, given by:

$$\begin{aligned} v_t &= w, \\ w_t &= v_{xx} + a v_{ttxx} + b v_{xxxx} + d v_{txx} + f'(v_x) v_{xx} + \beta v_{yy}. \end{aligned} \qquad (18)$$

When $d = 0$, the Lagrangian structure (11) yields a Hamiltonian formulation for the potential system (18), given by:

$$\begin{pmatrix} v \\ w \end{pmatrix}_t = J \begin{pmatrix} \delta H/\delta v \\ \delta H/\delta w \end{pmatrix}, \quad J = \begin{pmatrix} 0 & 1 \\ -1 & 0 \end{pmatrix} \tag{19}$$

with the Hamiltonian density given by:

$$H = \int_{\mathbb{R}^2} \left(\tfrac{1}{2}w^2 + \tfrac{1}{2}v_x^2 - \tfrac{1}{2}av_{tt}v_{xx} - \tfrac{1}{2}bv_{xx}^2 + F(v_x) + \tfrac{1}{2}\beta v_y^2 \right) dx\, dy, \tag{20}$$

where $F'(v_x) = f(v_x)$ and J is the Hamiltonian operator [15]. There exists an equivalent Hamiltonian structure in terms of the variable u, with the Hamiltonian given by a nonlocal expression in u and w. By noting that $u_t = v_{tx} = w_x$ and by applying the variational derivative identity $\delta H/\delta v = -D_x(\delta H/\delta u)$, the Hamiltonian formulation is then given by:

$$\begin{pmatrix} u \\ w \end{pmatrix}_t = \mathcal{D} \begin{pmatrix} \delta H/\delta u \\ \delta H/\delta w \end{pmatrix}, \quad \mathcal{D} = \begin{pmatrix} 0 & D_x \\ D_x & 0 \end{pmatrix} \tag{21}$$

where $\mathcal{D}$ is the Hamiltonian operator [15].

Note that when $a = 0$ and $f(v_x) = v_x^{p+1}$, the generalized $(2+1)$-dimensional Boussinesq with p-power nonlinearity is obtained. The Hamiltonian structure for this equation was obtained in [11]. In addition, if $\beta = 0$ and $p = 1$, this Hamiltonian formulation is one of the Hamiltonian structures [15] of the ordinary Boussinesq Equation (1). In this section, these Hamiltonian structures have been extended for a more general $(2+1)$-dimensional double dispersion equation depending on an arbitrary function.

3. Lie Symmetries

For nonlinear evolution equations, symmetries are important since they can be used to determine groups of transformations, which leave the solution space of the equation invariant, and also because they lead to reductions and exact invariant solutions. The Hamiltonian structure derived in the previous section motivates studying the symmetries of the 2D gDD equation in potential form (10). The Lie symmetries of Equation (10) consist of point symmetries and contact symmetries, since the equation involves only a single dependent variable v [16].

3.1. Point Symmetries

An infinitesimal point symmetry of the potential 2D gDD Equation (10) is a vector field of the form:

$$\mathbf{X} = \tau(t,x,y,v)\partial_t + \xi^x(t,x,y,v)\partial_x + \xi^y(t,x,y,v)\partial_y + \eta(t,x,y,v)\partial_v, \tag{22}$$

whose prolongation leaves invariant the whole solution space of the equation,

$$\text{pr}\mathbf{X}(v_{tt} - v_{xx} - av_{ttxx} - bv_{xxxx} - dv_{txx} - f'(v_x)v_{xx} - \beta v_{yy})|_{G=0} = 0. \tag{23}$$

A point symmetry (22) of the potential 2D gDD Equation (10) generates a one-parameter Lie group of point transformations acting on dependent and independent variables that carries solutions of the equation into other solutions. This point symmetry transformation is then given by:

$$\begin{aligned}
\tilde{t} &= t + \epsilon\tau(t,x,y,v) + O(\epsilon^2), \\
\tilde{x} &= x + \epsilon\xi^x(t,x,y,v) + O(\epsilon^2), \\
\tilde{y} &= y + \epsilon\xi^y(t,x,y,v) + O(\epsilon^2), \\
\tilde{v} &= v + \epsilon\eta(t,x,y,v) + O(\epsilon^2),
\end{aligned} \tag{24}$$

with ϵ being the group parameter. The action of a point symmetry on solutions of the potential 2D gDD Equation (10) yields:

$$v(t,x,y) \to \tilde{v}(t,x,y) = v(t,x,y) + \epsilon(\eta(t,x,y,v(t,x,y)) - \tau(t,x,y,v(t,x,y))v_t(t,x,y) \\ - \eta^x(t,x,y,v(t,x,y))v_x(t,x,y) - \eta^y(t,x,y,v(t,x,y))v_y(t,x,y)) + O(\epsilon^2), \quad (25)$$

corresponding to the characteristic form of the generator, given by:

$$\hat{X} = P\partial_v, \quad P = \eta - \tau v_t - \xi^x v_x - \xi^y v_y, \quad (26)$$

where P is the symmetry characteristic. The invariance condition is then equivalently expressed in terms of the Fréchet derivative (13) of the potential 2D gDD Equation (10) acting on the symmetry characteristic P as:

$$0 = \text{pr}\hat{X}(G)|_{G=0} = \delta_P G|_{G=0}, \quad (27)$$

yielding:

$$D_t^2 P - f''(v_x)v_{xx}D_x P - (1 + f'(v_x))D_x^2 P - \beta D_y^2 P - dD_t D_x^2 P - aD_t^2 D_x^2 P - bD_x^4 P = 0, \quad (28)$$

which holds for all solutions of the 2D gDD equation in potential form (10), and is called the symmetry-determining equation of PDE (10). The determining Equation (28) splits with respect to the differential consequences of v and leads to an overdetermined system of equations for P, $f(v_x)$, a, b, d, and β. Furthermore, we will also impose the classification conditions $f''(v_x) \neq 0$, which implies that the equation is nonlinear; $a^2 + b^2 \neq 0$ and $\beta \neq 0$, which respectively imply that the equation is a fourth-order PDE, and generalizes the 1D gDD Equation (4) to two spatial dimensions. We set up and solve the resulting determining system by using Maple, in particular the "rifsimp" and "pdsolve" commands. A summary of the steps followed for this computation is included in Appendix A. Therefore, we have the following classification result.

Theorem 1. *(i) The point symmetries admitted by the 2D generalized double dispersion potential Equation (10) for arbitrary a, b, d, β, and $f(v_x)$ with the conditions $a^2 + b^2 \neq 0$, $\beta \neq 0$, and $f''(v_x) \neq 0$ are generated by the transformations:*

$$X_1 = \partial_t, \quad (29a)$$
$$(\tilde{t}, \tilde{x}, \tilde{y}, \tilde{v})_1 = (t + \epsilon, x, y, v), \quad \text{time-translation.} \quad (29b)$$

$$X_2 = \partial_x, \quad (30a)$$
$$(\tilde{t}, \tilde{x}, \tilde{y}, \tilde{v})_2 = (t, x + \epsilon, y, v), \quad \text{space-translation.} \quad (30b)$$

$$X_3 = \partial_y, \quad (31a)$$
$$(\tilde{t}, \tilde{x}, \tilde{y}, \tilde{v})_3 = (t, x, y + \epsilon, v), \quad \text{space-translation.} \quad (31b)$$

$$X_{4,g,h} = (g(y + \sqrt{\beta}t) + h(y - \sqrt{\beta}t))\partial_v, \quad (32a)$$
$$(\tilde{t}, \tilde{x}, \tilde{y}, \tilde{v})_4 = (t, x, y, v + (g(y + \sqrt{\beta}t) + h(y - \sqrt{\beta}t))\epsilon). \quad (32b)$$

The last symmetry is a linear combination of two infinite-dimensional families, with $g(y + \sqrt{\beta}t) + h(y - \sqrt{\beta}t)$ being the general solution of the linear equation $P_{tt} - \beta P_{yy} = 0$ for $P = P(t,y)$.

(ii) The 2D generalized double dispersion potential Equation (10) admits additional point symmetries for special $f(v_x)$, a, b, or d, in the following cases:

(a) $a = 0, d = 0$, arbitrary $f(v_x)$, b, and β,

$$X_5 = y\partial_t + \beta t\partial_y, \tag{33a}$$

$$(\tilde{t}, \tilde{x}, \tilde{y}, \tilde{v})_5 = (\cosh(\epsilon\sqrt{\beta})t + \tfrac{1}{\sqrt{\beta}}\sinh(\epsilon\sqrt{\beta})y, x, \cosh(\epsilon\sqrt{\beta})y + \sqrt{\beta}\sinh(\epsilon\sqrt{\beta})t, v), \tag{33b}$$

boost in the plane (y, t).

(b) $f(v_x) = \alpha(v_x + c)^{p+1} - v_x$, $a = 0$, arbitrary b, d, and β,

$$X_6 = 2pt\partial_t + px\partial_x + 2py\partial_y + ((p-2)v - 2cx)\partial_v, \tag{34a}$$

$$(\tilde{t}, \tilde{x}, \tilde{y}, \tilde{v})_6 = (e^{2p\epsilon}t, e^{p\epsilon}x, e^{2p\epsilon}y, e^{(p-2)\epsilon}v - 2cx\epsilon), \qquad \text{scaling and shift.} \tag{34b}$$

(c) $f(v_x) = \tfrac{1}{\alpha}\ln(\alpha(v_x + c)) - v_x$, $a = 0$, arbitrary b, d, and β,

$$X_7 = 2t\partial_t + x\partial_x + 2y\partial_y + (3v + 2cx)\partial_v, \tag{35a}$$

$$(\tilde{t}, \tilde{x}, \tilde{y}, \tilde{v})_7 = (e^{2\epsilon}t, e^{\epsilon}x, e^{2\epsilon}y, e^{3\epsilon}v + 2cx\epsilon), \qquad \text{scaling and shift.} \tag{35b}$$

(d) $f(v_x) = \alpha(v_x + c)^{p+1} - v_x$, $b = 0, d = 0$, arbitrary a and β,

$$X_8 = pt\partial_t + py\partial_y - 2(v + cx)\partial_v, \tag{36a}$$

$$(\tilde{t}, \tilde{x}, \tilde{y}, \tilde{v})_8 = (e^{p\epsilon}t, x, e^{p\epsilon}y, e^{-2\epsilon}v - 2cx\epsilon), \qquad \text{scaling and shift.} \tag{36b}$$

(e) $f(v_x) = \alpha e^{pv_x} - v_x$, $b = 0, d = 0$, arbitrary a and β,

$$X_9 = pt\partial_t + py\partial_y - 2x\partial_v, \tag{37a}$$

$$(\tilde{t}, \tilde{x}, \tilde{y}, \tilde{v})_9 = (e^{p\epsilon}t, x, e^{p\epsilon}y, v - 2x\epsilon), \qquad \text{scaling and shift.} \tag{37b}$$

The classification of the maximal point symmetry Lie algebras for the 2D gDD potential Equation (10) is shown in the following theorem. For each case, the basis of generators and its non-zero Lie brackets are included.

Theorem 2. *The 2D generalized double dispersion potential Equation* (10) *admits the maximal symmetry algebras (with corresponding non-zero commutator structure) given by:*

(i) arbitrary $f(v_x)$, a, b, d, and β,

$$X_1, X_2, X_3, X_{4,g,h};$$

$$[X_1, X_{4,g,h}] = X_{4,\sqrt{\beta}g', -\sqrt{\beta}h'}; \quad [X_3, X_{4,g,h}] = X_{4,g',h'}.$$

(ii) $a = 0, d = 0$, arbitrary $f(v_x)$, b, and β,

$$X_1, X_2, X_3, X_{4,g,h}, X_5;$$

$$[X_1, X_5] = \beta X_3; \quad [X_3, X_5] = X_1; \quad [X_{4,g,h}, X_5] = X_{4,g_1,h_1},$$

where $g_1 = \sqrt{\beta}(y + \sqrt{\beta}t)g'$ and $h_1 = -\sqrt{\beta}(y - \sqrt{\beta}t)h'$.

(iii) $f(v_x) = \alpha(v_x + c)^{p+1} - v_x$, $a = 0$, arbitrary b, d, and β,

$$X_1, X_2, X_3, X_{4,g,h}, X_6;$$

$$[X_1, X_6] = 2pX_1; \quad [X_2, X_6] = pX_2 + X_{4,-2c,0}; \quad [X_3, X_6] = 2pX_3; \quad [X_{4,g,h}, X_6] = X_{4,g_2,h_2},$$

where $g_2 = (p-2)g - 2p(y + \sqrt{\beta}t)g'$ and $h_2 = (p-2)h - 2p(y - \sqrt{\beta}t)h'$.

(iv) $f(v_x) = \frac{1}{\alpha} \ln(\alpha(v_x + c)) - v_x$, $a = 0$, arbitrary b, d, and β,

$$\mathbf{X}_1, \mathbf{X}_2, \mathbf{X}_3, \mathbf{X}_{4,g,h}, \mathbf{X}_7;$$

$$[\mathbf{X}_1, \mathbf{X}_7] = 2\mathbf{X}_1; \quad [\mathbf{X}_2, \mathbf{X}_7] = \mathbf{X}_2 + \mathbf{X}_{4,2c,0}; \quad [\mathbf{X}_3, \mathbf{X}_7] = 2\mathbf{X}_3; \quad [\mathbf{X}_{4,g,h}, \mathbf{X}_7] = \mathbf{X}_{4,g_3,h_3},$$

where $g_3 = 3g - 2(y + \sqrt{\beta}t)g'$ and $h_3 = 3h - 2(y - \sqrt{\beta}t)h'$.

(v) $f(v_x) = \alpha(v_x + c)^{p+1} - v_x$, $b = 0$, $d = 0$, arbitrary a and β,

$$\mathbf{X}_1, \mathbf{X}_2, \mathbf{X}_3, \mathbf{X}_{4,g,h}, \mathbf{X}_8;$$

$$[\mathbf{X}_1, \mathbf{X}_8] = p\mathbf{X}_1; \quad [\mathbf{X}_2, \mathbf{X}_8] = \mathbf{X}_{4,-2c,0}; \quad [\mathbf{X}_3, \mathbf{X}_8] = p\mathbf{X}_3; \quad [\mathbf{X}_{4,g,h}, \mathbf{X}_8] = \mathbf{X}_{4,g_4,h_4},$$

where $g_4 = -2g - p(y + \sqrt{\beta}t)g'$ and $h_4 = -2h - p(y - \sqrt{\beta}t)h'$.

(vi) $f(v_x) = \alpha e^{pv_x} - v_x$, $b = 0$, $d = 0$, arbitrary a and β,

$$\mathbf{X}_1, \mathbf{X}_2, \mathbf{X}_3, \mathbf{X}_{4,g,h}, \mathbf{X}_9;$$

$$[\mathbf{X}_1, \mathbf{X}_9] = p\mathbf{X}_1; \quad [\mathbf{X}_2, \mathbf{X}_9] = \mathbf{X}_{4,-2,0}; \quad [\mathbf{X}_3, \mathbf{X}_9] = p\mathbf{X}_3; \quad [\mathbf{X}_{4,g,h}, \mathbf{X}_9] = \mathbf{X}_{4,g_5,h_5},$$

where $g_5 = -p(y + \sqrt{\beta}t)g'$ and $h_5 = -p(y - \sqrt{\beta}t)h'$.

(vii) $f(v_x) = \alpha(v_x + c)^{p+1} - v_x$, $a = 0$, $d = 0$, arbitrary b and β,

$$\mathbf{X}_1, \mathbf{X}_2, \mathbf{X}_3, \mathbf{X}_{4,g,h}, \mathbf{X}_5, \mathbf{X}_6.$$

(viii) $f(v_x) = \frac{1}{\alpha} \ln(\alpha(v_x + c)) - v_x$, $a = 0$, $d = 0$, arbitrary b and β,

$$\mathbf{X}_1, \mathbf{X}_2, \mathbf{X}_3, \mathbf{X}_{4,g,h}, \mathbf{X}_5, \mathbf{X}_7.$$

For arbitrary $f(v_x)$, a, b, d, and β, the 2D gDD potential Equation (10) admits a four-dimensional Lie algebra consisting of time-translation symmetry (29), space-translation symmetries (29) and (30), and infinite-dimensional symmetry families (32). When $a = 0$ and $d = 0$, Equation (10) becomes a 2D generalized Boussinesq equation in potential form and admits a five-dimensional algebra that includes the previous symmetries and also a boost in the plane (y, t) (33). The classification of point symmetries was already known for the 2D Boussinesq Equation (7) in potential form when $f(v_x) = v_x^{p+1}$ [11]. However, we remark that this maximal algebra is also admitted by Equation (10) with $a = 0$ and $d = 0$ for any $f(v_x)$.

We also note that for each of the cases (iii)–(viii) of Theorem 2, the specific forms of the function $f(v_x)$ will result in the dropping of the second-order term v_{xx} in the 2D gDD potential Equation (10). Therefore, the resulting equations are not properly Boussinesq-type equations, but some other fourth-order nonlinear dispersive PDEs in 2 + 1 dimensions that admit five-dimensional algebras consisting of time and space-translation symmetries, infinite symmetry families, plus a scaling and shift symmetry. In particular, when $a = 0$ and $d = 0$ and for the specific forms of $f(v_x)$ in Cases (vii) and (viii), the equations are respectively given by:

$$v_{tt} - bv_{xxxx} - \alpha(p+1)(v_x + c)^p v_{xx} - \beta v_{yy} = 0 \qquad (38)$$

and:

$$v_{tt} - bv_{xxxx} - \frac{1}{\alpha(v_x + c)} v_{xx} - \beta v_{yy} = 0. \qquad (39)$$

Both equations admit six-dimensional algebras consisting of time and space-translation symmetries, infinite symmetry families, boost symmetry, and scaling-shift symmetry.

3.2. Contact Symmetries

A contact symmetry of the potential 2D gDD Equation (10) is a one-parameter Lie group of transformations that leaves invariant the solution space of the equation and in which the transformation of (t, x, y, v) essentially depends on v_t, v_x, or v_y. The corresponding symmetry generator is given by:

$$\mathbf{X} = \tau \partial_t + \xi^x \partial_x + \xi^y \partial_y + \eta \partial_v + \eta^t \partial_{v_t} + \eta^x \partial_{v_x} + \eta^y \partial_{v_y}, \tag{40}$$

and the characteristic form of this generator is then given by:

$$\hat{\mathbf{X}} = P(t, x, y, v, v_t, v_x, v_y) \partial_v, \tag{41}$$

with:

$$\begin{aligned}
\tau &= -P_{v_t}, & \xi^x &= -P_{v_x}, & \xi^y &= -P_{v_y}, & \eta &= P - v_t P_{v_t} - v_x P_{v_x} - v_y P_{v_y}, \\
\eta^t &= P_t + v_t P_v, & \eta^x &= P_x + v_x P_v, & \eta^y &= P_y + v_y P_v.
\end{aligned} \tag{42}$$

A contact symmetry yields a prolonged point symmetry iff the symmetry characteristic P is at most linear in v_t, v_x, and v_y.

The invariance of the potential 2D gDD Equation (10) under the contact symmetry transformation is expressed by the determining Equation (28) holding for all solutions $v(t, x, y)$ of Equation (10). Then, the determining equation splits into an overdetermined system of equations for $P(t, x, y, v, v_t, v_x, v_y)$ together with the function $f(v_x)$ and the parameters of the equation a, b, d, and β. We impose again the classification conditions $f''(v_x) \neq 0$, $a^2 + b^2 \neq 0$ and $\beta \neq 0$. We set up and solve this system by using the Maple "rifsimp" and "pdsolve" commands. This computation is analogously done by using the steps outlined in Appendix A for the Lie point symmetries, but considering in this case the contact symmetry in characteristic form (41). Therefore, we obtain the following result.

Theorem 3. *The 2D generalized double dispersion potential Equation (10) does not admit any contact symmetry except for those that reduce to prolongations of point symmetries.*

4. Variational Symmetries

When $d = 0$, the 2D generalized double dispersion potential Equation (10) admits a local Lagrangian structure (11) in terms of a Lagrangian functional (17), which will be used in this section to determine which of the Lie symmetries of Equation (10) with $d = 0$ are variational symmetries.

A variational symmetry is an infinitesimal symmetry $\hat{\mathbf{X}} = P \partial_v$ that leaves invariant a Lagrangian functional L up to a total divergence,

$$\hat{\mathbf{X}} L = D_t \Psi^t + D_x \Psi^x + D_y \Psi^y, \tag{43}$$

where Ψ^t, Ψ^x, Ψ^y depend on t, x, y, v, and derivatives of v. The invariance condition is usually verified by computing the left-hand side of (43) and integrating by parts the resulting expression to obtain a total divergence expression. This invariance condition is equivalent to:

$$E_v(PE_v(L)) = 0 \tag{44}$$

in terms of the Euler operator (12) (i.e., the variational derivative), involving only the symmetry characteristic P and the Lagrangian L [14,15].

Therefore, for each of the Lie point symmetries admitted by the 2D gDD potential Equation (10) with $d = 0$, it is only necessary to check that the variational symmetry condition (44) is satisfied, where L is the Lagrangian (17). We next summarize the results.

Theorem 4. *The 2D generalized double dispersion equation in potential form (10) with $d = 0$ admits the variational point symmetries spanned by the time-translation symmetry (29), the space-translation symmetries*

(30) and (31), the infinite symmetry families (32), the boost symmetry (33), and the scaling-shift symmetry (37). The scaling-shift symmetry (34) is only variational for $p = \frac{4}{3}$, and the scaling-shift symmetries (35) and (36) are not variational.

5. Conservation Laws

A local conservation laws of the potential 2D gDD Equation (10) is a space-time divergence expression:

$$D_t T + D_x X + D_y Y|_{G=0} = 0 \qquad (45)$$

holding on the solution space of Equation (10), with T and (X, Y) being respectively the density and the spatial flux, which are functions of t, x, y, v, and derivatives of v. The expression (T, X, Y) is called the conserved current. For any given nonlinear evolution equation, local conservation laws are important since they describe physical quantities that do not change over time within an isolated physical process.

For the potential 2D gDD Equation (10), every non-trivial conservation law (45) is equivalent to the characteristic equation [14,15,17], given by:

$$D_t \tilde{T} + D_x \tilde{X} + D_y \tilde{Y} = (v_{tt} - v_{xx} - av_{ttxx} - bv_{xxxx} - dv_{txx} - (f(v_x))_x - \beta v_{yy})Q, \qquad (46)$$

holding off of the solution space of Equation (10), where Q, $\tilde{T}$, $\tilde{X}$, and $\tilde{Y}$ are functions of t, x, y, v, and derivatives of v. When restricted to the solution space of Equation (10), the conserved density $\tilde{T}$ and the spatial flux $(\tilde{X}, \tilde{Y})$ respectively yield T and (X, Y). The function Q in (46) is called the conservation law multiplier.

For a given equation admitting a Lagrangian structure (11), Noether's theorem states a one-to-one correspondence between variational symmetries and locally non-trivial conservation laws [15,17]. In terms of the variational symmetry characteristic P and the conservation law multiplier Q, the correspondence in Noether's theorem is equivalent to the condition:

$$P = Q. \qquad (47)$$

Given a variational symmetry characteristic P, it is straightforward to derive the corresponding conserved current $(\tilde{T}, \tilde{X}, \tilde{Y})$ from the characteristic Equation (46) by using several methods. One method consists of first splitting the characteristic equation $D_t \tilde{T} + D_x \tilde{X} + D_y \tilde{Y} = PE_v(L)$ with respect to v and its derivatives and then integrating the resulting linear system [17]. A second method consists of applying a repeated integration process [18] to the terms in the expression $PE_v(L)$ to obtain $\tilde{T}, \tilde{X}, \tilde{Y}$. A third method consists of inverting the Euler operator in the variational symmetry equation $E_v(PE_v(L)) = 0$ by means of a homotopy integral formula [14,15,17].

Since the 2D gDD equation in potential form (10) with $d = 0$ possesses a Lagrangian formulation, we now derive the conservation laws associated with the variational symmetries obtained in Theorem 4.

Theorem 5. *(i) The conservation laws admitted by the 2D generalized double dispersion equation in potential form* (10) *with $d = 0$, for arbitrary $f(v_x)$, a, b, β, arising from variational symmetries, are given by:*

$$T_1 = \tfrac{1}{2} a v_{tx}^2 - \tfrac{1}{2} b v_{xx}^2 + \tfrac{1}{2} v_t^2 + \tfrac{1}{2} v_x^2 + \tfrac{1}{2} \beta v_y^2 + F(v_x),$$
$$X_1 = - a v_{ttx} v_t - b v_{xxx} v_t + b v_{tx} v_{xx} - v_x f(v_x) - v_t v_x, \quad (48)$$
$$Y_1 = - \beta v_t v_y,$$

$$T_2 = a v_{tx} v_{xx} + v_t v_x,$$
$$X_2 = - a v_x v_{ttx} - b v_x v_{xxx} - \tfrac{1}{2} a v_{tx}^2 + \tfrac{1}{2} b v_{xx}^2 + F(v_x) - v_x f(v_x) - \tfrac{1}{2} v_t^2 - \tfrac{1}{2} v_x^2 + \tfrac{1}{2} \beta v_y^2, \quad (49)$$
$$Y_2 = - \beta v_x v_y,$$

$$T_3 = a v_{tx} v_{xy} + v_t v_y,$$
$$X_3 = - a v_{ttx} v_y - b v_{xxx} v_y + b v_{xx} v_{xy} - v_y f(v_x) - v_x v_y, \quad (50)$$
$$Y_3 = - \tfrac{1}{2} a v_{tx}^2 - \tfrac{1}{2} b v_{xx}^2 - \tfrac{1}{2} v_t^2 + \tfrac{1}{2} v_x^2 - \tfrac{1}{2} \beta v_y^2 + F(v_x),$$

$$T_4 = (g(y + \sqrt{\beta} t) + h(y - \sqrt{\beta} t)) v_t - \sqrt{\beta} (g'(y + \sqrt{\beta} t) - h'(y - \sqrt{\beta} t)) v,$$
$$X_4 = - (g(y + \sqrt{\beta} t) + h(y - \sqrt{\beta} t))(a v_{ttx} + b v_{xxx} + f(v_x) + v_x), \quad (51)$$
$$Y_4 = \sqrt{\beta} ((g'(y + \sqrt{\beta} t) + h'(y - \sqrt{\beta} t)) v - (g(y + \sqrt{\beta} t) + h(y - \sqrt{\beta} t)) v_y),$$

where $F'(v_x) = f(v_x)$.

(ii) The 2D generalized double dispersion equation in potential form (10) *with $d = 0$ admits additional conservation laws corresponding to variational symmetries arising for the following special $f(v_x)$, a, or b:*

(a) $a = 0$, arbitrary $f(v_x)$, b and β,

$$T_5 = -\tfrac{1}{2} b y v_{xx}^2 + \tfrac{1}{2} y v_t^2 + \tfrac{1}{2} y v_x^2 + \beta \tfrac{1}{2} y v_y^2 + \beta t v_t v_y + y F(v_x),$$
$$X_5 = - b \beta t v_{xxx} v_y - b y v_{xxx} v_t + b y v_{tx} v_{xx} + b \beta t v_{xx} v_{xy} - \beta t v_y f(v_x) - y v_t f(v_x) - \beta t v_x v_y - y v_t v_x,$$
$$Y_5 = -\tfrac{1}{2} b \beta t v_{xx}^2 - \beta y v_t v_y - \tfrac{1}{2} \beta t v_t^2 - \tfrac{1}{2} \beta t v_x^2 - \tfrac{1}{2} \beta^2 t v_y^2 + \beta t F(v_x),$$
$$(52)$$

where $F'(v_x) = f(v_x)$.

(b) $f(v_x) = \alpha e^{p v_x} - v_x$, $b = 0$, arbitrary a and β,

$$T_6 = \tfrac{1}{4} a p t v_{tx}^2 + \tfrac{1}{2} a p y v_{xx} v_{ty} + \tfrac{1}{2} a p v_{xx} v_t + \tfrac{1}{4} p t v_t^2 + \tfrac{1}{4} \beta p t v_y^2 + \tfrac{1}{2} p y v_y v_t + x v_t + \alpha t \tfrac{1}{2} e^{p v_x},$$
$$X_6 = - \tfrac{1}{2} a (p t v_t + p y v_y + 2x)(v_{ttx} + \alpha e^{p v_x}) + \tfrac{1}{2} a (p y v_{xy} + 2a) v_{tt} - \tfrac{1}{2} a (p v_t + p y v_{ty}) v_{tx}, \quad (53)$$
$$Y_6 = -\tfrac{1}{2} a p y v_{tt} v_{xx} + \tfrac{1}{4} a p y v_{tx}^2 - \tfrac{1}{2} \beta p t v_t v_y - \tfrac{1}{4} \beta p y v_y^2 - \beta x v_y - \tfrac{1}{4} p y v_t^2 + \tfrac{1}{2} \alpha y e^{p v_x}.$$

(c) $f(v_x) = \alpha (v_x + c)^{7/3} - v_x$, $a = 0$, arbitrary b and β,

$$T_7 = - 2 b t v_{xx}^2 + 2 t v_t^2 + 2 \beta t v_y^2 + 2 x v_t v_x + 4 y v_t v_y + v v_t + 3 x c v_t + \tfrac{6}{5} \alpha t (v_x + c)^{10/3},$$
$$X_7 = - b (3 c x + 4 t v_t + 2 x v_x + 4 y v_y + b v) v_{xxx} + b (4 t v_{tx} + x v_{xx} + 4 y v_{xy} + 3 c + 4 v_x) v_{xx}$$
$$\quad - \alpha (\tfrac{12}{5} c x + 4 t v_t + \tfrac{7}{5} x v_x + 4 y v_y + v)(v_x + c)^{7/3} + \beta x v_y^2 - x v_t^2, \quad (54)$$
$$Y_7 = - 2 b y v_{xx}^2 - 3 x c \beta v_y - 4 \beta t v_t v_y - 2 \beta x v_x v_y - 2 \beta y v_y^2 - \beta v v_y - 2 y v_t^2 + \tfrac{6}{5} \alpha y (v_x + c)^{10/3}.$$

Next, we look at the meaning of these conservation laws.

When we consider solutions $v(t, x, y)$ of the given equation in a spatial domain $\Omega \subseteq \mathbb{R}^2$, for every conservation law (45), there is an associated conserved integral, given by:

$$\mathcal{C}[v] = \int_{\Omega} T \, dx \, dy \tag{55}$$

satisfying:

$$\frac{d}{dt} \mathcal{C}[v] = - \int_{\partial \Omega} (X, Y) \cdot \hat{n} \, ds \tag{56}$$

where $\hat{n}$ represents the unit normal vector, which points outward to $\partial \Omega$, the boundary curve of Ω, whereas ds represents the arc length along this curve. The global Equation (56) physically means that there is a balance between the rate of change of the quantity (55) on Ω and the net outward flux through $\partial \Omega$.

For the 2D generalized double dispersion potential Equation (10), the conservation law (48) leads to the conserved quantity:

$$\mathcal{E}[v] = \int_{\Omega} \left(\tfrac{1}{2} a v_{tx}^2 - \tfrac{1}{2} b v_{xx}^2 + \tfrac{1}{2} v_t^2 + \tfrac{1}{2} v_x^2 + \tfrac{1}{2} \beta v_y^2 + F(v_x) \right) dx \, dy, \tag{57}$$

where $F'(v_x) = f(v_x)$, which is an energy arising from the time-translation symmetry (29). The conservation laws (49) and (50) yield, respectively, the conserved quantities:

$$\mathcal{P}^x[v] = \int_{\Omega} (a v_{tx} v_{xx} + v_t v_x) \, dx \, dy, \tag{58}$$

$$\mathcal{P}^y[v] = \int_{\Omega} (a v_{tx} v_{xy} + v_t v_y) \, dx \, dy, \tag{59}$$

which are momentum quantities arising from the space-translation symmetries (30) and (31). The infinite family of conservation laws (51) leads to an infinite family of conserved quantities:

$$\mathcal{T}[v] = \int_{\Omega} \sqrt{\beta} \left((g'(y + \sqrt{\beta} t) + h'(y - \sqrt{\beta} t)) v - (g(y + \sqrt{\beta} t) + h(y - \sqrt{\beta} t)) v_y \right) dx \, dy, \tag{60}$$

arising from the infinite symmetry families (32) and corresponding to the conserved quantities of transverse momenta of the linear wave equation $v_{tt} - \beta v_{yy} = 0$. The conservation law (52) yields the conserved quantity:

$$\mathcal{Q}[v] = \int_{\Omega} \left(- \tfrac{1}{2} b y v_{xx}^2 + \tfrac{1}{2} y v_t^2 + \tfrac{1}{2} y v_x^2 + \beta \tfrac{1}{2} y v_y^2 + \beta t v_t v_y + y F(v_x) \right) dx \, dy, \tag{61}$$

which is a boost-momentum arising from the boost symmetry (33). The conservation laws (53) and (54) yield, respectively, the conserved quantities:

$$\mathcal{E}_1[v] = \int_{\Omega} \left(\tfrac{1}{4} a p t v_{tx}^2 + \tfrac{1}{2} a p y v_{xx} v_{ty} + \tfrac{1}{2} a p v_{xx} v_t + \tfrac{1}{4} p t v_t^2 + \tfrac{1}{8} \beta p t v_y^2 + \tfrac{1}{2} p y v_y v_t + x v_t + \alpha t \tfrac{1}{2} e^{p v_x} \right) dx \, dy, \tag{62}$$

$$\mathcal{E}_2[v] = \int_{\Omega} \left(- 2 b t v_{xx}^2 + 2 t v_t^2 + 2 \beta t v_y^2 + 2 x v_t v_x + 4 y v_t v_y + v v_t + 3 x c v_t + \tfrac{6}{5} \alpha t (v_x + c)^{10/3} \right) dx \, dy, \tag{63}$$

which are dilational energy quantities arising from the scaling and shift symmetries (37) and (34).

6. Conclusions

For the 2D generalized double dispersion Equation (8), we first expressed this equation in potential form, and then, we obtained a condition for this equation to admit a Lagrangian formulation. We also gave the corresponding Hamiltonian structure. Next, we classified all Lie symmetries (point and contact) of the 2D gDD potential Equation (10). Finally, we constructed all conservation laws that arise from variational point symmetries of (10) with $d = 0$. We remark that all of the conservation laws (48)–(54) of the potential Equation (10) depended essentially on the potential v; therefore, for the 2D gDD Equation (8), the corresponding conservation laws are nonlocal. Furthermore, we gave the physical meaning of the corresponding conserved quantities.

In future work, we will look for exact group-invariant solutions of Equation (8) by using systematically all symmetries and conservation laws of the equation. Specifically, particular cases of the generalized Equation (8) have line solitons and lump solutions, so we plan to study the line solitons and lump solution and other kinds of solitary waves of the generalized Equation (8).

Author Contributions: Conceptualization, E.R., T.M.G., R.d.l.R. and M.S.B.; methodology, E.R., T.M.G., R.d.l.R. and M.S.B.; software, E.R., T.M.G., R.d.l.R. and M.S.B.; validation, E.R., T.M.G., R.d.l.R. and M.S.B.; formal analysis, E.R.; investigation, E.R.; writing–original draft preparation, E.R.; writing–review and editing, E.R., T.M.G., R.d.l.R. and M.S.B.; supervision, E.R.

Acknowledgments: The authors thank the referees for their comments and suggestions. Stephen Anco is also gratefully thanked for the valuable remarks.

Conflicts of Interest: The authors declare no conflict of interest.

Appendix A

We provide the computational steps followed to solve the determining Equation (28) for point symmetries (26) by using the software Maple.

Firstly, we set up the determining Equation (28), and we used the command "coeffs" to split this equation with respect to the variables $\partial^2 v$, $\partial^3 v$, and $\partial^4 v$. Note that a leading derivative of Equation (8) and its differential consequences can be previously substituted to simplify the determining Equation (28). From the leading derivatives v_{tt}, v_{xxxx} or v_{ttxx} of Equation (8), it is convenient to choose any of the fourth-order derivatives v_{xxxx} or v_{ttxx}. This procedure gives an overdetermined system of 207 equations for the symmetry characteristic P, $f(v_x)$, a, b, d, and β. We also consider the classification conditions $f''(v_x) \neq 0$, $a^2 + b^2 \neq 0$, and $\beta \neq 0$.

Secondly, we use the command "rifsimp" to obtain a tree containing all solution cases.

Thirdly, every solution case consists of a system of PDEs for P and, possibly, an ODE for the function $f(v_x)$ and conditions for the parameters of the equation a, b, d, and β. We solve the ODE for $f(v_x)$ by using the command "dsolve" and the system of PDEs for P by using the command "pdsolve". We check the solutions by substituting in the overdetermined system.

Finally, by applying the method detailed in the Appendix of [19], the overlapping cases are combined, giving a classification of all point symmetries admitted by the 2D generalized double dispersion Equation (8).

References

1. Boussinesq, J. Théorie de l' intumescence liquide appelée 'onde solitaire' ou 'de translation', se propageant dans un canal rectangulaire. *C. R. Acad. Sci.* **1871**, *72*, 755–759.
2. Boussinesq, J. Théorie des ondes et des remous qui se propagent le long d'un canal rectangulaire horizontal, en communiquant au liquide continu dans ce canal des vitesses sensiblement pareilles de la surface au fond. *J. Math. Pure Appl.* **1872**, *17*, 55–108.
3. Chen, G.; Wang, Y.; Wang, S. Initial boundary value problem of the generalized cubic double dispersion equation. *J. Math. Anal. Appl.* **2004**, *299*, 563–577. [CrossRef]
4. Samsonov, A.M.; Sokurinskaya, E.V. *Energy Exchange between Nonlinear Waves in Elastic Waveguides and External Media Nonlinear Waves in Active Media*; Springer: Berlin, Germany, 1989.
5. Wang, S.; Chen, G. Cauchy problem of the generalized double dispersion equation. *Nonlinear Anal.* **2006**, *64*, 159–173. [CrossRef]
6. Yacheng, L.; Runzhang, X. Potential well method for Cauchy problem of generalized double dispersion equations. *J. Math. Anal. Appl.* **2008**, *338*, 1169–1187.

7. Zhengde, D.; Boling, G. Long time behavior of nonlinear strain waves in elastic waveguides. *J. Part. Diff. Eq.* **1999**, *12*, 301–312.
8. Zhengde, D.; Xianyun, D. Global attractor for the nonlinear strain waves in elastic waveguides. *Acta Math. Sci.* **2000**, *17*, 260–270. [CrossRef]
9. Wang, S.; Esfahani, A. Global rough solutions to the sixth-order Boussinesq equation. *Nonlinear Anal.* **2014**, *102*, 97–104. [CrossRef]
10. Recio, E.; Gandarias, M.L.; Bruzón, M.S. Symmetries and conservation laws for a sixth-order Boussinesq equation. *Chaos Solitons Fractals* **2016**, *89*, 572–577. [CrossRef]
11. Anco, S.C.; Gandarias, M.L.; Recio, E. Conservation laws, symmetries, and line soliton solutions of generalized KP and Boussinesq equations with *p*-power nonlinearities in two dimensions. *Theor. Math. Phys.* **2018**, *197*, 1393–1411. [CrossRef]
12. Matsukawa, M.; Watanaba, S.; Tanaca, H. Soliton solutions of generalized 2D Boussinesq equation with quadratic and cubic nonlinearity. *J. Phys. Soc. Jpn.* **1989**, *58*, 827–830. [CrossRef]
13. Rui, W.; Zhao, P.; Zhang, Y. Invariant solutions and conservation laws of the (2 + 1)-dimensional Boussinesq equation. *Abstr. Appl. Anal.* **2014**, 840405. [CrossRef]
14. Anco, S.C. Generalization of Noether's theorem in modern form to non-variational partial differential equations. In *Fields Institute Communications: Recent progress and Modern Challenges in Applied Mathematics, Modeling and Computational Science*; Springer: New York, NY, USA, 2017; pp. 119–182.
15. Olver, P.J. *Applications of Lie Groups to Differential Equations*; Springer: New York, NY, USA, 1993.
16. Anco, S.C.; Bluman, G. Direct construction method for conservation laws of partial differential equations Part II: General treatment. *Eur. J. Appl. Math.* **2002**, *41*, 567–585. [CrossRef]
17. Bluman, G.W.; Cheviakov, A.; Anco, S.C. *Applications of Symmetry Methods to Partial Differential Equations*; Springer: New York, NY, USA, 2009.
18. Wolf, T. A comparison of four approaches to the calculation of conservation laws. *Eur. J. Appl. Math.* **2002**, *13*, 129–152. [CrossRef]
19. Recio, E.; Anco, S.C. Conservation laws and symmetries of radial generalized nonlinear *p*-Laplacian evolution equations. *J. Math. Anal. Appl.* **2017**, *452*, 1229–1261. [CrossRef]

© 2019 by the authors. Licensee MDPI, Basel, Switzerland. This article is an open access article distributed under the terms and conditions of the Creative Commons Attribution (CC BY) license (http://creativecommons.org/licenses/by/4.0/).

Article

Symmetry Analysis and Conservation Laws of a Generalization of the Kelvin-Voigt Viscoelasticity Equation

Almudena P. Márquez * and María S. Bruzón *

Department of Mathematics, University of Cadiz, Facultad de Ciencias Campus del Río San Pedro, 11510 Puerto Real, Spain
* Correspondence: almudena.marquez@uca.es (A.P.M.); m.bruzon@uca.es (M.S.B.)

Received: 30 May 2019; Accepted: 24 June 2019; Published: 28 June 2019

Abstract: In this paper, we study a generalization of the well-known Kelvin-Voigt viscoelasticity equation describing the mechanical behaviour of viscoelasticity. We perform a Lie symmetry analysis. Hence, we obtain the Lie point symmetries of the equation, allowing us to transform the partial differential equation into an ordinary differential equation by using the symmetry reductions. Furthermore, we determine the conservation laws of this equation by applying the multiplier method.

Keywords: viscoelasticity; Kelvin-Voigt equation; Lie symmetries; optimal system; group-invariant solutions; conservation laws; multiplier method

1. Introduction

The continuous development of mechanics and its engineering applications have increased remarkably the interest in non-linear phenomena, such as viscoelasticity. Viscoelastic materials are of interest in a wide variety of applications, from passive damping to aircraft tire construction. A good modelling of the material's behaviour is essential for the accurate design incorporating this material.

Viscoelastic behaviour appears in materials showing some sort of liquid-like elastic behaviour. However, a simple Hooke's law linear elastic constitutive relationship is not an accurate representation of viscoelastic material's behaviour. Viscoelastic materials are commonly said to have "memory" because of their rheological properties.

The rheological models, such as the Kelvin-Voigt model or the Maxwell model, are usually used to describe the viscoelastic behaviour. The Kelvin-Voigt model consists of a lumped parameter model similar to a spring and dashpot in parallel, while the Maxwell model describes a serially connected spring and dashpot. In addition, many papers have been published studying these models [1–4].

Many physical phenomena, as viscoelasticity, are described by non-linear partial differential equations (PDEs). In particular, the Kelvin-Voigt viscoelasticity equation is given by

$$u_{tt} - (C(x)\, u_x)_x - (B(x)\, u_{tx})_x = 0.$$

Nevertheless, in this paper we focus on a generalization of the Kelvin-Voigt viscoelasticity equation described by

$$u_{tt} - (C(x)\, f(u)_x)_x - (B(x)\, u_{tx})_x = 0, \quad (t,x) \in \mathbb{R} \times \Omega, \tag{1}$$

where Ω is an open subset of $\mathbb{R}$ and u a scalar real-valued function. Also, $f(u)$ is a smooth enough non-linear function, and $C(x) \neq 0$, $B(x) \neq 0$ are smooth enough functions too, depending on the variable $x \in \Omega$. Throughout the paper the subscripts denote partial derivatives.

There is no general theory for solving non-linear PDEs. Therefore, in this work, we use Lie theory to analyse Equation (1). Lie group analysis is a powerful tool to find general solutions for PDEs. This theory, originally defined by Sophus Lie at the end of the nineteenth century, develops solutions for PDEs by the transformation groups of Lie [5–8]. The fundamental basis of the Lie group method is that if a differential equation is invariant under a Lie group of transformations, then a reduction transformation exists. For instance, for PDEs with two independent variables like Equation (1), a single group reduction can transform the PDE into an ordinary differential equation (ODE), easier to solve.

Furthermore, a very important concept in the analysis of PDEs is the notion of conservation law. Conservation laws determine conserved quantities and constants of motion. They also detect integrability and check accuracy of numerical solutions method. Recently, Anco and Bluman [9,10] developed a method that does not need the existence of Lagrangians because it is based on adjoint equations for non-linear equations and avoids the integrals of functions. This method called the multiplier method allows finding all local conservation laws admitted by any evolution equation. Many papers have been published in the last few years using this method [11–19].

The paper is organized as follows: In Section 2 we determine the Lie point symmetries of Equation (1). Then, in Section 3 we use the Lie point symmetries admitted by Equation (1) to obtain an optimal system of one-dimensional subalgebras. Afterwards, in Section 4 we find symmetry reductions for the one-dimensional subalgebras calculated previously. These reductions allow us to transform Equation (1) into an ODE. In Section 5 we derive the conservation laws of Equation (1) by applying the multiplier method. Finally, in Section 6 some conclusions are presented.

2. Lie Point Symmetries

A one-parameter group of infinitesimal transformations in (x, t, u) is given by

$$\begin{aligned} x^* &= x + \varepsilon \xi(x, t, u) + \mathcal{O}(\varepsilon^2), \\ t^* &= t + \varepsilon \tau(x, t, u) + \mathcal{O}(\varepsilon^2), \\ u^* &= u + \varepsilon \phi(x, t, u) + \mathcal{O}(\varepsilon^2), \end{aligned}$$

where ε is the group parameter and $\xi(x, t, u)$, $\tau(x, t, u)$, and $\phi(x, t, u)$ are the infinitesimals.

Definition 1. *A vector field*

$$X = \xi(x, t, u)\frac{\partial}{\partial x} + \tau(x, t, u)\frac{\partial}{\partial t} + \phi(x, t, u)\frac{\partial}{\partial u}, \tag{2}$$

where $\xi(x, t, u)$, $\tau(x, t, u)$, and $\phi(x, t, u)$ are the infinitesimals, is a generator of a Lie point symmetry of Equation (1) if

$$X^{(3)}\left(u_{tt} - (C(x) f(u)_x)_x - (B(x) u_{tx})_x\right) = 0, \tag{3}$$

where $X^{(3)}$ is the third prolongation of the vector field (2) defined by

$$X^{(3)} = X + \zeta_x \frac{\partial}{\partial u_x} + \zeta_t \frac{\partial}{\partial u_t} + \zeta_{xx} \frac{\partial}{\partial u_{xx}} + \zeta_{xt} \frac{\partial}{\partial u_{xt}} + \zeta_{tt} \frac{\partial}{\partial u_{tt}} + \zeta_{xxt} \frac{\partial}{\partial u_{xxt}},$$

where the coefficients $\zeta_x, \zeta_t, \zeta_{xx}, \zeta_{xt}, \zeta_{tt}, \zeta_{xxt}$ are given by

$$\begin{aligned}
\zeta_x &= D_x\phi - u_t D_x\tau - u_x D_x\xi, \\
\zeta_t &= D_t\phi - u_t D_t\tau - u_x D_t\xi, \\
\zeta_{xx} &= D_x(\zeta_x) - u_{xt} D_x\tau - u_{xx} D_x\xi, \\
\zeta_{xt} &= D_t(\zeta_x) - u_{xt} D_x\tau - u_{xx} D_t\xi, \\
\zeta_{tt} &= D_t(\zeta_t) - u_{tt} D_t\tau - u_{xt} D_t\xi, \\
\zeta_{xxt} &= D_x(\zeta_{xt}) - u_{ttx} D_x\tau - u_{xxt} D_x\xi.
\end{aligned}$$

Here D_i stands for the total derivative operator.

Theorem 1. *The Lie point symmetries of the generalization of the Kelvin-Voigt Equation (1), with $f(u)$ non-linear function, and $C(x) \neq 0$, $B(x) \neq 0$ arbitrary functions, are generated by the operator*

$$X_1 = \partial_t.$$

For some particular functions of $f(u)$, $C(x)$, $B(x)$, there are additional generators given below.

1. If $f(u)$ is an arbitrary function, $C(x) = c_1$ and $B(x) = b_1$, with c_1, b_1 arbitrary constants,

$$X_2^1 = \partial_x.$$

2. If $f(u)$ is an arbitrary function, $C(x) = c_1$ and $B(x) = b_1 x + b_2$, with $c_1, b_1 \neq 0$, b_2 arbitrary constants,

$$X_2^2 = (b_1 x + b_2)\partial_x + b_1 t\, \partial_t.$$

3. If $f(u)$ is an arbitrary function, $C(x) = \frac{4}{(nx+c_1)^2}$ and $B(x) = b_1$, with c_1, b_1 arbitrary constants, and n a positive integer,

$$X_2^3 = (nx + c_1)\partial_x + 2nt\, \partial_t.$$

4. If $f(u)$ is an arbitrary function, $C(x) = c_2(c_1 - x)^n$ and $B(x) = \frac{(-c_1+x)c_1\sqrt{(c_1-x)^n c_2}}{n}$, with c_1, c_2 arbitrary constants, and n a positive integer,

$$X_2^4 = (c_1 - x)\partial_x + \frac{1}{2}(n-2)t\, \partial_t.$$

5. If $f(u) = \dfrac{f_0^2 e^{\frac{-ke^k}{e^k-1}}\left(\frac{(e^k-1)u}{f_0} + \frac{f_1 e^k - f_1}{f_0}\right)^{\frac{e^k}{e^k-1}+1} + 2f_2 e^k - f_2}{2e^k - 1}$, $C(x) = c_1$ and $B(x) = b_1$, with k a positive integer, f_0, f_1, f_2 positive constants, and c_1, b_1 arbitrary constants, besides $X_2^5 = X_2^1$,

$$X_3^5 = x\partial_x + 2t\, \partial_t + 2(-1 + e^{-k})(u + f_1)\partial_u.$$

6. If $f(u) = \dfrac{f_1^2 e^{\frac{-f_0 e^{f_0}}{e^{f_0}-1}}\left(\frac{(e^{f_0}-1)u}{f_1} + \frac{f_2 e^{f_0} - f_2}{f_1}\right)^{\frac{e^{f_0}}{e^{f_0}-1}+1}}{(e^{f_0}-1)\left(\frac{e^{f_0}}{e^{f_0}-1}+1\right)} + f_3$, $C(x) = c_2(c_1 - x)^n$ and $B(x) = b_1$, with f_0, f_1, f_2, f_3 positive constants, c_1, c_2, b_1 arbitrary constants, and n a positive integer,

$$X_2^6 = (c_1 - x)\partial_x - 2t\, \partial_t - (1 + e^{-f_0})(u + f_2)(n+2)\partial_u.$$

7. If $f(u) = \dfrac{f_0 e^{\frac{-ke^k}{e^k-1}} \left(\frac{(e^k-1)(u+f_1)}{f_0} \right)^{\frac{2e^k-1}{e^k-1}} + 2f_2 e^k - f_2}{2e^k - 1}$, $C(x) = c_2(b_2 - x)^{-m}$ and $B(x) = b_3(b_2 - x)^n$, with k a positive integer, f_0, f_1, f_2 positive constants, b_2, b_3 arbitrary constants, and n, m positive integers,

$$X_2^7 = (b_2 - x)\partial_x + t(n-2)\partial_t + (-1 + e^{-k})(u + f_1)(m + 2n - 2)\partial_u.$$

Proof of Theorem 1. Expanding (3), we obtain an overdetermined system satisfying the determining equations for the symmetry group. From $B^2(\tau_u) = 0$ and $B^2(\tau_x) = 0$, we find that τ depends only on t. The equations $B^2(\xi_t) = 0$ and $B^2(\xi_u) = 0$ reveal that ξ is a function of x alone. The remaining determining equations are

$$B^2(\xi_{uu}) = 0,\ B^2(\xi_{tu}) = 0,\ B^2(\tau_{uu}) = 0,\ B(\tau_{uu}) = 0,\ B^2(\tau_{uuu}) = 0,\ B^2(\xi_{uuu}) = 0,\ = 0,$$
$$B\left(2(\tau_{ux})B - \xi_u\right) = 0,$$
$$B\left(2(\tau_{uux})B - \xi_{uu}\right) = 0,$$
$$B\left((\tau_{ux})B + \xi_u\right) = 0,$$
$$B\left((\tau_{uxx})B + \phi_{uu} - 2(\tau_{tu})\right) = 0,$$
$$B\left((\xi_u)(f_{uu})C - (\xi_{uu})(f_u)C - (\xi_{tuu})B\right) = 0,$$
$$B\left(f_u(\phi_{xx})C + (\phi_{txx})B - \phi_{tt}\right) = 0,$$
$$B\left((\tau_u)(f_u)C + \phi_{uu})B - (\tau_{tu})B - 2(\xi_{ux})B\right) = 0,$$
$$B\left((\tau_u)(f_u)C - \phi_{uu})B + (\tau_{tu})B + 2(\xi_{ux})B\right) = 0,$$
$$B\left((\tau_{uu})(f_u)C - \phi_{uuu})B + (\tau_{tuu})B + 2(\xi_{uux})B\right) = 0,$$
$$\xi\left(B'\right) - (\tau_{xx})B^2 + (\tau_t)B - 2(\xi_x B) = 0,$$
$$B\left(2(\tau_x)(f_u)C - 2(\phi_{ux})B + 2(\tau_{tx})B + (\xi_{xx})B - 2(\xi_t)\right) = 0,$$
$$B\left((\tau_{xx})(f_u)C - (\phi_{uxx})B + (\tau_{txx})B + 2(\phi_{tu}) - \tau_{tt}\right) = 0,$$
$$\xi(f_u)B(C') - \xi f_u(B')C + ((f_{uu}))\phi BC + (\tau_t)(f_u)BC + (\phi_{tu})B^2 - 2(\xi_{tx})B^2 = 0,$$
$$B\left(2(\tau_x)(f_{uu})C + 2(\tau_{ux})(f_u)C - 2(\phi_{uux})B + 2(\tau_{tux})B + (\xi_{uxx})B - 2(\xi_{tu})\right) = 0,$$
$$B\left(2(f_{uu})(\phi_x)C + 2f_u(\phi_{ux})C - (\xi_{xx})(f_u)C + 2(\phi_{tux})B - (\xi_{txx})B + \xi_{tt}\right) = 0,$$
$$\xi(f_{uu})B(C') - \xi(f_{uu})(B')C + f_u(\phi_{uu})BC + (f_{uu})(\phi_u)BC + ((f_{uuu}))\phi BC + (\tau_t)(f_{uu})BC$$
$$-2(\xi_{ux})(f_u)BC + (\phi_{tuu})B^2 - 2(\xi_{tux})B^2 = 0.$$

Solving this system of equations we find the infinitesimals ξ, τ and ϕ of (2). $\square$

3. Optimal Systems

It is important to classify invariant solutions according to the classification of the associated symmetry generators. Then, one generator from each class is used to determine the desired set of invariant solutions. An optimal system of generators is defined as a set consisting of exactly one generator from each class [20].

The problem of obtaining an optimal system of subgroups is equivalent to that of obtaining an optimal system of subalgebras, and so we concentrate on the latter. For one-dimensional subalgebras, this classification problem is essentially equal to classifying the orbits of the adjoint representation [5].

The most important operator on vector fields is their Lie bracket or commutator. If X_i and X_j are vector fields, then their Lie bracket $[X_i, X_j]$ is the unique vector field satisfying

$$[X_i, X_j] = X_i(X_j) - X_j(X_i).$$

The commutator table for the Lie algebra of Case 5 of Theorem (1) is shown in Table 1. The (i, j)-th entry of the table expresses the Lie bracket $[X_i, X_j]$, for $i, j = 1, 2, 3$.

Table 1. The commutator table for Case 5 of Theorem (1).

$[X_i, X_j]$	X_1	X_2^5	X_3^5
X_1	0	0	$2X_1$
X_2^5	0	0	X_2^5
X_3^5	$-2X_1$	$-X_2^5$	0

The adjoint representation can be constructed by summing the Lie series

$$Ad(\exp(\epsilon X_i))X_j = \sum_{n=0}^{\infty} \frac{\epsilon^n}{n!}(ad\, X_i)^n(X_j),$$

$$= X_j - \epsilon[X_i, X_j] + \frac{\epsilon^2}{2}[X_i, [X_i, X_j]] - \cdots . \quad (4)$$

To compute the adjoint representation, we use the Lie series (4) in conjunction with the commutator table in Table 1. The adjoint table of this Lie algebra is shown in Table 2, with the (i, j)-th entry indicating $Ad(\exp(\epsilon X_i))X_j$.

Table 2. The adjoint table for Case 5 of Theorem (1).

$Ad(\exp(\epsilon X_i))X_j$	X_1	X_2^5	X_3^5
X_1	X_1	X_2^5	$X_3^5 - 2\epsilon X_1$
X_2^5	X_1	X_2^5	$X_3^5 - \epsilon X_2^5$
X_3^5	$e^{2\epsilon}X_1$	$e^{\epsilon}X_2^5$	X_3^5

Theorem 2. *A one-dimensional optimal system for the generalization of the Kelvin-Voigt Equation (1) is given by*

$$\lambda X_1 + \mu X_2^1, \quad X_2^2, \quad X_2^3, \quad X_2^4, \quad X_3^5, \quad X_2^6, \quad X_2^7,$$

where λ and μ are arbitrary constants.

Proof of Theorem 2. Let $\mathcal{G}$ be the symmetry algebra of Equation (1), with the adjoint representation for Case 5 in Theorem (1) determined in Table 2. Let $v = a_1v_1 + a_2v_2 + a_3v_3$ be a non-zero vector field of $\mathcal{G}$. For each case, we simplify the coefficients a_i, $i = 1, 2, 3$, as much as possible through proper adjoints applications on v. □

4. Symmetry Reductions

In this section, we use the optimal system of one-dimensional subalgebras of Theorem (2) to determine the symmetry reductions of Equation (1).

The symmetry variables are found by solving the invariant surface condition

$$\Phi \equiv \zeta \partial_x + \tau \partial_t - \phi = 0.$$

Reduction 1. From $\lambda X_1 + \mu X_2^1$, we obtain the travelling wave reduction

$$z = \mu x - \lambda t, \quad u = h(z),$$

where $h(z)$ satisfies

$$\lambda \mu^2 b_1 h''' - \mu^2 c_1 f' h'' + \lambda^2 h'' - \mu^2 c_1 f'' (h')^2 = 0.$$

Reduction 2. From X_2^2, we obtain the invariant solution

$$z = \frac{b_1 x + b_2}{t}, \quad u = h(z),$$

where $h(z)$ satisfies
$$b_1^2 h''' z^2 + h'' z^2 + 3 b_1^2 h'' z + 2 h' z + b_1^2 h' = 0.$$

Reduction 3. From X_2^3, we obtain the invariant solution
$$z = \frac{nx + c_1}{\sqrt{t}}, \qquad u = h(z),$$

where $h(z)$ satisfies
$$2 b_1 h''' n^2 z + h'' z^2 + 4 b_1 h'' n^2 + 3 h' z = 0.$$

Reduction 4. From X_2^4, we obtain the invariant solution
$$z = -t^{\frac{2}{n-2}}(c_1 - x), \qquad u = h(z),$$

where $h(z)$ satisfies
$$2 c_1 \sqrt{c_2} h''' n (-z)^{\frac{n}{2}} z^2 - 4 c_1 \sqrt{c_2} h''' (-z)^{\frac{n}{2}} z^2 - 4 h'' n z^2 + c_1 \sqrt{c_2} h'' n^2 (-z)^{\frac{n}{2}} z$$
$$+ 4 c_1 \sqrt{c_2} h'' n (-z)^{\frac{n}{2}} z - 12 c_1 \sqrt{c_2} h'' (-z)^{\frac{n}{2}} z + 2 h' n^2 z - 8 h' n z + c_1 \sqrt{c_2} h' n^2 (-z)^{\frac{n}{2}}$$
$$- 4 c_1 \sqrt{c_2} h' (-z)^{\frac{n}{2}} = 0.$$

Reduction 5. From X_3^5, we obtain the invariant solution
$$z = \frac{x}{\sqrt{t}}, \qquad u = \frac{t^{-1+e^{-k}}}{h(z)} - f_1,$$

where $h(z)$ must satisfy a non-autonomous equation.

Reduction 6. From X_2^6, we obtain the invariant solution
$$z = -(c_1 - x)\sqrt{t}, \qquad u = \frac{t^{\frac{(-1+e^{-f_0})(n+2)}{2}}}{h(z)} - f_2,$$

where $h(z)$ must satisfy a non-autonomous equation.

Reduction 7. From X_2^7, we obtain the invariant solution
$$z = -\frac{(b_2 - x)}{t^{\frac{1}{n-2}}}, \qquad u = \frac{t^{\frac{-(-1+e^{-k})(m+2n-2)}{n-2}}}{h(z)} - f_1,$$

where $h(z)$ must satisfy a non-autonomous equation.

The expressions of the reduced equations for X_3^5, X_2^6, and X_2^7 are omitted here to save space.

5. Conservation Laws

A local conservation law for the generalization of the Kelvin-Voigt viscoelasticity Equation (1) is a continuity equation
$$D_t T + D_x X = 0,$$
holding for all solutions of Equation (1), where the conserved density T and the spatial flux X are functions of x, t, u, and derivatives of u. Here D_t and D_x denote total derivatives with respect to t and x, respectively. The pair (T, X) is called a conserved current.

Two local conservation laws are considered to be locally equivalent [5,21] if they differ by a locally trivial conservation law $T = D_x \Theta$, $X = -D_t \Theta$, where T and X are evaluated on the set of solutions of Equation (1) and Θ is a function of x, t, u, and derivatives of u.

A non-trivial conservation law can be written in a general form as

$$\frac{d}{dt}\int_\Omega T\,dx = -X\Big|_{\partial\Omega},$$

where $\Omega \subseteq \mathbb{R}$ is any fixed spatial domain.

Any local conservation law can be stated by using the characteristic form arising from a divergence identity

$$D_t \tilde{T} + D_x \tilde{X} = (u_{tt} - (C(x)f(u)_x)_x - (B(x)u_{tx})_x)Q, \quad (5)$$

where $\tilde{T} = T + D_x\Theta$ and $\tilde{X} = X - D_t\Theta$ are locally equivalent to T and X. The function Q is called a multiplier. It satisfies

$$Q = E_u(\tilde{T}),$$

where E_u represents the Euler operator with respect to u [5], that is

$$E_u = \partial_u - D_x\partial_{u_x} - D_t\partial_{u_t} + D_x D_t \partial_{u_{xt}} + D_x^2 \partial_{u_{xx}} + \cdots.$$

For evolution equations, there is a one-to-one relationship between non-zero multipliers and non-trivial conserved current vectors up to local equivalence [5,9]. In general, a function $Q(x,t,u,u_t,u_x,\ldots)$ is a multiplier if it verifies that $(u_{tt} - (C(x)f(u)_x)_x - (B(x)u_{tx})_x)Q$ is a divergence expression for all function $u(x,t)$. Given a multiplier Q, the conserved density can be determined by

$$T = \int_0^1 u\, Q(x,t,\lambda u, \lambda u_x, \lambda u_{xx},\ldots)\,d\lambda.$$

The divergence condition yields to the determining equation

$$E_u\Big((u_{tt} - (C(x)f(u)_x)_x - (B(x)u_{tx})_x)Q\Big) = 0. \quad (6)$$

In order to give a complete classification of multipliers, we write and split the determining Equation (6) with respect to the variables $u_{tt}, u_{ttt}, u_{ttx}, u_{txx}, u_{xxx}, u_{tttt}, u_{tttx}, u_{txxx}, u_{xxxx}$. Thus, we get a linear determining system for $Q(x,t,u,u_t,u_x,\ldots)$. The multipliers are found by solving the system with the same algorithmic method used for the determining equations for infinitesimal symmetries. Then, integrating the characteristic Equation (5) for each multiplier, we find the conserved current.

Theorem 3. *The multipliers admitted by the generalization of the Kelvin-Voigt Equation (1), with $f(u)$ a smooth enough non-linear function, and $C(x) \neq 0$, $B(x) \neq 0$ smooth enough arbitrary functions, are given by*

$$Q_1 = 1, \qquad Q_2 = t, \qquad Q_3 = \int \frac{1}{C(x)}dx.$$

Theorem 4. *All non-trivial local conservation laws admitted by the generalization of the Kelvin-Voigt Equation (1), with $f(u)$ a smooth enough non-linear function, and $C(x) \neq 0$, $B(x) \neq 0$ smooth enough arbitrary functions, are given by*

1. *For the multiplier $Q_1 = 1$, the conserved density and the spatial flux are*

$$\begin{aligned} T_1 &= u_t, \\ X_1 &= -B(x)u_{tx} - C(x)f(u)_x. \end{aligned}$$

2. *For the multiplier $Q_2 = t$, the conserved density and the spatial flux are*

$$\begin{aligned} T_2 &= t u_t - u, \\ X_2 &= -t B(x) u_{tx} - t C(x) f(u)_x. \end{aligned}$$

3. For the multiplier $Q_3 = \int \frac{1}{C(x)} dx$, the conserved density and the spatial flux are

$$T_3 = \left(\int \frac{1}{C(x)} dx\right) u_t + \left(\frac{B(x) C'(x)}{C(x)^2} - \frac{B'(x)}{C(x)}\right) u,$$

$$X_3 = -\left(\int \frac{1}{C(x)} dx\right) B(x) u_{tx} - \left(\int \frac{1}{C(x)} dx\right) C(x) f(u)_x + \frac{B(x) u_t}{C(x)} + f(u).$$

6. Conclusions

In this paper, we studied a generalization of the Kelvin-Voigt viscoelasticity equation given by the partial differential Equation (1). Firstly, we determined a complete Lie group classification. Then, we constructed the optimal system of one-dimensional subalgebras. These one-dimensional subalgebras have been used to find the symmetry reductions, allowing us to transform the partial differential equation into an ordinary differential equation. Moreover, we analysed all conservation laws for this equation by applying the multiplier method.

Author Contributions: A.P.M. and M.S.B. worked together in the derivation of the mathematical results. Both authors provided critical feedback and helped shape the research, analysis and manuscript.

Funding: This research received no external funding.

Acknowledgments: The authors express their sincere gratitude to the financial support of *Junta de Andalucía* FQM-201 group. We gratefully thank the reviewers for their assistance.

Conflicts of Interest: The authors declare no conflict of interest.

Abbreviations

The following abbreviations are used in this manuscript:

PDE   Partial Differential Equation
ODE   Ordinary Differential Equation

References

1. Atallah-Baraket, A.; Trabelsi, M. Analysis of the energy decay of a viscoelasticity type equation. *Analele Stiintifice ale Universitatii Ovidius Constanta* **2016**, *24*, 21–45. [CrossRef]
2. Schiessel, H.; Metzler, R.; Blumen, A.; Nonnenmacher, T.F. Generalized viscoelastic models: Their fractional equations with solutions. *J. Phys. A: Math. Gen.* **1995**, *28*, 6567. [CrossRef]
3. Lei, Y.; Adhikari, S.; Friswell, M.I. Vibration of nonlocal Kelvin-Voigt viscoelastic damped Timoshenko beams. *Int. J. Eng. Sci.* **2013**, *66*, 1–13. [CrossRef]
4. Lewandowski, R.; Chorazyczewski, B. Identification of the parameters of the Kelvin-Voigt and the Maxwell fractional models, used to modeling of viscoelastic dampers. *Comput. Struct.* **2010**, *88*, 1–17. [CrossRef]
5. Olver, P.J. *Applications of Lie Groups to Differential Equations*; Springer-Verlag: New York, NY, USA, 1986.
6. Bluman, G.W.; Anco, S.C. *Symmetry and Integration Methods for Differential Equations*; Springer-Verlag: New York, NY, USA, 2002.
7. Bluman, G.W.; Cheviakov, A.F.; Anco, S.C. *Applications of Symmetry Methods to Partial Differential Equations*; Springer-Verlag: New York, NY, USA, 2010.
8. Bluman, G.W.; Kumei, S. *Symmetries and Differential Equations*; Springer-Verlag: New York, NY, USA, 1989.
9. Anco, S.C.; Bluman, G.W. Direct construtcion method for conservation laws of partial differential equations Part I: Examples of conservation law classifications. *Eur. J. Appl. Math.* **2002**, *5*, 545–566. [CrossRef]
10. Anco, S.C.; Bluman, G.W. Direct construtcion method for conservation laws of partial differential equations Part 2: General treatment. *Eur. J. Appl. Math.* **2002**, *5*, 545–566. [CrossRef]
11. Bruzón, M.S.; Márquez, A.P. Conservation laws of one-dimensional strain-limiting viscoelasticity model. *AIP Conf. Proc.* **2017**, *1836*, 020081.
12. Bruzón, M.S.; Recio, E.; Garrido, T.M.; Márquez, A.P. Conservation laws, classical symmetries and exact solutions of the generalized KdV-Burguers-Kuramoto equation. *Open Phys.* **2017**, *15*, 433–439. [CrossRef]

13. Bruzón, M.S.; Recio, E.; Garrido, T.M.; Márquez, A.P.; de la Rosa, R. On the similarity solutions and conservation laws of the Cooper-Shepard-Sodano equation. *Math. Meth. Appl. Sci.* **2018**, *41*, 7325–7332. [CrossRef]
14. Motsepa, T.; Khalique, C.M.; Gandarias, M.L. Symmetry Analysis and Conservation laws of the Zoomeron Equation. *Symmetry* **2017**, *9*, 27. [CrossRef]
15. Mothibi, D.M.; Khalique, C.M. Conservation laws and Exact Solutions of a Generalized Zakharov-Kuznetsov Equation. *Symmetry* **2015**, *7*, 949–961. [CrossRef]
16. Anco, S.C.; Rosa, M.; Gandarias, M.L. Conservation laws and symmetries of time-dependent generalized KdV equations. *Discret. Contin. Dyn. Syst. Ser. S* **2018**, *11*, 607–615. [CrossRef]
17. Gandarias, M.L.; Khalique, M. Symmetries, solutions and conservation laws of a class of nonlinear dispersive wave equations. *Commun. Nonlinear Sci. Numer. Simul.* **2016**, *32*, 114–131. [CrossRef]
18. de la Rosa, R.; Gandarias, M.L.; Bruzón, M.S. On symmetries and conservation laws of a Gardner equation involving arbitrary functions. *Appl. Math. Comput.* **2016**, *290*, 125–134. [CrossRef]
19. Bruzón, M.S.; Márquez, A.P.; Garrido, T.M.; Recio, E.; de la Rosa, R. Conservation laws for a generalized seventh order KdV equation. *J. Comput. Appl. Math.* **2019**, *354*, 682–688. [CrossRef]
20. Hydon, P.E. *Symmetry Methods for Differential Equations: A Beginner's Guide*; Cambridge University Press: Cambridge, UK, 2000.
21. Anco, S.C. Generalization of Noether's theorem in modern form to non-variational partial differential equations. In *Recent Progress and Modern Challenges in Applied Mathematics, Modeling and Computational Science*; Springer: New York, NY, USA, 2017; pp. 119–182.

© 2019 by the authors. Licensee MDPI, Basel, Switzerland. This article is an open access article distributed under the terms and conditions of the Creative Commons Attribution (CC BY) license (http://creativecommons.org/licenses/by/4.0/).

Article

Invariant Solutions of the Wave Equation on Static Spherically Symmetric Spacetimes Admitting G_7 Isometry Algebra

Hassan Azad [1], Khaleel Anaya [2], Ahmad Y. Al-Dweik [3] and M. T. Mustafa [3,*]

1. Abdus Salam School of Mathematical Sciences, GC University, Lahore 54600, Pakistan; hassan.azad@sms.edu.pk
2. Department of Mathematics and Statistics, King Fahd University of Petroleum and Minerals, Dhahran 31261, Saudi Arabia; khalil.anaya1@gmail.com
3. Department of Mathematics, Statistics and Physics, Qatar University, Doha 2713, Qatar; aydweik@qu.edu.qa
* Correspondence: tahir.mustafa@qu.edu.qa

Received: 30 October 2018; Accepted: 19 November 2018; Published: 23 November 2018

Abstract: Algorithms to construct the optimal systems of dimension of at most three of Lie algebras are given. These algorithms are applied to determine the Lie algebra structure and optimal systems of the symmetries of the wave equation on static spherically symmetric spacetimes admitting G_7 as an isometry algebra. Joint invariants and invariant solutions corresponding to three-dimensional optimal systems are also determined.

Keywords: wave equation; spherically symmetric spacetimes; lie symmetries; roots; optimal systems; invariant solutions

1. Introduction

It was shown in [1–3] that spherically symmetric spacetimes belong to one of the following four classes according to their isometries and metrics:

- G_{10} corresponding to the static spacetimes Minkowski, de Sitter and anti de Sitter.
- G_7 corresponding to the static spacetimes Einstein and the anti Einstein universe, and one non-static spacetime.
- G_6 corresponding to the static spacetimes Bertotti–Robinson and two other metrics of Petrov type D, and six non-static spacetimes.
- G_4 is a class of metrics involving one or two arbitrary functions of one variable.

Azad et al. [4] applied Lie group analysis to study the wave equation on the classes of static spherically symmetric spacetimes admitting the isometry groups G_{10} or G_7 or G_6. The Iwasawa decomposition for the symmetry algebras was obtained to partially classify non-conjugate solvable algebras. The optimal system of subalgebras was not given in this previous study.

The G_7 spacetimes admit either $so(4) \oplus \mathbb{R}$ or $so(1,3) \oplus \mathbb{R}$ as isometry algebras as shown in [3]. In this paper, we continue the investigation started in [4] by finding the optimal system of subalgebras of dimension of at most three and the corresponding invariant solutions for spacetimes admitting G_7 as isometry algebras. We expect these solutions to be of interest to mathematical physicists.

As regards optimal systems, we can always construct a family of group invariant solutions obtained by using a subgroup of a symmetry group admitted by a given differential equation, as explained in [5]. Since there are infinitely many subgroups of a symmetry group admitted by a given differential equation, listing of all the group invariant solutions is impossible. However, obtaining optimal systems-meaning conjugacy classes- of s-dimensional subgroups of the symmetry group and applying the optimal systems leads to an effective and systematic mechanism of classifying the group invariant solutions. This leads to non-similar invariant solutions under symmetry transformations.

Classifying the group invariant solutions by utilizing optimal systems is a significant application of Lie group and Lie symmetry methods to differential equations. The method was first introduced by Ovsiannikov [6]. He applied this method in classifying the invariant solutions of the one-dimensional gasdynamic equation [7]. Ibragimov extended this work to the two-dimensional adiabatic gas motions in his master thesis [8] by applying the expansion method for solvable Lie algebra. The main idea behind the method is discussed in detail in Ibragimov [5,9], Olver [10] and Hydon [11].

The symmetry Lie algebra of the equations under study is non-solvable, but finding the optimal systems for non-solvable Lie algebras is more challenging. In this paper, improved algorithms are introduced and applied to construct the optimal systems of dimension of at most three of Lie algebras. The reason is that a PDE with four independent variables can be reduced to an ordinary differential equation (ODE) using three-dimensional subalgebras satisfying the transversality condition with rank three [10]. This provides the non-trivial invariant solutions under a maximum number of symmetries.

The paper is organized as follows: in Section 2, algorithms to construct the optimal systems of dimension of at most three of Lie algebras are introduced. In Section 3, Lie point symmetry transformations of the wave equation on the metrics considered in this paper are found. In Section 4, the algorithms are applied to determine the Lie algebra structure and optimal systems of the symmetries. In Section 5, joint invariants and invariant solutions corresponding to three-dimensional optimal systems are determined.

2. Algorithms to Construct the Optimal Systems of Dimension of at Most Three of Non-Solvable Lie Algebras

In this paper, we are interested only in finding an optimal system of subalgebras of dimension of at most three as explained in the introduction. This is achieved by using the algorithms explained below. These algorithms are based on a combination of the expansion method and algorithms for determining maximal solvable subalgebras of semi-simple Lie algebras.

If X is a solvable, then either X is abelian or it can be obtained from its commutator X' by a sequence of one-dimensional ideals . Thus, in any case, by using normalizers or centralizers, one can reach X from lower dimensional subalgebras. In more detail, the expansion method is revised and improved to a systematic method by using the normalizers and their associated quotient algebras as follows:

Let Θ_r be the optimal systems of r-dimensional solvable subalgebras of the solvable algebra $\mathcal{L}$. For every $X \in \Theta_{t-1}$, find the normalizer $\mathcal{N}(X)$. In case the quotient algebra $\mathcal{N}(X)/X$ is non-zero, we find a one-dimensional optimal system in $\mathcal{N}(X)/X$ for every $X \in \Theta_{t-1}$ by considering the invariants of the adjoint representation of $\mathcal{N}(X)/X$.

Among the constructed optimal systems of $\mathcal{N}(X)/X$ for every $X \in \Theta_{t-1}$, we may still have repetitions in their preimages in $\mathcal{L}$. Removing the repetitions provides an optimal system Θ_t. Enumeration of all non-conjugate solvable subalgebras of $\mathcal{L}$ can finally be done through consecutive choice of the values of t from 1 till $dim(\mathcal{L})$.

The Expansion method can be used to find optimal systems of solvable subalgebras in solvable or non-solvable Lie algebras. However, dealing with the general adjoint action of the group once the Lie algebra is non-solvable is very difficult. Therefore, in order to find the optimal systems of solvable subalgebras in a non-solvable Lie algebra, we proceed as follows:

For a general Lie algebra with Levi decomposition $\mathcal{L} = \mathcal{S} \oplus_s \mathcal{R}(\mathcal{L})$, where $\mathcal{S}$ is a semisimple subalgebra of $\mathcal{L}$ and $\mathcal{R}(\mathcal{L})$ is the radical of $\mathcal{L}$, every maximal solvable subalgebra is of the form $\mathcal{M} \oplus_s \mathcal{R}(\mathcal{L})$, where $\mathcal{M}$ is maximal solvable in $\mathcal{S}$. The maximal solvable subalgebras can be determined using the algorithms given in [12] or more efficiently using the method detailed in Section 2.1. For a semisimple algebra $\mathcal{S}$, there is a subalgebra $\mathcal{N}$ in which all elements are ad-nilpotent and which contains—up to conjugacy—all the commutators of solvable subalgebras of $\mathcal{S}$. All the maximal solvable subalgebras that are not compact tori can be constructed from the normalizers of conjugacy classes in $\mathcal{N}$—as detailed in Section 2.1. Then, the task of finding the optimal systems of solvable subalgebras of the Lie algebra $\mathcal{L}$ is reduced to finding the optimal systems of solvable subalgebras in each of these maximal solvable subalgebras using the expansion method. Finally, the repetitions in the obtained rough classification of subalgebras are removed using the adjoint representation of $\mathcal{L}$.

As a special case, if the radical is the center, then the calculations are greatly simplified. The reason is that it is enough to find the optimal systems of solvable subalgebras in each conjugacy class of maximal solvable subalgebras in the semisimple part of $\mathcal{L}$. Then, the repetitions in the obtained rough classification of subalgebras are removed using the adjoint representation of the semisimple part of $\mathcal{L}$. Finally, adjoining the subalgebras of the radical gives the optimal systems of solvable subalgebras in $\mathcal{L}$.

In order to find the general adjoint action of the semisimple part of $\mathcal{L}$, we need to make a suitable change of basis depending on the root space decomposition or the Iwasawa decomposition of the semisimple part $\mathscr{L}$ based on the signature of the Killing form.

2.1. Algorithm for Finding the Conjugacy Classes of Maximal Solvable Subalgebras

For the convenience of the reader who is not a specialist in Lie theory, we first recall how to construct Cartan algebras and roots algorithmically from a knowledge of the commutator table of a given Lie algebra.

The structure of a semisimple Lie algebra is determined by its roots. For more details, the reader is referred to [13]; see also [14–16].

Definition 1. *A Lie subalgebra H of a Lie algebra $\mathcal{L}$ is said to be a **Cartan subalgebra** if H is abelian and every element $h \in H$ is semisimple: by a semisimple element, we mean an element that is diagonalizable in the adjoint representation. Moreover, H is maximal with these properties.*

Definition 2. *Let C be a Cartan subalgebra of a semisimple Lie algebra $\mathcal{L}$. A non-zero vector $v \in \mathcal{L}^{\mathbb{C}} := \mathcal{L} + i\mathcal{L}$ such that $[h,v] = \lambda(h)v$ for all $h \in C$ is called a **root vector** and the corresponding linear function λ is called a **root of the Cartan algebra** C.*

In general, the roots will be complex-valued. In the following argument, we will use the notion of positive roots, so one needs to define what it means for a complex valued root to be positive.

Definition 3. *A complex number $z = a + ib$, $a,b \in \mathbb{R}$ is **positive** if either its real part a is positive or $a = 0$, but its imaginary part b is positive.*

Fix a basis $h_1, \ldots, h_r$ of a Cartan algebra C. A non-zero root λ is **positive** if the first non-zero number $\lambda(h_i)$ is a complex positive number. Otherwise, it is called a **negative root**. Positive roots which are not a sum of two positive roots are called **simple roots**.

The well known software Maple is able to find the root space decompositions of Lie algebras of fairly high dimensions by using the command "RootSpaceDecomposition(C)", where C is a list of vectors in a Lie algebra, defining a Cartan subalgebra.

The Cartan algebra is picked up using an algorithm due to de Graaf [15]. However, one gets better coordinates for computation if one chooses a Cartan algebra by enlarging a given diagonalizable subalgebra to a Cartan subalgebra following the algorithms given in [13]. We need in this paper only a special case of these algorithms to compute the Cartan subalgebras. We first compute the Killing form of the Lie algebra. If it is negative definite, pick any non-zero element X and compute its centralizer. By a negative definite matrix, we mean a matrix which is equal to its conjugate transpose and its eigenvalues are strictly negative. If the centralizer of X, $\mathcal{C}(X)$, is self centralizing, i.e., $\mathcal{C}(\mathcal{C}(X)) = \mathcal{C}(X)$, then $\mathcal{C}(X)$ is the Cartan subalgebra. Otherwise, we can find a linearly independent element Y in the centralizer of X. Continue this procedure with the abelian algebra $\langle X, Y \rangle$ until a self centralizing subalgebra is reached. The obtained algebra is the Cartan algebra because it is abelian and every element is diagonalizable.

On the other hand, if the Killing form is not negative definite and a maximal compact subalgebra is known, say K, then computing a Cartan subalgebra C of K using the procedure explained in the previous paragraph for compact algebras and the centralizer of C in the full Lie algebra gives us the required Cartan algebra.

The main use of Cartan algebras is to find all the maximal solvable subalgebras [17]. In case the Lie algebra $\mathcal{L}$ is compact, a Cartan algebra is, up to conjugacy, the only maximal solvable subalgebra. This follows from Lie's theorem on solvable algebras [14].

There is a solvable subalgebra B with real eigenvalues in the adjoint representation of $\mathcal{L}$ with the property that any other solvable algebra with real eignvalues in the adjoint representation is conjugate to a subalgebra of B.

In [12], it is found that the algebra B can be constructed algorithmically by using positive roots of a given maximally real Cartan subalgebra; by maximally real Cartan subalgebra, we mean a Cartan algebra whose real part has maximal possible dimension. In case the Killing form is not negative definite, any Cartan algebra is a sum of two subalgebras such that one of them has all real eigenvalues in the adjoint representation in $\mathcal{L}$ and the other has all purely imaginary eigenvalues in the adjoint representation in $\mathcal{L}$. We call the first subalgebra the real part of the Cartan subalgebra and the second subalgebra the compact part of the Cartan subalgebra. Let N be the algebra consisting of the real and imaginary parts of the positive root vectors for the given maximally real Cartan subalgebra. Then, the algebra $B = A + N$ where A is the real part of the maximally real Cartan subalgebra has the property that every solvable algebra with real eigenvalues in the adjoint representation is conjugate to subalgebra of B. Moreover, all maximal solvable algebras which are non-abelian can be obtained by computing normalizers of subalgebras of N. In more detail, we consider conjugacy classes of subalgebras of N. If X is a representative of such a class, we compute the normalizer of X and its Levi decomposition. We keep only those X in which the normalizer of X has Levi decomposition $\mathcal{N}(X) = S + \mathcal{R}(\mathcal{N}(X))$, $\mathcal{R}(\mathcal{N}(X))/X$ a torus and where the semisimple part has a compact Cartan subalgebra. If T is this compact Cartan subalgebra, then $T + \mathcal{R}(\mathcal{N}(X))$ is a maximal solvable subalgebra and all such, apart from compact maximal tori-if any- are obtained in this way.

2.2. Algorithm for Finding Three-Dimensional Optimal System of Non-Solvable Subalgebras of a Lie Algebra

- It is a classical fact that any non-solvable three-dimensional subalgebra is isomorphic to either $sl(2,\mathbb{R})$ or $so(3)$ copies in $\mathscr{S}$ up to conjugacy where $\mathscr{S}$ is a semisimple subalgebra of the given Lie algebra $\mathcal{L}$. Therefore, one can construct the three-dimensional optimal system of non-solvable subalgebras by finding copies of $so(3)$ and $sl(2,\mathbb{R})$ in $\mathscr{S}$.
- In order to find such copies in the semisimple Lie algebra S, we have developed the following algorithms which are based on the canonical relations for $so(3)$:

$$[X, A] = Y, \quad [A, Y] = X, \quad [Y, X] = A, \tag{1}$$

and $sl(2,\mathbb{R})$

$$[A, B] = 2B, \quad [A, Y] = -2Y, \quad [B, Y] = A. \tag{2}$$

To find the non-conjugate copies of $so(3)$:

- We start with an element A of the one-dimensional optimal system of S whose non-zero eigenvalues in the adjoint representation are purely imaginary.
- By scaling, we may assume that this eigenvalue is i. Let $X + iY$ be the eigenvector of A corresponding to the eigenvalue i. If $[X, Y] = \lambda A$ for some negative constant λ, then the algebra $\langle A, X, Y \rangle$ forms a copy of $so(3)$.

- Applying this algorithm for all elements in the one-dimensional optimal system gives us the copies of $so(3)$.
- Removing the repetitions using invariant tools gives the non-conjugate copies of $so(3)$.

To find the copies of $sl(2,\mathbb{R})$:

- We start with an element of the two-dimensional optimal system of non-abelian subalgebras.
- If $\langle A, B \rangle$ is such algebra with $[A, B] = cB$ for some non-zero constant c, find the eigenvectors of $\operatorname{ad} A$, if any corresponding to the eigenvalue $-c$. We reject $\langle A, B \rangle$ if there is no such eigenvalue. Otherwise, let Y be an eigenvector of $\operatorname{ad}(A)$ with eigenvalue $-c$. If the commutator $[B, Y]$ is a nonzero multiple of A, then $\langle A, B, Y \rangle$ is a copy of $sl(2,\mathbb{R})$.
- Removing the repetitions using invariant tools gives the non-conjugate copies of $sl(2,\mathbb{R})$.

3. Lie Point Symmetry Transformations of the Wave Equation

The wave equation on a spacetime is given by $\Box_g u = 0$, where $\Box_g = \frac{\partial}{\partial x_i}(\sqrt{|g|}\, g^{ik} \frac{\partial}{\partial x_k})$ is called the Laplace–Beltrami operator for the metric given by

$$ds^2 = e^{\nu(r,t)} dt^2 - e^{\lambda(r,t)} dr^2 - e^{\mu(r,t)} d\theta^2 - e^{\mu(r,t)} \sin^2\theta\, d\varphi^2. \tag{3}$$

Hence, the wave equation $\Box_g u = 0$ on the metric (3) can be written as

$$\frac{\partial}{\partial t}\left(e^{\left(\mu - \frac{\nu}{2} + \frac{\lambda}{2}\right)} \sin\theta \frac{\partial u}{\partial t}\right) - \frac{\partial}{\partial r}\left(e^{\left(\mu + \frac{\nu}{2} - \frac{\lambda}{2}\right)} \sin\theta \frac{\partial u}{\partial r}\right) - \frac{\partial}{\partial \theta}\left(e^{\left(\frac{\nu}{2} + \frac{\lambda}{2}\right)} \sin\theta \frac{\partial u}{\partial \theta}\right) - \frac{\partial}{\partial \varphi}\left(\frac{e^{\left(\frac{\nu}{2} + \frac{\lambda}{2}\right)}}{\sin\theta} \frac{\partial u}{\partial \varphi}\right) = 0. \tag{4}$$

The approach to find the symmetries of the wave equation using the conformal Killing vector field of the underlying spacetimes metric is due to Yuri Bozhkov and Igor Leite Freire [18].

Theorem 1 ([18]). *Let M^n be a Lorentzian manifold of dimension $n \geq 3$ with the metric g given in local coordinates $\{x_1, x_2, ..., x_n\}$. The Lie symmetries of wave equation $\Box_g u = 0$ on M^n have the form*

$$X = \xi^i(x) \frac{\partial}{\partial x_i} + \left(\left(\frac{2-n}{4}\mu(x) + c\right) u + b(x)\right) \frac{\partial}{\partial u}, \tag{5}$$

where c is an arbitrary constant,

$$\Box_g b(x) = 0, \quad \Box_g \mu(x) = 0, \tag{6}$$

$Y = \xi^i(x) \frac{\partial}{\partial x_i}$ *is a conformal Killing vector field of the metric g such that*

$$(\pounds_Y g)_{ab} = \xi^c \partial_c g_{ab} + g_{cb} \partial_a \xi^c + g_{ca} \partial_b \xi^c = \mu(x) g_{ab} \tag{7}$$

and $\pounds_Y$ denotes the Lie derivative with respect to vector field Y, where $b(x)$ and $\mu(x)$ satisfy (6).

3.1. Lie Point Symmetry Transformations of the Wave Equation on Einstein Spacetime

The wave equation $\Box_g u = 0$ on the spherically symmetric space admitting $so(4) \oplus \mathbb{R}$ as isometry algebra can be obtained from Equation (4) by substituting $\nu = 0$, $\lambda = -\ln(\alpha r^2 + 1)$ and $\mu = \ln r^2$, $\alpha = -c^2 < 0$ as shown in [3].

From now on, we will work with Cartesian coordinates as their introduction simplifies many comutations. The wave equation under study can be written in Cartesian coordinates $x = r\cos\varphi \sin\theta$, $y = r\sin\varphi \sin\theta$, $z = r\cos\theta$ as:

$$u_{tt} + (c^2 x^2 - 1) u_{xx} + (c^2 y^2 - 1) u_{yy} + (c^2 z^2 - 1) u_{zz} + 3c^2 z u_z + 2c^2 xz u_{zx} \\ + 3c^2 x u_x + 2c^2 xy u_{xy} + 2c^2 yz u_{yz} + 3c^2 y u_y = 0. \tag{8}$$

By using Theorem 1 and the isometries of the metric given in [3], the Lie symmetry algebra of the wave Equation (8) consists of the eight-dimensional subalgebra spanned by

$$X_1 = B\frac{\partial}{\partial y}, \quad X_2 = B\frac{\partial}{\partial x}, \quad X_3 = B\frac{\partial}{\partial z}, \quad X_4 = x\frac{\partial}{\partial z} - z\frac{\partial}{\partial x},$$
$$X_5 = z\frac{\partial}{\partial y} - y\frac{\partial}{\partial z}, \quad X_6 = x\frac{\partial}{\partial y} - y\frac{\partial}{\partial x}, \quad X_7 = \frac{\partial}{\partial t}, \quad X_8 = u\frac{\partial}{\partial u}, \tag{9}$$

and the infinite-dimensional ideal consisting of the operators

$$X_\tau = \tau(t,x,y,z)\frac{\partial}{\partial u}, \tag{10}$$

where $\tau(t,x,y,z)$ is an arbitrary solution of the wave Equation (8) and $B = \sqrt{1 - c^2(x^2 + y^2 + z^2)}$.

Moreover, the one-parameter groups $G_i(\varepsilon) = \{e^{\varepsilon X_i}, \varepsilon \in \mathbb{R}\}$ generated by (9) are given as follows:

$$\begin{aligned}
G_1(\varepsilon_1) &: (t,x,y,z,u) \mapsto (t,x,\tfrac{1}{c}\sqrt{1-c^2(x^2+z^2)}\sin(\arctan(\tfrac{cy}{B})+c\,\varepsilon_1),z,u),\\
G_2(\varepsilon_2) &: (t,x,y,z,u) \mapsto (t,\tfrac{1}{c}\sqrt{1-c^2(y^2+z^2)}\sin(\arctan(\tfrac{cx}{B})+c\,\varepsilon_2),y,z,u),\\
G_3(\varepsilon_3) &: (t,x,y,z,u) \mapsto (t,x,y,\tfrac{1}{c}\sqrt{1-c^2(x^2+y^2)}\sin(\arctan(\tfrac{cz}{B})+c\,\varepsilon_3),u),\\
G_4(\varepsilon_4) &: (t,x,y,z,u) \mapsto (t,x\sin\varepsilon_4 - z\cos\varepsilon_4, y, x\sin\varepsilon_4 + z\cos\varepsilon_4, u),\\
G_5(\varepsilon_5) &: (t,x,y,z,u) \mapsto (t,x,-z\sin\varepsilon_5 - y\cos\varepsilon_5, y\sin\varepsilon_5 - z\cos\varepsilon_5, u),\\
G_6(\varepsilon_6) &: (t,x,y,z,u) \mapsto (t,-y\sin\varepsilon_6 + x\cos\varepsilon_5, x\sin\varepsilon_6 + y\cos\varepsilon_6, z, u),\\
G_7(\varepsilon_7) &: (t,x,y,z,u) \mapsto (t+\varepsilon_7,y,z,u),\\
G_8(\varepsilon_8) &: (t,x,y,z,u) \mapsto (t,x,y,z,u+\varepsilon_8).
\end{aligned} \tag{11}$$

3.2. Lie Point Symmetry Transformations of the Wave Equation on Anti-Einstein Spacetime

The wave Equation $\Box_g u = 0$ on the spherically symmetric space admitting $so(1,3) \oplus \mathbb{R}$ as isometry algebra can be obtained from Equation (4) by substituting $\nu = 0$, $\lambda = -\ln(\alpha r^2 + 1)$ and $\mu = \ln r^2$, $\alpha = c^2 > 0$ as shown in [3].

As before, we will work with Cartesian coordinates as their introduction simplifies many comutations. The wave equation under study can be written in Cartesian coordinates $x = r\cos\varphi\sin\theta$, $y = r\sin\varphi\sin\theta$, $z = r\cos\theta$ as:

$$(c^2x^2 + 1)u_{xx} + (c^2y^2 + 1)u_{yy} + (c^2z^2 + 1)u_{zz} + 2c^2yzu_{yz},$$
$$+3xc^2u_x + 2c^2xyu_{xy} + 2c^2xzu_{xz} + 3c^2zu_z + 3c^2yu_y - u_{tt} = 0. \tag{12}$$

By using Theorem 1 and the isometries of the metric given in [3], the Lie symmetry algebra of the wave Equation (12) consists of the eight-dimensional subalgebra spanned by

$$X_1 = B\frac{\partial}{\partial y}, \quad X_2 = B\frac{\partial}{\partial x}, \quad X_3 = B\frac{\partial}{\partial z}, \quad X_4 = x\frac{\partial}{\partial z} - z\frac{\partial}{\partial x},$$
$$X_5 = z\frac{\partial}{\partial y} - y\frac{\partial}{\partial z}, \quad X_6 = x\frac{\partial}{\partial y} - y\frac{\partial}{\partial x}, \quad X_7 = \frac{\partial}{\partial t}, \quad X_8 = u\frac{\partial}{\partial u}, \tag{13}$$

and the infinite-dimensional ideal consisting of the operators

$$X_\tau = \tau(t,x,y,z)\frac{\partial}{\partial u}, \tag{14}$$

where $\tau(t,x,y,z)$ is an arbitrary solution of the wave Equation (12) and $B = \sqrt{1 + c^2(x^2 + y^2 + z^2)}$.

Moreover, the one-parameter groups $G_i(\varepsilon) = \{e^{\varepsilon X_i}, \varepsilon \in \mathbb{R}\}$ generated by (13) are given as follows:

$$G_1(\varepsilon_1) : (t,x,y,z,u) \mapsto \left(t, x, \frac{e^{-c\varepsilon_1}\left(2cBe^{2c\varepsilon_1}y + (B^2+c^2y^2)e^{2c\varepsilon_1} - 1 - c^2(x^2+z^2)\right)}{2c^2y+2cB}, z, u\right),$$

$$G_2(\varepsilon_2) : (t,x,y,z,u) \mapsto \left(t, \frac{(2e^{2c\varepsilon_2}(B^2-1) + 2cBe^{2c\varepsilon_2}x - c^2(y^2+z^2) + e^{2c\varepsilon_2} - 1)e^{-c\varepsilon_2}}{2c^2x+2cB}, y, z, u\right),$$

$$G_3(\varepsilon_3) : (t,x,y,z,u) \mapsto \left(t, x, y, \frac{e^{-c\varepsilon_3}\left(2cBe^{2c\varepsilon_3}z + B^2 e^{2c\varepsilon_3} - 1 - c^2(x^2+y^2)\right)}{2c^2z+2cB}, u\right),$$

$$G_4(\varepsilon_4) : (t,x,y,z,u) \mapsto (t, x\sin\varepsilon_4 - z\cos\varepsilon_4, y, x\sin\varepsilon_4 + z\cos\varepsilon_4, u),$$

$$G_5(\varepsilon_5) : (t,x,y,z,u) \mapsto (t, x, -z\sin\varepsilon_5 - y\cos\varepsilon_5, y\sin\varepsilon_5 - z\cos\varepsilon_5, u),$$

$$G_6(\varepsilon_6) : (t,x,y,z,u) \mapsto (t, -y\sin\varepsilon_6 + x\cos\varepsilon_5, x\sin\varepsilon_6 + y\cos\varepsilon_6, z, u),$$

$$G_7(\varepsilon_7) : (t,x,y,z,u) \mapsto (t + \varepsilon_7, x, y, z, u),$$

$$G_8(\varepsilon_8) : (t,x,y,z,u) \mapsto (t, x, y, z, u + \varepsilon_8).$$

(15)

4. Lie Algebra Structure and Optimal Systems

4.1. Lie Point Symmetry Algebra of the Wave Equation on Einstein Spacetime

The non-zero Lie brackets of (9) are:

$$\begin{array}{llll}
[X_1, X_2] = c^2 X_6, & [X_1, X_3] = c^2 X_5, & [X_1, X_5] = -X_3, & [X_1, X_6] = -X_2, \\
[X_2, X_3] = -c^2 X_4, & [X_2, X_4] = X_3, & [X_2, X_6] = X_1, & [X_3, X_4] = -X_2, \\
[X_3, X_5] = X_1, & [X_4, X_5] = X_6, & [X_4, X_6] = -X_5, & [X_5, X_6] = X_4.
\end{array}$$

(16)

The Levi-Decomposition of this algebra is $\mathcal{L} = \{X_1, X_2, X_3, X_4, X_5, X_6\} \oplus \{X_7, X_8\}$. Let S be the semisimple part. To identify the semisimple part, we need to find a Cartan algebra and the corresponding root space decomposition.

First of all, after computing the Killing form, we see that it is negative definite. Thus, to determine a Cartan algebra, choose any non-zero element in the semisimple part S. We choose, for example, the element X_3 and compute its centralizer. The centralizer turns out to be $\{X_3, X_6\}$ and the subalgebra $\{X_3, X_6\}$ is self centralizing. Thus, $C = \{X_3, X_6\}$ is a Cartan subalgebra which is itself the only maximal solvable subalgebra up to the conjugacy as mentioned in Section 2.1. The roots for this Cartan subalgebra are $\{(ci, i), (-ci, i), (-ci, -i), (ci, -i)\}, i = \sqrt{-1}$. Therefore, the positive roots are $\{(ci, i), (ci, -i)\}$. The root vectors for the positive roots are $\{X_1 + cX_4 + i(X_2 + cX_5), X_1 - cX_4 + i(X_2 - cX_5)\}$. Since the negative roots are conjugates of the positive roots, the real and the imaginary parts of the positive root vectors must generate, as a Lie algebra, the full Lie algebra.

This gives us the change of basis which gives the general adjoint action of the group of symmetry transformations:

$$\begin{array}{llll}
V_1 = X_1 + cX_4, & V_2 = X_2 + cX_5, & V_3 = X_3 - cX_6, & V_4 = X_1 - cX_4, \\
V_5 = X_2 - cX_5, & V_6 = X_3 + cX_6, & V_7 = X_7, & V_8 = X_8.
\end{array}$$

(17)

The corresponding non-zero Lie brackets of this subalgebra are:

$$\begin{array}{llll}
[V_1, V_2] = -2cV_3 & [V_1, V_3] = 2cV_2, & [V_2, V_3] = -2cV_1, & [V_4, V_5] = 2cV_6, \\
& [V_4, V_6] = -2cV_5, & [V_5, V_6] = 2cV_4.
\end{array}$$

(18)

It is obvious from (18) that the subalgebra $\langle V_1, V_4 \rangle$ is Cartan since it is abelian and it is self centralizing. Since our Lie algebra is compact, therefore, $\langle V_1, V_4 \rangle$ is the only maximal solvable algebra up to the conjugacy. The subalgebra $\langle V_1, V_4 \rangle$ is conjugate to $\langle V_3, V_6 \rangle$.

As we will see later, $\langle V_1, V_2, V_3 \rangle$ and $\langle V_4, V_5, V_6 \rangle$ form two copies of $so(3)$ which commute with each other. Since $so(4)$, which is the set of all skew symmetric 4×4 matrices, is also isomorphic to $so(3) \oplus so(3)$ [19], we see that the semisimple part is isomorphic to $so(4)$. This decomposition can be obtained by working with a Cartan subalgebra of $so(4)$ and determining its root space decomposition as was done above.

4.1.1. Optimal Systems of Solvable Subalgebras of $so(4)$

To find the optimal systems of $so(4)$, we first find the one-dimensional optimal system and a rough classification of higher order subalgebras inside the maximal solvable subalgebra $\langle V_1, V_4 \rangle$ which is abelian. Secondly, we obtain the higher-dimensional optimal system by removing the repetitions from the obtained rough classification of subalgebras using the adjoint representation of $so(4)$.

Theorem 2. *The optimal systems Θ_i up to order three of solvable subalgebras of $so(4)$ with the non-zero Lie brackets (18) are the following:*

- *The one-dimensional optimal system Θ_1 is $\{\langle V_1 + \alpha V_4 \rangle, \langle V_1 \rangle, \langle V_4 \rangle, \alpha \neq 0\}$,*
- *The two-dimensional optimal system Θ_2 is $\{\langle V_1, V_4 \rangle\}$.*

There is no three-dimensional optimal system.

Proof. Clearly, it is enough to deal with the one-dimensional optimal system only. The one-dimensional optimal system of the maximal solvable subalgebra of $so(4)$ is itself the one-dimensional optimal system of $so(4)$. This is because the representative elements are non-conjugate under the adjoint representation of $so(4)$ given by

$$A(\varepsilon_1, \ldots, \varepsilon_6) = e^{\varepsilon_1 C(1)} \ldots e^{\varepsilon_6 C(6)}, \qquad (19)$$

where $C(j)$ is the matrix whose $(i,k)^{th}$ entries are given as c_{ij}^k: here, the constants c_{ij}^k are the **structure constants** relative to the basis $V_1, \ldots, V_6$. □

4.1.2. Optimal Systems of Solvable Subalgebras of $\mathcal{L} = so(4) \oplus \mathbb{R}^2$

First, note the general fact that if $\mathcal{L} = S \oplus R$ where S is the semisimple part and the radical R is the center, then the conjugacy classes of S can be joined with elements of the center to obtain conjugacy classes of $\mathcal{L}$, as follows:

Let $\pi : S \oplus R \to S$ be the projection defined by $\pi(x,y) = x$. This is a homomorphism because R is an ideal. Therefore, it will map conjugate classes to conjugate classes.

Every k-dimensional subalgebra of $\mathcal{L}$ is of the form $\langle x_1 + y_1, x_2 + y_2, \ldots, x_k + y_k \rangle$, where $x_i \in S$, $y_i \in R$. Its projection is $\langle x_1, \ldots, x_k \rangle$ of dimension less than or equal to k. Moreover, if $\langle x_1 + y_1, x_2 + y_2, \ldots, x_k + y_k \rangle$ is conjugate to $\langle \tilde{x}_1 + \tilde{y}_1, \tilde{x}_2 + \tilde{y}_2, \ldots, \tilde{x}_k + \tilde{y}_k \rangle$, then as the radical R is the center, $y_i = \tilde{y}_i$ and $\langle x_1, x_2, \ldots, x_k \rangle$ is conjugate to $\langle \tilde{x}_1, \tilde{x}_2, \ldots, \tilde{x}_k \rangle$. However, the dimension of the image algebra of $\langle \tilde{x}_1 + \tilde{y}_1, \tilde{x}_2 + \tilde{y}_2, \ldots, \tilde{x}_k + \tilde{y}_k \rangle$ can go down. Thus, to get all conjugacy classes of the full algebra, we start with the elements of the optimal systems of S and add to each one of them arbitrary elements of the center and keep those that form a subalgebra. The classes of the center correspond to the zero subspace of S. This will give all the conjugacy classes of the full algebra. Applying this to $\mathcal{L} = so(4) \oplus \mathbb{R}^2$, we obtain the following classes.

Clearly, the one-dimensional optimal system $\tilde{\Theta}_1$ of $\mathbb{R}^2$ is $\{\langle V_7 \rangle, \langle V_8 \rangle, \langle V_7 + \alpha V_8 \rangle, \alpha \neq 0\}$ and the only two-dimensional optimal system $\tilde{\Theta}_2$ of $\mathbb{R}^2$ is $\{\langle V_7, V_8 \rangle\}$.

In order to get the optimal systems of the full Lie algebra up to order three, we use the optimal systems of $so(4)$ constructed in Theorem 2. We join it with the optimal system of the abelian algebra $\mathbb{R}^2$ as explained above.

- To get the one-dimensional optimal system of $\mathcal{L}$, we have the cases:

    1. We add an arbitrary element from $\mathbb{R}^2$ to every element in Θ_1; in this case, we get $\{\langle V_1 + \alpha V_4 + Z_1 \rangle, \langle V_1 + Z_1 \rangle, \langle V_4 + Z_1 \rangle, \alpha \neq 0\}$.

2. We take $\tilde{\Theta}_1$ itself; in this case, we get $\{\langle V_7\rangle, \langle V_8\rangle, \langle V_7+\beta V_8\rangle, \beta\neq 0\}$.

- To get the two-dimensional optimal system of $\mathcal{L}$, we have the cases:

  1. We add an arbitrary element from $\mathbb{R}^2$ to every element in Θ_2; in this case, we get $\{\langle V_1+Z_1, V_4+Z_2\rangle\}$.
  2. We add an arbitrary element from $\mathbb{R}^2$ to every element in Θ_1 and combine the result with an element from $\tilde{\Theta}_1$; in this case, we get $\{\langle V_1+\alpha V_4+Z_1, Z_3\rangle, \langle V_1+Z_1, Z_3\rangle, \langle V_4+Z_1, Z_3\rangle, \alpha\neq 0\}$.
  3. Take $\tilde{\Theta}_2$ itself; in this case, we get $\{\langle V_7, V_8\rangle\}$.

- To get the three-dimensional optimal system of $\mathcal{L}$,

  1. Either we add an arbitrary element from $\mathbb{R}^2$ to every element in Θ_2 and combine the result with an element from $\tilde{\Theta}_1$; in this case, we get $\{\langle V_1+Z_1, V_4+Z_2, Z_3\rangle\}$;
  2. or we add an arbitrary element from $\mathbb{R}^2$ to every element in Θ_1 and combine the result with an element from $\tilde{\Theta}_2$; in this case, we get $\{\langle V_1+\alpha V_4+Z_1, V_7, V_8\rangle, \langle V_1+Z_1, V_7, V_8\rangle, \langle V_4+Z_1, V_7, V_8\rangle, \alpha\neq 0\}$.

Finally, we check that the obtained class is a subalgebra by taking the wedge product of it with its commutator and equate by zero and see if we can kill some constants. This leads to the following theorem:

Theorem 3. *The optimal systems of solvable subalgebra of $\mathcal{L}$ with the non-zero Lie brackets (18) are as follows:*

- *The one-dimensional optimal system is $\{\langle V_1+\alpha V_4+Z_1\rangle, \langle V_1+Z_1\rangle, \langle V_4+Z_1\rangle, \langle V_7\rangle, \langle V_8\rangle, \langle V_7+\beta V_8\rangle, \alpha, \beta\neq 0\}$.*
- *The two-dimensional optimal system is $\{\langle V_1+Z_1, V_4+Z_2\rangle, \langle V_1+\alpha V_4+Z_1, Z_3\rangle, \langle V_1+Z_1, Z_3\rangle, \langle V_4+Z_1, Z_3\rangle, \langle V_7, V_8\rangle, \alpha\neq 0\}$.*
- *The three-dimensional optimal system is $\{\langle V_1+Z_1, V_4+Z_2, Z_3\rangle, \langle V_1+\alpha V_4+Z_1, V_7, V_8\rangle, \langle V_1+Z_1, V_7, V_8\rangle, \langle V_4+Z_1, V_7, V_8\rangle, \alpha\neq 0\}$.*

Here, $Z_1 = \alpha_1 V_7 + \beta_1 V_8$, $Z_2 = \alpha_2 V_7 + \beta_2 V_8$ are arbitrary elements of $\mathbb{R}^2$ and $Z_3 = V_7 + \alpha_3 V_8$ or $Z_3 = V_8$ represent a one-dimensional optimal system of $\mathbb{R}^2$.

4.1.3. Three-dimesional Optimal System of Non-solvable Subalgebras of $\mathcal{L} = so(4) \oplus \mathbb{R}^2$

If H is a three-dimensional non-solvable algebra, then H equals its commutator. As the commutator of $\mathcal{L}$ is $so(4)$, all such subalgebras of $\mathcal{L}$ are subalgebras of $so(4)$. We need to construct the copies of $so(3)$ and $sl(2,\mathbb{R})$, if any, by following the algorithm given in Section 2.2:

- First, construct the copies of $so(3)$:

  1. The element V_1 has the eigenvector $V_2 + iV_3$ corresponding to the eigenvalue $2ci$. Therefore, $\tilde{V}_1 = \frac{V_1}{2c}$ has the same eigenvector with the eigenvalue i. Moreover, $[V_2, V_3] = -(2c)^2 \tilde{V}_1$. Hence, $\langle V_1, V_2, V_3\rangle$ forms a copy of $so(3)$.
  2. The element V_4 has the eigenvector $V_5 + iV_6$ corresponding to the eigenvalue $-2ci$. Therefore, $\tilde{V}_4 = \frac{V_4}{-2c}$ has the same eigenvector with the eigenvalue i. Moreover, $[V_5, V_6] = -(2c)^2 \tilde{V}_4$. Hence, $\langle V_4, V_5, V_6\rangle$ forms a copy of $so(3)$.
  3. The element $V_1 + \alpha V_4$ has the eigenvector $V_2 + V_5 + i(V_3 - V_6)$ corresponding to the eigenvalue $2ci$. Therefore, $\tilde{V} = \frac{V_1+\alpha V_4}{2c}$ has the same eigenvector with the eigenvalue i. Moreover, and $[V_2+V_5, V_3-V_6] = -2c\tilde{V}$. Hence, $\langle V_1+V_4, V_2+V_5, V_3-V_6\rangle$ forms a copy of $so(3)$. Note that here α must be equal to one to ensure that $\langle V_1+\alpha V_4, V_2+V_5, V_3-V_6\rangle$ is a subalgebra.

- The Lie algebra $so(4)$ does not contain any copy of $sl(2,\mathbb{R})$, since it does not contain any non-abelian two-dimensional subalgebra.

This proves the following theorem:

Theorem 4. *The three-dimensional optimal system of non-solvable subalgebras of $\mathcal{L}$ is $\{\langle V_1, V_2, V_3\rangle, \langle V_4, V_5, V_6\rangle, \langle V_1 + V_4, V_2 + V_5, V_3 - V_6\rangle\}$, where the $V_i, i = 1, \ldots, 6$ form a basis of so(4) given in (17).*

4.2. Lie Point Symmetry Algebra of the Wave Equation on Anti-Einstein Spacetime

The non-zero Lie brackets of (13) are:

$$\begin{array}{llll}
[X_1, X_2] = -c^2 X_6, & [X_1, X_3] = -c^2 X_5, & [X_1, X_5] = -X_3, & [X_1, X_6] = -X_2, \\
[X_2, X_3] = c^2 X_4, & [X_2, X_4] = X_3, & [X_2, X_6] = X_1, & [X_3, X_4] = -X_2, \\
[X_3, X_5] = X_1, & [X_4, X_5] = X_6, & [X_4, X_6] = -X_5, & [X_5, X_6] = X_4.
\end{array} \quad (20)$$

The Levi-Decomposition of this algebra is $\mathcal{L} = \{X_1, X_2, X_3, X_4, X_5, X_6\} \oplus \{X_7, X_8\}$. Let S be the semisimple part.

To determine the structure of the semisimple part, we need to find a Cartan algebra and the root space decomposition with respect to the Cartan algebra. In this case, the Killing form is not negative definite and it has exactly three negative eigenvalues. This means that the maximal compact algebra is three-dimensional.

The reason is that, if K is a maximal compact subalgebra of the Lie algebra $\mathcal{L}$, then any compact subalgebra of $\mathcal{L}$ is conjugate to a subalgebra of K. Moreover, every one-dimensional subalgebra of K is conjugate to a subalgebra of a fixed Cartan subalgebra of K [14–16].

As we will explain later, the subalgebra $\langle X_4, X_5, X_6 \rangle$ is a copy of so(3) in the given Lie algebra. Thus, $K = \langle X_4, X_5, X_6 \rangle$ is a maximal compact subalgebra of the algebra S. A Cartan subalgebra of S can be obtained by choosing any element of K and computing its centralizer. We choose, for example, X_6 as a representative of a Cartan algebra of K. Computing the centralizer of X_6, we find that it is $\langle X_3, X_6 \rangle$. In addition, as the centralizer of $\langle X_3, X_6 \rangle$ is itself, $C = \langle X_3, X_6 \rangle$ is a Cartan subalgebra of S. Moreover, computing the eigenvalues of X_3, we find that all eigenvalues of adX_3 are real and X_3 is diagonalizable. Moreover, the centralizer of X_3 is C and the centralizer of X_6 is also C; this means that C is the maximally real Cartan subalgebra.

We find roots of C in S. The roots are $(c, i), (c, -i), (-c, i), (-c, -i)$, the positive roots are $(c, i), (c, -i)$ and clearly the sum of these positive roots is not a root. The root spaces for the positive roots (c, i) and $(c, -i)$ are $\langle X_1 + cX_5 + i(cX_4 - X_2)\rangle$. Let $N = \langle X_1 + cX_5, X_2 - cX_4\rangle$. The algebra $B = A \oplus N$, where $A = \langle X_3 \rangle$ is the real part of C, has the property that every solvable algebra with real eigenvalues in the adjoint representation is conjugate to a subalgebra of B. We compute the normalizers of each conjugacy class of N. The normalizer of each representative element of the one-dimensional optimal system of N does not contain a Cartan algebra. Therefore, we keep only N because its normalizer $\mathcal{N}(N)$ is solvable and contains a Cartan algebra. Thus, there is only one maximal solvable subalgebra, namely $\mathcal{N}(N) = \langle X_6, X_3 \rangle \oplus \langle X_1 + cX_5, X_2 - cX_4 \rangle$. Therfore, the Iwasawa decomposition of S is $K \oplus A \oplus N = \langle X_4, X_5, X_6 \rangle \oplus \langle X_3 \rangle \oplus \langle X_1 + cX_5, X_2 - cX_4 \rangle$ [14,16].

This gives us the following change of basis which makes the computations easier:

$$\begin{array}{llll}
V_1 = X_4, & V_2 = X_5, & V_3 = X_6, & V_4 = X_3, \\
V_5 = X_1 + cX_5, & V_6 = X_2 - cX_4, & V_7 = X_7, & V_8 = X_8.
\end{array} \quad (21)$$

The non-zero Lie brackets of (21) are

$$\begin{array}{llll}
[V_1, V_2] = V_3, & [V_1, V_3] = -V_2, & [V_1, V_4] = cV_1 + V_6, & [V_1, V_5] = cV_3, \\
[V_1, V_6] = -V_4, & [V_2, V_3] = V_1, & [V_2, V_4] = cV_2 - V_5, & [V_2, V_5] = V_4, \\
[V_2, V_6] = cV_3 & [V_3, V_5] = V_6, & [V_3, V_6] = -V_5, & [V_4, V_5] = cV_5, \\
[V_4, V_6] = cV_6.
\end{array} \quad (22)$$

In fact, the semisimple part S is isomorphic to $so(1,3)$ as can be seen by working with its Cartan algebra and the associated root space decompositions. The algebra $\langle V_7, V_8 \rangle$ is the center of the Lie algebra $\mathcal{L}$.

4.2.1. Optimal Systems of Solvable Subalgebras of $so(1,3)$

To find the optimal system of $so(1,3)$, we first find the one-dimensional optimal system and a rough classification of higher order subalgebras inside the maximal solvable subalgebra spanned by $E_1 := V_3, E_2 := V_4, E_3 := V_5, E_4 := V_6$. The corresponding non-zero Lie brackets of this subalgebra are:

$$[E_1, E_3] = E_4, \quad [E_1, E_4] = -E_3, \quad [E_2, E_3] = cE_3, \quad [E_2, E_4] = cE_4. \tag{23}$$

Secondly, we obtain the higher-dimensional optimal system by removing the repetitions from the obtained rough classification of subalgebras using the adjoint action of $so(1,3)$.

Theorem 5. *The optimal systems Θ_i up to order three of the solvable subalgebras of $so(1,3)$ with the non-zero Lie brackets (22) are the following:*

- *The one-dimensional solvable optimal system Θ_1 is*
  $\{\langle V_4 \rangle, \langle V_5 \rangle, \langle V_3 + \alpha V_4 \rangle : \alpha \in \mathbb{R}\}$.
- *The two-dimensional solvable optimal system Θ_2 is $\{\langle V_3, V_4 \rangle, \langle V_4, V_5 \rangle, \langle V_5, V_6 \rangle\}$.*
- *The three-dimensional solvable optimal system Θ_3 is*
  $\{\langle V_4, V_5, V_6 \rangle, \langle V_3 + \alpha V_4, V_5, V_6 \rangle : \alpha \in \mathbb{R}\}$.

Proof. To remove the repetitions in the obtained one-dimensional optimal system and the higher-dimensional rough classification of the maximal solvable subalgebra of $\mathcal{L}$, we use their normalizers in $so(1,3)$ as follows:

- The one-dimensional optimal system of the maximal solvable subalgebra of $so(1,3)$ is itself the one-dimensional optimal system of $so(1,3)$. This is because the representative elements are non-conjugate under the adjoint action of $so(1,3)$, as can be seen using the action of corresponding adjoint group given as in (19).
- The two-dimensional abelian subalgebras are $\langle V_3, V_4 \rangle, \langle V_5, V_6 \rangle$. The non-abelian subalgebra $\langle V_4, V_5 \rangle$ is clearly non-conjugate with both of them. Moreover, since the normalizers of the two-dimensional abelian subalgebras are $\mathcal{N}(\langle V_3, V_4 \rangle)/\langle V_3, V_4 \rangle = 0$, $\mathcal{N}(\langle V_5, V_6 \rangle)/\langle V_5, V_6 \rangle = \langle \tilde{V}_3, \tilde{V}_4 \rangle$. As their dimensions are different, they are non-conjugate.
- All the three-dimensional subalgebras given in the rough classification have the same normalizers, centralizers and commutators, namely the abelian subalgebra $\langle V_5, V_6 \rangle$.

Let X be one of these algebras. We find that the eigenvalues of X/X' are repeated real in one case, purely imaginary in one case and complex conjugates but not purely imaginary in the third case. Therefore, they are non-conjugate. □

4.2.2. Optimal Systems of Solvable Subalgebras of $\mathcal{L} = so(1,3) \oplus \mathbb{R}^2$

Clearly, the one-dimensional optimal system $\tilde{\Theta}_1$ of $\mathbb{R}^2$ is $\{\langle V_8 \rangle, \langle V_7 + \alpha V_8 \rangle : \alpha \in \mathbb{R}\}$ and the only two-dimensional optimal system $\tilde{\Theta}_2$ of $\mathbb{R}^2$ is $\{\langle V_7, V_8 \rangle\}$.

In order to get the optimal systems of the full Lie algebra up to order three, we use the optimal systems of $so(1,3)$ constructed in Theorem 5 and join each one of them with the optimal systems of the abelian algebra $\mathbb{R}^2$.

- To get the one-dimensional optimal system of $\mathcal{L}$,

  1. either we take $\tilde{\Theta}_1$ itself; in this case, we get $\{\langle V_8 \rangle, \langle V_7 + \beta V_8 \rangle : \beta \in \mathbb{R}\}$;

2. or we add an arbitrary element from $\mathbb{R}^2$ to every representative element in Θ_1; in this case, we get $\{\langle V_4 + Z_1\rangle, \langle V_5 + Z_1\rangle, \langle V_3 + \alpha V_4 + Z_1\rangle : \alpha \in \mathbb{R}\}$.

- To get the two-dimensional optimal system of $\mathcal{L}$,

    1. either we add an arbitrary element from $\mathbb{R}^2$ to every element in Θ_2; in this case, we get $\{\langle V_3 + Z_1, V_4 + Z_2\rangle, \langle V_4 + Z_1, V_5 + Z_2\rangle, \langle V_5 + Z_1, V_6 + Z_2\rangle\}$,
    2. or we add an arbitrary element from $\mathbb{R}^2$ to every element in Θ_1 and combine the result with an element from $\tilde{\Theta}_1$; in this case, we get $\{\langle V_4 + Z_1, Z_3\rangle, \langle V_5 + Z_1, Z_3\rangle, \langle V_3 + \alpha V_4 + Z_1, Z_3\rangle : \alpha \in \mathbb{R}\}$.
    3. or take $\tilde{\Theta}_2$ itself; in this case, we get $\{\langle V_7, V_8\rangle\}$.

- To get the three-dimensional optimal system of $\mathcal{L}$,

    1. either we add an arbitrary element from $\mathbb{R}^2$ to every element in Θ_3; in this case, we get $\{\langle V_4 + Z_1, V_5 + Z_2, V_6 + Z_3\rangle, \langle V_3 + \alpha V_4 + Z_1, V_5 + Z_2, V_6 + Z_3\rangle : \alpha \in \mathbb{R}\}$,
    2. or we add an arbitrary element from $\mathbb{R}^2$ to every element in Θ_2 and combine the result with an element from $\tilde{\Theta}_1$; in this case, we get $\{\langle V_3 + Z_1, V_4 + Z_2, Z_3\rangle, \langle V_4 + Z_1, V_5 + Z_2, Z_3\rangle, \langle V_5 + Z_1, V_6 + Z_2, Z_3\rangle\}$,
    3. or we add an arbitrary element from $\mathbb{R}^2$ to every element in Θ_1 and combine the result with an element from $\tilde{\Theta}_2$; in this case, we get $\{\langle V_4 + Z_1, V_7, V_8\rangle, \langle V_5 + Z_1, V_7, V_8\rangle, \langle V_3 + \alpha V_4 + Z_1, V_7, V_8\rangle : \alpha \in \mathbb{R}\}$,

where $Z_1 = \alpha_1 V_7 + \beta_1 V_8, Z_2 = \alpha_2 V_7 + \beta_2 V_8$ are arbitrary elements of $\mathbb{R}^2$ and $Z_3 = V_7 + \alpha_3 V_8$ or $Z_3 = V_8$ represents a one-dimensional optimal system of $\mathbb{R}^2$ and $\alpha_1, \alpha_2, \alpha_3, \beta_1, \beta_2 \in \mathbb{R}$.

Finally, we check that the obtained class is a subalgebra by taking the wedge product of its commutator with each element in the class and make these wedges equal to zero. Therefore, we have the following theorem.

Theorem 6. *The optimal systems of solvable subalgebras of $\mathcal{L}$ with the non-zero Lie brackets* (22) *are as follows:*

- *The one-dimensional solvable optimal system is* $\{\langle V_4 + Z_1\rangle, \langle V_5 + Z_1\rangle, \langle V_8\rangle, \langle V_3 + \alpha V_4 + Z_1\rangle, \langle V_7 + \beta V_8\rangle : \alpha, \beta \in \mathbb{R}\}$.
- *The two-dimensional solvable optimal system is*
$\{\langle V_3 + Z_1, V_4 + Z_2\rangle, \langle V_4 + Z_1, V_5\rangle, \langle V_5 + Z_1, V_6 + Z_2\rangle, \langle V_4 + Z_1, Z_3\rangle, \langle V_5 + Z_1, Z_3\rangle, \langle V_7, V_8\rangle, \langle V_3 + \alpha V_4 + Z_1, Z_3\rangle : \alpha \in \mathbb{R}\}$.
- *The three-dimensional solvable optimal system is* $\{\langle V_4 + Z_1, V_5, V_6\rangle, \langle V_3 + Z_1, V_4 + Z_2, Z_3\rangle, \langle V_4 + Z_1, V_5, Z_3\rangle, \langle V_5 + Z_1, V_6 + Z_2, Z_3\rangle, \langle V_4 + Z_1, V_7, V_8\rangle, \langle V_5 + Z_1, V_7, V_8\rangle, \langle V_3 + \alpha V_4 + Z_1, V_5, V_6\rangle, \langle V_3 + \alpha V_4 + Z_1, V_7, V_8\rangle : \alpha \in \mathbb{R}\}$.

Here, $Z_1 = \alpha_1 V_7 + \beta_1 V_8, Z_2 = \alpha_2 V_7 + \beta_2 V_8$ are arbitrary elements of $\mathbb{R}^2$ and $Z_3 = V_7 + \alpha_3 V_8$ or $Z_3 = V_8$ represents a one-dimensional optimal system of $\mathbb{R}^2$.

4.2.3. Three-Dimensional Optimal System of Non-Solvable Subalgebras of $\mathcal{L} = so(1,3) \oplus \mathbb{R}^2$

If H is a three-dimensional non-solvable algebra, then H equals its commutator. As the commutator of $\mathcal{L}$ is $so(1,3)$, all such subalgebras of $\mathcal{L}$ are subalgebras of $so(1,3)$. To find such subalgebras, we follow the algorithm given in Section 2.2.

We need to construct the copies of $so(3)$ and $sl(2,\mathbb{R})$, if any, by following the algorithm that is given in Section 2.2:

- First, construct the copies of $so(3)$: the element V_3 has the eigenvector $V_1 + iV_2$ corresponding to the eigenvalue $-i$. Therefore, $\tilde{V}_3 = -V_3$ has the same eigenvector with the eigenvalue i. Moreover, $[V_1, V_2] = -\tilde{V}_3$. Therefore, $\langle V_1, V_2, V_3\rangle$ forms a copy of $so(3)$.

- The only non-abelian two-dimensional subalgebra in $so(1,3)$ is $\langle V_4, V_5 \rangle$ with $[V_4, V_5] = cV_5$. Moreover, the eigenvector of $ad V_4$ corresponding to the eigenvalue $-c$ is $V_5 - 2cV_2$ and $[V_5 - 2cV_2, V_5] = -2cV_4$. Hence, the subalgebra $\langle V_4, V_5, V_5 - 2cV_2 \rangle$ forms a copy of $sl(2, \mathbb{R})$.

This proves the theorem.

Theorem 7. *The three-dimensional non-solvable optimal system is $\{\langle V_1, V_2, V_3 \rangle, \langle V_4, V_5, V_5 - 2cV_2 \rangle\}$, where the V_i ($i = 1, \ldots, 6$) forms a basis of $so(1,3)$ given in (21).*

5. Joint Invariants and Invariant Solutions Corresponding to Three-Dimensional Optimal Systems of $\mathcal{L}$

The invariant solutions can be obtained through symmetry reductions carried out by implementing the well-known procedure of utilizing the joint invariants of the subalgebras obtained by three-dimensional optimal system, see, e.g., [6,11,20] for details.

Remark 1 ([10]). Let $\langle X_1, \ldots, X_n \rangle$ be a Lie algebra with basis

$$X_i = \xi_i^1 \frac{\partial}{\partial t} + \xi_i^2 \frac{\partial}{\partial r} + \xi_i^3 \frac{\partial}{\partial \theta} + \xi_i^4 \frac{\partial}{\partial \varphi} + \eta_i \frac{\partial}{\partial u}, i = 1, \ldots, n.$$

A necessary condition for the existence of an invariant solution under the Lie algebra $\langle X_1, \ldots, X_n \rangle$ is the following **transversality condition**:

$$\text{rank}\{E_1\} = \text{rank}\{E_2\}, \tag{24}$$

where

$$E_1 = \begin{pmatrix} \xi_1^1 & \xi_1^2 & \xi_1^3 & \xi_1^4 \\ \vdots & \vdots & \vdots & \vdots \\ \xi_n^1 & \xi_n^2 & \xi_n^3 & \xi_n^4 \end{pmatrix}, \quad E_2 = \begin{pmatrix} \xi_1^1 & \xi_1^2 & \xi_1^3 & \xi_1^4 & \eta_1 \\ \vdots & \vdots & \vdots & \vdots & \vdots \\ \xi_n^1 & \xi_n^2 & \xi_n^3 & \xi_n^4 & \eta_n \end{pmatrix}.$$

Before giving the formal definition of equivalent invariant solutions, let us note the following general fact:

Whenever a transformation group G operates on a set S and U is a subset of S and H is the stabilizer of U, then the stabilizer of $a.U$, $a \in G$ is aHa^{-1}. We will apply this where the set S is the set of solutions of a differential equation, U is the set of invariant solutions and the group G is the local group whose Lie algebra is the symmetry algebra of the differential equation.

Definition 4. *Consider the differential equation admitting the group of transformations G. Let $\mathcal{L}$ be the Lie algebra corresponding the group G. If $u = \Theta_1(x)$ and $u = \Theta_2(x)$ are two invariant solutions of the given differential equation under the subalgebras H_1 and H_2 of $\mathcal{L}$, respectively, then we call $u = \Theta_1(x)$ and $u = \Theta_2(x)$ **equivalent invariant solutions** with respect to the group G if one can find some transformation in G that transforms $u = \Theta_1(x)$ to $u = \Theta_2(x)$.*

Let H_1 be conjugate to H_2 with respect to the group of transformations G. Define U to be the set of invariant surfaces under H_1. Then, H_1 belongs to the stabilizer of U and H_2 belongs to the stabilizer of $a.U$ for some $a \in G$. The set of invariant surfaces under H_2 should be of the form $a.U$.

Therefore, the problem of classifying the invariant solutions is reduced to classifying the corresponding conjugacy classes of subalgebras of the symmetry algebra $\mathcal{L}$ [6].

In this section, we compute the invariant solutions corresponding to three-dimensional subalgebras of $\mathcal{L}$.

5.1. Invariant Solutions of the Wave Equation on Einstein Spacetime

5.1.1. Solvable Subalgebras of of $\mathcal{L}$

Example 1. *In the case of* $\mathcal{L}_1 = \langle V_1 + Z_1, V_4 + Z_2, Z_3 \rangle$, $Z_3 = V_7 + \alpha_3 V_8$. *The generators of* $\mathcal{L}_1$ *in Cartesian coordinates are as follows:*

$$V_1 + Z_1 = \alpha_1 \tfrac{\partial}{\partial t} - cz \tfrac{\partial}{\partial x} + \sqrt{1 - c^2(x^2 + y^2 + z^2)} \tfrac{\partial}{\partial y} + cx \tfrac{\partial}{\partial z} + \beta_1 u \tfrac{\partial}{\partial u},$$
$$V_4 + Z_2 = \alpha_2 \tfrac{\partial}{\partial t} + cz \tfrac{\partial}{\partial x} + \sqrt{1 - c^2(x^2 + y^2 + z^2)} \tfrac{\partial}{\partial y} - cx \tfrac{\partial}{\partial z} + \beta_2 u \tfrac{\partial}{\partial u}, \qquad (25)$$
$$Z_3 = \tfrac{\partial}{\partial t} + \alpha_3 u \tfrac{\partial}{\partial u}.$$

The transversality condition (24) of (25) with rank three is always satisfied. Since the Lie algebra $\mathcal{L}_1$ is abelian, one can find the invariant functions, we call them also invariants, of $\mathcal{L}_1$ in any order. The invariants of Z_3 are:

$$m_1 = x, m_2 = y, m_3 = z, m_4 = u e^{-\alpha_3 t}. \qquad (26)$$

The remaining operators can be given in terms of the variables $m_i, i = 1, \ldots, 4$ as

$$V_1 + Z_1 = -cm_3 \tfrac{\partial}{\partial m_1} + \sqrt{1 - c^2(m_1^2 + m_2^2 + m_3^2)} \tfrac{\partial}{\partial m_2} + cm_1 \tfrac{\partial}{\partial m_3} - m_4 (\alpha_3 \alpha_1 - \beta_1) \tfrac{\partial}{\partial m_4},$$
$$V_4 + Z_2 = cm_3 \tfrac{\partial}{\partial m_1} + \sqrt{1 - c^2(m_1^2 + m_2^2 + m_3^2)} \tfrac{\partial}{\partial m_2} - cm_1 \tfrac{\partial}{\partial m_3} - m_4 (\alpha_3 \alpha_2 - \beta_2) \tfrac{\partial}{\partial m_4}. \qquad (27)$$

Next, the invariants of $V_1 + Z_1$ are

$$n_1 = m_1^2 + m_3^2, \; n_2 = \arctan\left(\tfrac{c\, m_2}{\sqrt{1 - c^2(m_1^2 + m_2^2 + m_3^2)}} \right) - \arctan\left(\tfrac{m_3}{m_1} \right),$$
$$n_3 = m_4\, e^{\tfrac{(\alpha_3 \alpha_1 - \beta_1)}{c} \arctan\left(\tfrac{m_3}{m_1} \right)}. \qquad (28)$$

In terms of the variables $n_i, i = 1, \ldots, 3$, the remaining operator is

$$V_4 + Z_2 = -2c \tfrac{\partial}{\partial n_2} - n_3 \left((\alpha_1 + \alpha_2) \alpha_3 - \beta_2 - \beta_1 \right) \tfrac{\partial}{\partial n_3}. \qquad (29)$$

Finally, the invariants of $V_4 + Z_2$ are

$$n_1, \; n_3\, e^{\tfrac{(\beta_1 + \beta_2 - \alpha_3 \alpha_1 - \alpha_3 \alpha_2)}{2c} n_2}. \qquad (30)$$

Writing the invariants (30) in terms of the original variables gives the joint invariants of $\mathcal{L}_1$ as

$$x^2 + z^2, \; u\, e^{\left(A_1 \arctan\left(\tfrac{c\, y}{\sqrt{1 - c^2(x^2 + y^2 + z^2)}} \right) + A_2 \arctan\left(\tfrac{z}{x} \right) - \alpha_3 t \right)}, \qquad (31)$$

where $A_1 = \tfrac{(\alpha_1 + \alpha_2)\alpha_3 - \beta_1 - \beta_2}{2c}$, $A_2 = \tfrac{(3\alpha_1 + \alpha_2)\alpha_3 - 3\beta_1 - \beta_2}{2c}$.

Note that $A_1 = A_2 = 0$ when $Z_1 = Z_2 = 0$. Therefore, for simplicity, let us discuss the invariant solution for this case.

The invariant transformations in this case are:

$$w = x^2 + z^2, Z(w) = u\, e^{-\alpha_3 t}. \qquad (32)$$

Thus, using (32), Equation (8) can be reduced to the ODE:

$$4w\left(c^2 w - 1\right) Z'' + 4(2c^2 w - 1) Z' + \alpha_3^2 Z = 0. \qquad (33)$$

It was found that the transformation

$$w = \frac{r}{c^2}, \quad Z(w) = R(r) \tag{34}$$

reduces Equation (33) to the hypergeometric differential equation

$$r(r-1)R'' + \left((\nu + \mu + 1)r - \gamma\right)R' + \nu\mu R = 0 \tag{35}$$

with $\nu = \frac{c - \sqrt{c^2 - \alpha_3^2}}{2c}$, $\mu = \frac{c + \sqrt{c^2 - \alpha_3^2}}{2c}$, $\gamma = 1$. The solution of (35) is given in terms of the hypergeometric function $F(\nu, \mu; \gamma; r)$ as

$$R(r) = c_1 F(\mu, \nu; \nu + \mu; 1 - r) + c_2 (r-1)^{1-\nu-\mu} F(1-\nu, 1-\mu; 2-\nu-\mu; 1-r). \tag{36}$$

Therefore, the solution of (33) is

$$Z(w) = R(c^2 w). \tag{37}$$

Thus, the invariant solution of (8) is

$$u(t, x, z) = e^{\alpha_3 t}\left(c_1 F(\mu, \nu; \nu + \mu; 1 - c^2(x^2 + z^2)) + c_2(c^2(x^2 + z^2) - 1)^{1-\nu-\mu} F(1-\nu, 1-\mu; 2-\nu-\mu; 1 - c^2(x^2 + z^2))\right). \tag{38}$$

Another interesting special case when $\alpha_3 = c$, the solution of Equation (35) becomes

$$R(r) = c_1 EllipticK(\sqrt{r}) + c_2 EllipticCK(\sqrt{r}), \tag{39}$$

where *EllipticK* and *EllipticCK* are respectively the complete and the complementary Elliptic integrals of the first kind.

Thus, the invariant solution of (8) is

$$u(t, x, z) = e^{ct}\left(c_1 EllipticK\left(\sqrt{c^2(x^2 + z^2)}\right) + c_2 EllipticCK\left(\sqrt{c^2(x^2 + z^2)}\right)\right). \tag{40}$$

5.1.2. Non-Solvable Subalgebras of of $\mathcal{L}$

As is well known, all three-dimensional non-solvable subalgebras are simple. As they have no non-trivial ideal, we use the method of reduced row echelon form of operators in any convenient basis. As shown in [21], the operators of the three-dimensional non-solvable subalgebra in the reduced row echelon form always form an abelian algebra. Clearly, the joint invariants of the three-dimensional non-solvable subalgebra are the same as those of this abelian algebra. Using this, we find that the joint invariants for $\mathcal{L}$ as follows:

Example 2. *In the case of $\mathcal{L}_1 = \langle V_1, V_2, V_3 \rangle$, by writing $\mathcal{L}_1$ in the reduced row echelon form, the fundamental set of the invariants can be obtained by solving the following system:*

$$\begin{pmatrix} 0 & 1 & 0 & 0 & 0 \\ 0 & 0 & 1 & 0 & 0 \\ 0 & 0 & 0 & 1 & 0 \end{pmatrix} \begin{pmatrix} I_t \\ I_x \\ I_y \\ I_z \\ I_u \end{pmatrix} = \begin{pmatrix} 0 \\ 0 \\ 0 \\ 0 \\ 0 \end{pmatrix}. \tag{41}$$

Clearly, the joint invariants are t, u. Therefore, the invariant transformations are:

$$w = t, Z(w) = u. \tag{42}$$

Thus, using (125), Equation (8) can be reduced to the ODE:

$$Z'' = 0, \tag{43}$$

which has the solution

$$Z(w) = c_1 + c_2\, w. \tag{44}$$

Thus, the invariant solution of (8) is

$$u(t) = c_1 + c_2\, t. \tag{45}$$

Example 3. *In the case of $\mathcal{L}_2 = \langle V_4, V_5, V_6 \rangle$, since the reduced row echelon form of the operators of $\mathcal{L}_2$ coincides with that in (124), it follows that they have the same solution.*

5.2. Invariant Solutions of the Wave Equation on Anti-Einstein Spacetime

5.2.1. Solvable Subalgebras of $\mathcal{L}$

Example 4. *Case $\mathcal{L}_1 = \langle V_3 + \alpha V_4 + Z_1, V_5, V_6 \rangle$, $\alpha \neq 0$. The generators of $\mathcal{L}_1$ in Cartesian coordinates are as follows:*

$$\begin{aligned}
V_3 + \alpha V_4 + Z_1 &= \alpha_1 \tfrac{\partial}{\partial t} - y \tfrac{\partial}{\partial x} + x \tfrac{\partial}{\partial y} + \alpha\, \sqrt{1 + c^2(x^2 + y^2 + z^2)}\, \tfrac{\partial}{\partial z}, \beta_1 u \tfrac{\partial}{\partial u}, \\
V_5 &= (\sqrt{1 + c^2(x^2 + y^2 + z^2)} + cz) \tfrac{\partial}{\partial y} - cy \tfrac{\partial}{\partial z}, \\
V_6 &= (\sqrt{1 + c^2(x^2 + y^2 + z^2)} + cz) \tfrac{\partial}{\partial x} - cx \tfrac{\partial}{\partial z}.
\end{aligned} \tag{46}$$

The transversality condition (24) of (46) with rank three is always satisfied. Since the derived Lie algebra generated by $\mathcal{L}_1$ is $\langle V_5, V_6 \rangle$ which is abelian, one can find the invariants of $\mathcal{L}_1$ by starting with V_5 or V_6. The invariants of V_5 are:

$$m_1 = t,\ m_2 = x,\ m_3 = \frac{-cz + \sqrt{1 + c^2(x^2 + y^2 + z^2)}}{1 + c^2(x^2 + y^2)},\ m_4 = u. \tag{47}$$

The operator V_6 can be given in terms of the variables $m_i, i = 1, \ldots, 4$ as

$$V_6 = \tfrac{1}{m_3} \tfrac{\partial}{\partial m_2}. \tag{48}$$

Next, the invariants of V_6 are

$$n_1 = m_1,\ n_2 = m_3,\ n_3 = m_4. \tag{49}$$

In terms of the variables $n_i, i = 1, 2, 3$, the remaining operator is

$$V_3 + \alpha V_4 + Z_1 = \alpha_1 \tfrac{\partial}{\partial m_1} - \alpha\, cn_2 \tfrac{\partial}{\partial n_2} + \beta_1 n_3 \tfrac{\partial}{\partial n_3}. \tag{50}$$

We have to study the following two cases:

- Case 1: If $\alpha_1 \neq 0$, the invariants of $V_3 + \alpha V_4 + Z_1$ are

$$n_2\, e^{\frac{c\alpha}{\alpha_1} n_1},\ n_3\, e^{-\frac{\beta_1}{\alpha_1} n_1}. \tag{51}$$

Writing the invariants (51) in terms of the original variables gives the joint invariants of $\mathcal{L}_1$ as

$$\frac{-cz + \sqrt{1+c^2(x^2+y^2+z^2)}}{1+c^2(x^2+y^2)} e^{\frac{ac}{\alpha_1}t}, \quad u e^{-\frac{\beta_1}{\alpha_1}t}.$$

Therefore, the invariant transformations are:

$$w = \frac{-cz + \sqrt{1+c^2(x^2+y^2+z^2)}}{1+c^2(x^2+y^2)} e^{\frac{ac}{\alpha_1}t}, \quad Z(w) = u\, e^{-\frac{\beta_1}{\alpha_1}t}. \tag{52}$$

Thus, using (52), Equation (12) can be reduced to the ODE:

$$c\left(\left(\alpha^2 + \alpha_1^2\right)c + 2\alpha\beta_1\right) w\, Z' + c^2\left(\alpha^2 - \alpha_1^2\right) w^2\, Z'' + \beta_1^2\, Z = 0, \tag{53}$$

which has the non-trivial solution for the following cases:

1. $\alpha^2 - \alpha_1^2 \neq 0$:

$$Z(w) = c_1\, w^{\frac{-c\alpha_1^2 - \alpha\beta_1 + \alpha_1\sqrt{c^2\alpha_1^2 + 2c\alpha\beta_1 + \beta_1^2}}{(\alpha^2 - \alpha_1^2)c}} + c_2\, w^{\frac{-c\alpha_1^2 - \alpha\beta_1 - \alpha_1\sqrt{c^2\alpha_1^2 + 2c\alpha\beta_1 + \beta_1^2}}{(\alpha^2 - \alpha_1^2)c}}. \tag{54}$$

2. $\alpha^2 - \alpha_1^2 = 0$, $c\alpha_1 + \beta_1 \neq 0$:

$$Z(w) = c_1\, w^{-\frac{\beta_1^2}{2c\alpha_1(\alpha_1 c + \beta_1)}}. \tag{55}$$

Thus, the invariant solution of (12) is

$$u(t,x,y,z) = Z\left(\frac{-cz + \sqrt{1+c^2(x^2+y^2+z^2)}}{1+c^2(x^2+y^2)} e^{\frac{ac}{\alpha_1}t}\right) e^{\frac{\beta_1}{\alpha_1}t} \tag{56}$$

- Case 2: If $\alpha_1 = 0$, the invariants of $V_3 + \alpha V_4 + Z_1$ are

$$n_1,\ n_3 n_2^{\frac{\beta_1}{ac}}. \tag{57}$$

Writing the invariants (57) in terms of the original variables gives the joint invariants of $\mathcal{L}_1$ as

$$t,\ u\left(\frac{-cz + \sqrt{1+c^2(x^2+y^2+z^2)}}{1+c^2(x^2+y^2)}\right)^{\frac{\beta_1}{ac}}.$$

Therefore, the invariant transformations are:

$$w = t,\quad Z(w) = u\left(\frac{-cz + \sqrt{1+c^2(x^2+y^2+z^2)}}{1+c^2(x^2+y^2)}\right)^{\frac{\beta_1}{ac}}. \tag{58}$$

Thus, using (58), Equation (12) can be reduced to the ODE:

$$\alpha^2 Z'' - \beta_1\,(\beta_1 + 2\alpha\, c)\, Z = 0, \tag{59}$$

which has the solution

$$Z(w) = c_1\, e^{\frac{\sqrt{\beta_1^2 + 2\beta_1 \alpha c}}{\alpha} w} + c_2\, e^{-\frac{\sqrt{\beta_1^2 + 2\beta_1 \alpha c}}{\alpha} w}. \tag{60}$$

Thus, the invariant solution of (12) is

$$u(t,x,y,z) = Z(t)\left(\frac{-cz + \sqrt{1+c^2(x^2+y^2+z^2)}}{1+c^2(x^2+y^2)}\right)^{-\frac{\beta_1}{ac}}. \tag{61}$$

Example 5. *Case $\mathcal{L}_2 = \langle V_3 + Z_1, V_5, V_6 \rangle$. The generators of $\mathcal{L}_2$ in Cartesian coordinates are as follows:*

$$\begin{aligned} V_3 + Z_1 &= \alpha_1 \frac{\partial}{\partial t} - y\frac{\partial}{\partial x} + x\frac{\partial}{\partial y} + \beta_1 u \frac{\partial}{\partial u}, \\ V_5 &= (\sqrt{1+c^2(x^2+y^2+z^2)}+cz)\frac{\partial}{\partial y} - cy\frac{\partial}{\partial z}, \\ V_6 &= (\sqrt{1+c^2(x^2+y^2+z^2)}+cz)\frac{\partial}{\partial x} - cx\frac{\partial}{\partial z}. \end{aligned} \tag{62}$$

The transversality condition (24) of (62) with rank three is satisfied for $\alpha_1 \neq 0$. Since the derived Lie algebra generated by $\mathcal{L}_2$ is $\langle V_5, V_6 \rangle$ which is abelian, one can find the invariants of $\mathcal{L}_2$ by starting with the invariants of $\langle V_5, V_6 \rangle$ which are given by (49) as

$$n_1 = t, n_2 = \frac{-cz + \sqrt{1+c^2(x^2+y^2+z^2)}}{1+c^2(x^2+y^2)}, n_3 = u. \tag{63}$$

In terms of the variables $n_i, i = 1, 2, 3$, the remaining operator is

$$V_3 + Z_1 = \alpha_1 \frac{\partial}{\partial n_1} + \beta_1 n_3 \frac{\partial}{\partial n_3}. \tag{64}$$

Finally, the invariants of $V_3 + Z_1$ are

$$n_2, n_3 \, e^{-\frac{\beta_1}{\alpha_1} n_1}. \tag{65}$$

Writing the invariants (65) in terms of the original variables gives the joint invariants of $\mathcal{L}_2$ as

$$\frac{-cz + \sqrt{1+c^2(x^2+y^2+z^2)}}{1+c^2(x^2+y^2)}, ue^{-\frac{\beta_1}{\alpha_1} t}.$$

Therefore, the invariant transformations are:

$$w = \frac{-cz + \sqrt{1+c^2(x^2+y^2+z^2)}}{1+c^2(x^2+y^2)}, \; Z(w) = ue^{-\frac{\beta_1}{\alpha_1} t}. \tag{66}$$

Thus, using (66), Equation (12) can be reduced to the Cauchy–Euler ODE:

$$w^2 Z'' - wZ' - \left(\frac{\beta_1}{c\alpha_1}\right)^2 Z = 0, \tag{67}$$

which has the solution

$$Z(w) = c_1 w^{\frac{c\alpha_1 + \sqrt{c^2\alpha_1^2 + \beta_1^2}}{c\alpha_1}} + c_2 w^{\frac{c\alpha_1 - \sqrt{c^2\alpha_1^2 + \beta_1^2}}{c\alpha_1}}. \tag{68}$$

Thus, the invariant solution of (12) is

$$u(t,x,y,z) = Z\left(\frac{-cz + \sqrt{1+c^2(x^2+y^2+z^2)}}{1+c^2(x^2+y^2)}\right) e^{\frac{\beta_1}{\alpha_1} t}. \tag{69}$$

Example 6. Case $\mathcal{L}_3 = \langle V_4 + Z_1, V_5, V_6 \rangle$. The generators of $\mathcal{L}_3$ in Cartesian coordinates are as follows:

$$\begin{aligned} V_4 + Z_1 &= \alpha_1 \tfrac{\partial}{\partial t} + \sqrt{1+c^2(x^2+y^2+z^2)} \tfrac{\partial}{\partial z} + \beta_1 u \tfrac{\partial}{\partial u}, \\ V_5 &= (\sqrt{1+c^2(x^2+y^2+z^2)} + cz) \tfrac{\partial}{\partial y} - cy \tfrac{\partial}{\partial z}, \\ V_6 &= (\sqrt{1+c^2(x^2+y^2+z^2)} + cz) \tfrac{\partial}{\partial x} - cx \tfrac{\partial}{\partial z}. \end{aligned} \quad (70)$$

The transversality condition (24) of (70) with rank three is always satisfied. Since the derived Lie algebra generated by $\mathcal{L}_3$ is $\langle V_5, V_6 \rangle$ which is abelian, one can find the invariants of $\mathcal{L}_3$ by starting with the invariants of $\langle V_5, V_6 \rangle$ which are given by (49) as

$$n_1 = t, n_2 = \frac{-cz + \sqrt{1+c^2(x^2+y^2+z^2)}}{1+c^2(x^2+y^2)}, n_3 = u. \quad (71)$$

In terms of the variables $n_i, i = 1, 2, 3$, the remaining operator is

$$V_4 + Z_1 = \alpha_1 \tfrac{\partial}{\partial n_1} - c\, n_2 \tfrac{\partial}{\partial n_2} + \beta_1 n_3 \tfrac{\partial}{\partial n_3}. \quad (72)$$

We have to consider the following two cases:

- **Case 1**: If $\alpha_1 \neq 0$, the invariants of $V_4 + Z_1$ are

$$n_2 e^{\frac{c}{\alpha_1} n_1}, n_3 e^{-\frac{\beta_1}{\alpha_1} n_1}. \quad (73)$$

Writing the invariants (73) in terms of the original variables gives the joint invariants of $\mathcal{L}_3$ as

$$\frac{-cz + \sqrt{1+c^2(x^2+y^2+z^2)}}{1+c^2(x^2+y^2)} e^{\frac{ct}{\alpha_1}}, u e^{-\frac{\beta_1}{\alpha_1} t}. \quad (74)$$

Therefore, the invariant transformations are:

$$w = \frac{-cz + \sqrt{1+c^2(x^2+y^2+z^2)}}{1+c^2(x^2+y^2)} e^{\frac{ct}{\alpha_1}}, \quad Z(w) = u\, e^{-\frac{\beta_1}{\alpha_1} t}. \quad (75)$$

Thus, using (75), Equation (12) can be reduced to the ODE:

$$c^2 \left(\alpha_1^2 - 1\right) w^2 Z'' - \left(\left(1+\alpha_1^2\right)c + 2\beta_1\right) cw Z' - \beta_1^2 Z = 0, \quad (76)$$

which has the non-trivial solution for the following cases:

1. $\alpha_1^2 - 1 \neq 0$:

$$Z(w) = c_1 w^{\frac{c\alpha_1^2 + \beta_1 + \alpha_1 \sqrt{c^2 \alpha_1^2 + 2c\beta_1 + \beta_1^2}}{c(\alpha_1^2 - 1)}} + c_2 w^{\frac{c\alpha_1^2 + \beta_1 - \alpha_1 \sqrt{c^2 \alpha_1^2 + 2c\beta_1 + \beta_1^2}}{c(\alpha_1^2 - 1)}}, \quad (77)$$

2. $\alpha_1^2 - 1 = 0, c + \beta_1 \neq 0$:

$$Z(w) = c_1 w^{-\frac{\beta_1^2}{2c(c+\beta_1)}}. \quad (78)$$

Thus, the invariant solution of (12) is

$$u(t,x,y,z) = Z\left(\frac{-cz + \sqrt{1+c^2(x^2+y^2+z^2)}}{1+c^2(x^2+y^2)} e^{\frac{ct}{\alpha_1}}\right) e^{\frac{\beta_1}{\alpha_1}t}. \tag{79}$$

- **Case 2:** If $\alpha_1 = 0$, the invariants of $V_4 + Z_1$ are

$$n_1, n_3 \, n_2^{\frac{\beta_1}{c}}. \tag{80}$$

Writing the invariants (80) in terms of the original variables gives the joint invariants of $\mathcal{L}_3$ as

$$t, u \left(\frac{-cz + \sqrt{1+c^2(x^2+y^2+z^2)}}{1+c^2(x^2+y^2)}\right)^{\frac{\beta_1}{c}}. \tag{81}$$

Therefore, the invariant transformations are:

$$w = t, \ Z(w) = u\left(\frac{-cz + \sqrt{1+c^2(x^2+y^2+z^2)}}{1+c^2(x^2+y^2)}\right)^{\frac{\beta_1}{c}}. \tag{82}$$

Thus, using (82), Equation (12) can be reduced to the Cauchy–Euler ODE:

$$Z'' - \left(2\beta_1 c + \beta_1^2\right) Z = 0, \tag{83}$$

which has the solution

$$Z(w) = c_1 e^{\sqrt{\beta_1^2 + 2c\beta_1} w} + c_2 e^{-\sqrt{\beta_1^2 + 2c\beta_1} w}. \tag{84}$$

Thus, the invariant solution of (12) is

$$u(t,x,y,z) = Z(t)\left(\frac{-cz + \sqrt{1+c^2(x^2+y^2+z^2)}}{1+c^2(x^2+y^2)}\right)^{-\frac{\beta_1}{c}}. \tag{85}$$

Example 7. *Case* $\mathcal{L}_4 = \langle V_3 + Z_1, V_4 + Z_2, Z_3 \rangle$, $Z_3 = V_7 + \alpha_3 V_8$. *The generators of* $\mathcal{L}_4$ *in Cartesian coordinates are as follows:*

$$\begin{aligned}V_3 + Z_1 &= \alpha_1 \tfrac{\partial}{\partial t} - y\tfrac{\partial}{\partial x} + x\tfrac{\partial}{\partial y} + \beta_1 \, u\tfrac{\partial}{\partial u}, \\ V_4 + Z_2 &= \alpha_2 \tfrac{\partial}{\partial t} + \sqrt{1+c^2(x^2+y^2+z^2)}\tfrac{\partial}{\partial z} + \beta_2 \, u\tfrac{\partial}{\partial u}, \\ Z_3 &= \tfrac{\partial}{\partial t} + \alpha_3 \, u\tfrac{\partial}{\partial u}.\end{aligned} \tag{86}$$

The transversality condition (24) of (86) with rank three is always satisfied. Since the Lie algebra generated by $\mathcal{L}_4$ is abelian, one can find the invariants of $\mathcal{L}_4$ in any order. The invariants of Z_3 are:

$$m_1 = x, \, m_2 = y, \, m_3 = z, \, m_4 = u e^{-\alpha_3 t}. \tag{87}$$

The operators can be given in terms of the variables $m_i, i = 1, \ldots, 4$ as

$$\begin{aligned}V_3 + Z_1 &= -m_2 \tfrac{\partial}{\partial m_1} + m_1 \tfrac{\partial}{\partial m_2} + (\beta_1 - \alpha_3 \alpha_1) \, m_4 \tfrac{\partial}{\partial m_4}, \\ V_4 + Z_2 &= \sqrt{1+c^2(m_1^2 + m_2^2 + m_3^2)}\tfrac{\partial}{\partial m_3} + (\beta_2 - \alpha_2 \alpha_3) \, m_4 \tfrac{\partial}{\partial m_4}.\end{aligned} \tag{88}$$

Next, the invariants of $V_3 + Z_1$ are

$$n_1 = m_1^2 + m_2^2, n_2 = m_3, n_3 = m_4 e^{(\beta_1 - \alpha_3 \alpha_1) \arctan\left(\frac{m_1}{m_2}\right)}. \tag{89}$$

In terms of the variables $n_i, i = 1, 2, 3$, the remaining operator is

$$V_4 + Z_2 = \sqrt{1 + c^2(n_1 + n_2^2)}\frac{\partial}{\partial n_1} + (\beta_2 - \alpha_2 \alpha_3) n_3 \frac{\partial}{\partial n_3}. \tag{90}$$

Finally, the invariants of $V_4 + Z_2$ are

$$n_1, n_3 \left(c n_2 + \sqrt{1 + c^2(n_1 + n_2^2)}\right)^{\frac{\alpha_2 \alpha_3 - \beta_2}{c}}. \tag{91}$$

Writing the invariants (91) in terms of the original variables gives the joint invariants of $\mathcal{L}_4$ as

$$x^2 + y^2, u e^{-\alpha_3 t} e^{A_1 \arctan\left(\frac{x}{y}\right)} \left(cz + \sqrt{1 + c^2 (x^2 + y^2 + z^2)}\right)^{A_2}, \tag{92}$$

where $A_1 = (\beta_1 - \alpha_3 \alpha_1)$, $A_2 = \frac{\alpha_2 \alpha_3 - \beta_2}{c}$.

Therefore, the invariant transformations are:

$$w = x^2 + y^2, \; Z(w) = u e^{-\alpha_3 t} e^{A_1 \arctan\left(\frac{x}{y}\right)} \left(cz + \sqrt{1 + c^2 (x^2 + y^2 + z^2)}\right)^{A_2}. \tag{93}$$

Thus, using (93), Equation (12) can be reduced to the ODE:

$$4(w^2 + c^2 w^3) Z'' + 4(w - c^2 (A_2 - 2) w^2) Z' + \left(\left(A_2^2 - 2 A_2\right) c^2 - a^2\right) w + A_1^2 \right) Z = 0, \tag{94}$$

which can be transformed using the transformation $w = -\frac{r}{c^2}$.
$Z(w) = r^{\frac{1}{2} i A_1} R(r)$ to the hypergeometric differential equation

$$r(r-1) R'' + \left((\nu + \mu + 1) r - \gamma\right) R' + \nu \mu R = 0 \tag{95}$$

with $\nu = \frac{(1+iA_1 - A_2)c + \sqrt{c^2 + \alpha_3^2}}{2c}, \mu = \frac{(1+iA_1 - A_2)c - \sqrt{c^2 + \alpha_3^2}}{2c}, \gamma = 1 + iA_1$. The solution of (95) is given in terms of the hypergeometric function $F(\nu, \mu; \gamma; r)$ as

$$R(r) = c_1 F(\nu, \mu; \gamma; r) + c_2 r^{1-\gamma} F(\nu - \gamma + 1, \mu - \gamma + 1; 2 - \gamma; r). \tag{96}$$

Therefore, the solution of (94) is

$$Z(w) = (-c^2 w)^{\frac{1}{2}(\gamma - 1)} R(-c^2 w). \tag{97}$$

Thus, the invariant solution of (12) is

$$u(t, x, y, z) = Z(x^2 + y^2) \, e^{\alpha_3 t} e^{i(\gamma - 1) \arctan\left(\frac{x}{y}\right)} \left(cz + \sqrt{1 + c^2 (x^2 + y^2 + z^2)}\right)^{3\nu - \gamma - \mu}, \tag{98}$$

where $\alpha_3^2 = c^2(\nu - \mu)^2 - c^2$.

Example 8. Case $\mathcal{L}_5 = \langle V_4 + Z_1, V_5, Z_3 \rangle$, $Z_3 = V_7 + \alpha_3 V_8$. The generators of $\mathcal{L}_5$ in Cartesian coordinates are as follows:

$$V_4 + Z_1 = \alpha_1 \tfrac{\partial}{\partial t} + \sqrt{1 + c^2(x^2 + y^2 + z^2)} \tfrac{\partial}{\partial z} + \beta_1 u \tfrac{\partial}{\partial u},$$
$$V_5 = (\sqrt{1 + c^2(x^2 + y^2 + z^2)} + cz) \tfrac{\partial}{\partial y} - cy \tfrac{\partial}{\partial z}, \qquad (99)$$
$$Z_3 = \tfrac{\partial}{\partial t} + \alpha_3 u \tfrac{\partial}{\partial u}.$$

The transversality condition (24) of (99) with rank three is always satisfied. Since the derived Lie algebra generated by $\mathcal{L}_5$ is $\langle V_5 \rangle$, one can find the invariants of $\mathcal{L}_5$ by starting with V_5. The invariants of V_5 are:

$$m_1 = t, \; m_2 = x, \; m_3 = u, \; m_4 = \frac{-cz + \sqrt{1 + c^2(x^2 + y^2 + z^2)}}{1 + c^2(x^2 + y^2)}. \qquad (100)$$

The operators can be given in terms of the variables $m_i, i = 1, \ldots, 4$ as

$$V_4 + Z_1 = \alpha_1 \tfrac{\partial}{\partial m_1} + \beta_1 m_3 \tfrac{\partial}{\partial m_3} - c m_4 \tfrac{\partial}{\partial m_4}, \qquad (101)$$
$$Z_3 = \tfrac{\partial}{\partial t} + \alpha_3 m_3 \tfrac{\partial}{\partial m_3}.$$

Next, the invariants of $V_4 + Z_1$ are

$$n_1 = m_2, \; n_2 = m_3 e^{-\frac{\beta_1}{\alpha_1} m_1}, \; n_3 = m_4 e^{\frac{c}{\alpha_1} m_1}. \qquad (102)$$

In terms of the variables $n_i, i = 1, 2, 3$, the remaining operator is

$$Z_3 = \tfrac{(\alpha_3 \alpha_1 - \beta_1)}{\alpha_1} n_2 \tfrac{\partial}{\partial n_2} + \tfrac{c}{\alpha_1} n_3 \tfrac{\partial}{\partial n_3}. \qquad (103)$$

Finally, the invariants of Z_3 are

$$n_1, \; n_2 \, n_3^{\frac{\beta_1 - \alpha_3 \alpha_1}{c}}. \qquad (104)$$

Writing the invariants (104) in terms of the original variables gives the joint invariants of $\mathcal{L}_5$ as

$$x, \; u \, e^{-\alpha_3 t} \left(\frac{1 + c^2(x^2 + y^2)}{-cz + \sqrt{1 + c^2(x^2 + y^2 + z^2)}} \right)^{A_1}, \qquad (105)$$

where $A_1 = \tfrac{\beta_1 - \alpha_1 \alpha_3}{c}$.
Therefore, the invariant transformations are:

$$w = x, \; Z(w) = u \, e^{-\alpha_3 t} \left(\frac{1 + c^2(x^2 + y^2)}{-cz + \sqrt{1 + c^2(x^2 + y^2 + z^2)}} \right)^{A_1}. \qquad (106)$$

Thus, using (106), Equation (12) can be reduced to the ODE:

$$\left(c^2 w^2 + 1\right) Z'' - c^2 (2A_1 - 3) w Z' + \left(c^2 (A_1^2 - 2A_1) - \alpha_3^2\right) Z = 0. \qquad (107)$$

It was found using Maple software (Maple 13.0, Waterloo Maple Inc., Waterloo, ON, Canada) that the transformation

$$w = \frac{r}{ic}, \; Z(w) = \left(r^2 - 1\right)^{\frac{1}{2} A_1 - \frac{1}{4}} R(r) \qquad (108)$$

reduces Equation (107) to the associated Legendre equation

$$(1 - r^2) R'' - 2r R' + \left(\nu(\nu + 1) - \frac{\mu^2}{1 - r^2}\right) R = 0 \qquad (109)$$

with $\nu = \frac{2\sqrt{c^2+\alpha_3^2}-c}{2c}$, $\mu = A_1 - \frac{1}{2}$. Therefore, the solution of (107) is

$$Z(w) = \left(-c^2 w^2 - 1\right)^{\frac{\mu}{2}} \left(c_1 P_\nu^\mu(icw) + c_2 Q_\nu^\mu(icw)\right), \tag{110}$$

where P_ν^μ and Q_ν^μ are the associated Legendre functions of the first and second kinds respectively. Thus, the invariant solution of (12) is

$$u(t, x, y, z) = Z(x) e^{\alpha_3 t} \left(\frac{-cz + \sqrt{1 + c^2(x^2 + y^2 + z^2)}}{1 + c^2(x^2 + y^2)}\right)^{\mu + \frac{1}{2}}, \tag{111}$$

where $\alpha_3^2 = c^2(\nu + \frac{1}{2})^2 - c^2$.

Example 9. *Case* $\mathcal{L}_6 = \langle V_5 + Z_1, V_6 + Z_2, Z_3 \rangle$, $Z_3 = V_7 + \alpha_3 V_8$. *The generators of* $\mathcal{L}_6$ *in Cartesian coordinates are as follows:*

$$\begin{aligned} V_5 + Z_1 &= \alpha_1 \frac{\partial}{\partial t} + (\sqrt{1 + c^2(x^2 + y^2 + z^2)} + cz)\frac{\partial}{\partial y} - cy\frac{\partial}{\partial z} + \beta_1 u \frac{\partial}{\partial u}, \\ V_6 + Z_2 &= \alpha_2 \frac{\partial}{\partial t} + (\sqrt{1 + c^2(x^2 + y^2 + z^2)} + cz)\frac{\partial}{\partial x} - cx\frac{\partial}{\partial z} + \beta_2 u \frac{\partial}{\partial u}, \\ Z_3 &= \frac{\partial}{\partial t} + \alpha_3 u \frac{\partial}{\partial u}. \end{aligned} \tag{112}$$

The transversality condition (24) of (112) with rank three is always satisfied. Since the Lie algebra generated by $\mathcal{L}_6$ is abelian, one can find the invariants of $\mathcal{L}_6$ in any order. The invariants of Z_3 are:

$$m_1 = x, m_2 = y, m_3 = z, m_4 = u e^{-\alpha_3 t}. \tag{113}$$

The operators can be given in terms of the variables $m_i, i = 1, \ldots, 4$ as

$$\begin{aligned} V_5 + Z_1 &= (\sqrt{1 + c^2(m_1^2 + m_2^2 + m_3^2)} + cm_3)\frac{\partial}{\partial m_2} - cm_2 \frac{\partial}{\partial m_3} + (\beta_1 - \alpha_3 \alpha_1) m_4 \frac{\partial}{\partial m_4}, \\ V_6 + Z_2 &= (\sqrt{1 + c^2(m_1^2 + m_2^2 + m_3^2)} + c\, m_3)\frac{\partial}{\partial m_1} - c\, m_1 \frac{\partial}{\partial m_3} + (\beta_2 - \alpha_2 \alpha_3) m_4 \frac{\partial}{\partial m_4}. \end{aligned} \tag{114}$$

Next, the invariants of $V_5 + Z_1$ are

$$\begin{aligned} n_1 &= m_1, \\ n_2 &= \frac{-cm_3 + \sqrt{1 + c^2(m_1^2 + m_2^2 + m_3^2)}}{1 + c^2(m_1^2 + m_2^2)}, \\ n_3 &= m_4 e^{\frac{(\alpha_3 \alpha_1 - \beta_1) m_2 \left(-cm_3 + \sqrt{1 + c^2(m_1^2 + m_2^2 + m_3^2)}\right)}{1 + c^2(m_1^2 + m_2^2)}}. \end{aligned} \tag{115}$$

In terms of the variables $n_i, i = 1, 2, 3$, the remaining operator is

$$V_6 + Z_2 = \frac{1}{n_2}\frac{\partial}{\partial n_1} + (\beta_2 - \alpha_2 \alpha_3) n_3 \frac{\partial}{\partial n_3}. \tag{116}$$

Finally, the invariants of $V_6 + Z_2$ are

$$n_2, n_3 \, e^{(\alpha_2 \alpha_3 - \beta_2) n_1 n_2}. \tag{117}$$

Writing the invariants (117) in terms of the original variables gives the joint invariants of $\mathcal{L}_6$ as

$$\frac{-cz + \sqrt{1 + c^2(x^2 + y^2 + z^2)}}{1 + c^2(x^2 + y^2)}, \, u e^{-\alpha t} e^{\left(A_1 x + A_2 y\right)\left(\frac{-cz + \sqrt{1 + c^2(x^2+y^2+z^2)}}{1+c^2(x^2+y^2)}\right)}, \tag{118}$$

where $A_1 = \alpha_2 \alpha_3 - \beta_2$, $A_2 = \alpha_1 \alpha_3 - \beta_1$.

Therefore, the invariant transformations are:

$$w = \frac{-cz + \sqrt{1 + c^2(x^2 + y^2 + z^2)}}{1 + c^2(x^2 + y^2)}, \quad Z(w) = ue^{-\alpha t}e^{\left(A_1 x + A_2 y\right)\left(\frac{-cz + \sqrt{1+c^2(x^2+y^2+z^2)}}{1+c^2(x^2+y^2)}\right)}. \tag{119}$$

Thus, using (119), Equation (12) can be reduced to the ODE:

$$c^2 w^2 Z'' - c^2 w Z' + \left((A_1^2 + A_2^2)w^2 - \alpha_3^2\right)Z = 0, \tag{120}$$

which can be transformed using the transformation $w = r$, $Z(w) = rR(r)$ to the parametric Bessel equation:

$$r^2 R'' + rR' + \left(\alpha^2 r^2 - v^2\right)R = 0 \tag{121}$$

with $\alpha = \frac{\sqrt{A_1^2 + A_2^2}}{c}$, $v = \frac{\sqrt{c^2 + \alpha^2}}{c}$.
Therefore, the solution of (120) is

$$Z(w) = c_1 w J_v(\alpha w) + c_2 w Y_v(\alpha w), \tag{122}$$

where J_v and Y_v are the Bessel functions of the first and second kind, respectively. Thus, the invariant solution of (12) is

$$u(t, x, y, z) = Z\left(\frac{-cz + \sqrt{1 + c^2(x^2 + y^2 + z^2)}}{1 + c^2(x^2 + y^2)}\right) e^{\alpha_3 t} e^{\left(A_1 x + A_2 y\right)\left(\frac{cz - \sqrt{1+c^2(x^2+y^2+z^2)}}{1+c^2(x^2+y^2)}\right)}. \tag{123}$$

5.2.2. Non-Solvable Subalgebras of $\mathcal{L}$

Example 10. *Case* $\mathcal{L}_1 = \langle V_4, V_5, V_5 - 2cV_2 \rangle$. *By writing* $\mathcal{L}_1$ *in the reduced row echelon form, the fundamental set of the invariants can be obtained by solving the following system:*

$$\begin{pmatrix} 0 & 1 & 0 & 0 & 0 \\ 0 & 0 & 1 & 0 & 0 \\ 0 & 0 & 0 & 1 & 0 \end{pmatrix} \begin{pmatrix} I_t \\ I_x \\ I_y \\ I_z \\ I_u \end{pmatrix} = \begin{pmatrix} 0 \\ 0 \\ 0 \\ 0 \\ 0 \end{pmatrix}. \tag{124}$$

Clearly, the joint invariants are t, u. Therefore, the invariant transformations are:

$$w = t, Z(w) = u. \tag{125}$$

Thus, using (125), Equation (12) can be reduced to the ODE:

$$Z'' = 0, \tag{126}$$

which has the solution

$$Z(w) = c_1 + c_2 \, w. \tag{127}$$

Thus, the invariant solution of (12) is

$$u(t) = c_1 + c_2 \, t. \tag{128}$$

6. Concluding Remarks and Future Research

Improved algorithms of the expansion method introduced by Ovsiannikov [6] are introduced to construct the optimal systems of dimension of at most three of non-solvable Lie algebra. The algorithms are then applied to determine the Lie algebra structure and optimal systems of the symmetries of the wave equation on static spherically symmetric spacetimes admitting G_7 as an isometry algebra, while joint invariants and invariant solutions corresponding to three-dimensional optimal systems are also found. The energy density $e(u) = \frac{1}{2}g^{ij}u_iu_j$ and the corresponding energy of the solutions can be investigated for physical significance of the wave functions obtained in the examples.

It would be of interest to complete and extend this study by applying the algorithms in this paper to equations of physical interest on all static and non-static spherically symmetric spacetimes and to find the corresponding invariant solutions.

Author Contributions: Conceptualization, H.A., A.Y.A.-D. and M.T.M.; Formal analysis, A.Y.A.-D. and M.T.M.; Investigation, H.A., K.A., A.Y.A.-D. and M.T.M.; Methodology, K.A. and A.Y.A.-D.; Supervision, A.Y.A.-D. and M.T.M.; Writing—original draft, H.A. and A.Y.A.-D.; Writing—review & editing, K.A., A.Y.A.-D. and M.T.M.

Acknowledgments: The publication of this article was funded by the Qatar National Library. The authors are thankful to Qatar University and King Fahd university of petroleum and minerals for their continuous support and excellent research facilities.

Conflicts of Interest: The authors declare no conflict of interest.

References

1. Azad, H.; Ziad, M. Spherically symmetric manifolds which admit five isometries. *J. Math. Phys.* **1995**, *36*, 1908–1911. [CrossRef]
2. Bokhari, A.H.; Qadir, A. Symmetries of static, spherically symmetric space-times. *J. Math. Phys.* **1987**, *28*, 1019–1022. [CrossRef]
3. Qadir, A.; Ziad, M. The classification of spherically symmetric space-times. *Il Nuovo Cimento B (1971–1996)* **1995**, *110*, 317–334. [CrossRef]
4. Azad, H.; Al-Dweik, A.Y.; Ghanam, R.; Mustafa, M.T. Symmetry analysis of wave equation on static spherically symmetric spacetimes with higher symmetries. *J. Math. Phys.* **2013**, *54*, 063509. [CrossRef]
5. Ibragimov, N.H. *Transformation Groups Applied to Mathematical Physics*; Springer Science & Business Media: New York, NY, USA, 2001; Volume 3.
6. Ovsiannikov, L.V. *Group Analysis of Differential Equations*; Academic Press: New York, NY, USA, 1982.
7. Ovsiannikov, L.V. *Group Properties of Differential Equations*; Siberian Branch, USSR Academy of Sciences: Novosibirsk, Russia, 1962.
8. Ibragimov, N.H. Optimal Systems of Subgroups and Classification of Invariant Solutions of Equations for Planar Non-stationary Gas Flows. Master of Science Thesis in Mathematics, Institute of Hydrodynamics, USSR Academy Science, Novosibirsk State University, Novosibirsk, Rrussia, 1965.
9. Ibragimov, N.H. *Selected Works*; ALGA Publication: Karlskrona, Sweden, 2006; Volume II.
10. Olver, P.J. *Applications of Lie Groups to Differential Equations*; Springer Science and Business Media: New York, NY, USA, 2000; Volume 107.
11. Hydon, P.E. *Symmetry Methods for Differential Equations: A Beginner's Guide*; Cambridge University Press: Cambridge, UK, 2000; Volume 22.
12. Azad, H.; Biswas, I.; Chatterjee, P. On the maximal solvable subgroups of semisimple algebraic groups. *J. Lie Theory Vol.* **2012**, *22*, 1169–1179.
13. Ali, S.; Azad, H.; Biswas, I.; Ghanam, R.; Mustafa, M.T. Embedding algorithms and applications to differential equations. *J. Symb. Comput.* **2018**, *86*, 166–188. [CrossRef]
14. Hilgert, J.; Neeb, K.H. *Structure and Geometry of Lie Groups*; Springer: New York, NY, USA, 2011.
15. De Graaf, W.A. *Lie Algebras: Theory and Algorithms*, North-Holland Mathematical Library; Elsevier Science: Amsterdam, The Netherlands, 2000.
16. Knapp, A.W. *Lie Groups beyond an Introduction*, 2nd ed.; Progress in Mathematics; Birkhäuser Boston, Inc.: Boston, MA, USA, 2002.
17. Azad, H.; Biswas, I. A note on real algebraic groups. *Forum Math.* **2016**, *28*, 539–543. [CrossRef]

18. Yuri, B.; Freire, I.L. Special conformal groups of a Riemannian manifold and Lie point symmetries of the nonlinear Poisson equation. *J. Differ. Equat.* **2010**, *249*, 872–913.
19. Libor, S.; Winternitz, P. *Classification and Identification of Lie Algebras*; American Mathematical Society: Provence, RI, USA, 2014; Volume 33.
20. Bluman, G.; Stephen, C.A. *Symmetry and Integration Methods for Differential Equations*; Springer Science, Business Media: New York, NY, USA, 2002; Volume 154.
21. Azad, H.; Biswas, I.; Ghanam, R.; Mustafa, M.T. On computing joint invariants of vector fields. *J. Geom. Phys.* **2015**, *97*, 69–76. [CrossRef]

© 2018 by the authors. Licensee MDPI, Basel, Switzerland. This article is an open access article distributed under the terms and conditions of the Creative Commons Attribution (CC BY) license (http://creativecommons.org/licenses/by/4.0/).

Article

Positive Energy Condition and Conservation Laws in Kantowski-Sachs Spacetime via Noether Symmetries

Sumaira Saleem Akhtar [1], Tahir Hussain [1] and Ashfaque H. Bokhari [2,*]

1. Department of Mathematics, University of Peshawar, Peshawar 25000, Khyber Pakhtunkhwa, Pakistan; sumairamaths@gmail.com (S.S.A.); tahirhussain@uop.edu.pk (T.H.)
2. Department of Mathematics and Statistics, King Fahd University of Petroleum and Minerals, Dhahran 31261, Saudi Arabia
* Correspondence: abokhari@kfupm.edu.sa

Received: 16 November 2018; Accepted: 30 November 2018; Published: 4 December 2018

Abstract: In this paper, we have investigated Noether symmetries of the Lagrangian of Kantowski–Sachs spacetime. The associated Lagrangian of the Kantowski–Sachs metric is used to derive the set of determining equations. Solving the determining equations for several values of the metric functions, it is observed that the Kantowski–Sachs spacetime admits the Noether algebra of dimensions 5, 6, 7, 8, 9, and 11. A comparison of the obtained Noether symmetries with Killing and homothetic vectors is also presented. With the help of Noether's theorem, we have presented the expressions for conservation laws corresponding to all Noether symmetries. It is observed that the positive energy condition is satisfied for most of the obtained metrics.

Keywords: Noether symmetries; conservation laws; Kantowski–Sachs metric

PACS: 04.20.Jb

1. Introduction

Einstein's field equations (EFEs), $G_{ab} = kT_{ab}$, are ten-tensor equations, which describe the gravitational effects produced by a given mass in a spacetime. In these equations, the stress-energy tensor T_{ab} gives the distribution of energy and momentum, k represents the gravitational constant, while G_{ab} expresses the spacetime curvature and is known as the Einstein tensor. Moreover, the Einstein tensor G_{ab} can be expressed as $G_{ab} = R_{ab} - \frac{R}{2} g_{ab}$, where R_{ab} = the Ricci tensor, g_{ab} = the metric tensor, and R = the Ricci scalar.

The quantities R_{ab} and R appearing in the EFEs are built up from g_{ab} and its partial derivatives. In this way, the EFEs form a system of partial differential equations. A Lorentz metric g_{ab} is regarded as an exact solution of EFEs if it is obtained by solving the EFEs exactly in closed form and is conformable to a physically realistic T_{ab}. Finding the exact solutions of EFEs is not an easy task unless some simplifying assumptions are employed. Therefore, only a few physically meaningful solutions of these equations are found in the literature [1]. Among the approaches followed for obtaining the exact solutions of EFEs, the most popular is to use some symmetry restrictions on the tensor g_{ab}. These restrictions are mathematically expressed as $\mathcal{L}_X g_{ab} = 0$, where X is called a Killing vector (KV) and $\mathcal{L}$ denotes the Lie derivative operator. It is well known that every KV corresponds to a conservation law. A large body of the literature is devoted to the investigation of KVs and corresponding conservation laws in spacetimes [1–3].

The symmetries of tensors other than the metric tensor are usually referred to as collineations. Some examples of collineations include curvature collineations ($\mathcal{L}_X R^a_{bcd} = 0$), Ricci collineations ($\mathcal{L}_X R_{ab} = 0$), and matter collineations ($\mathcal{L}_X T_{ab} = 0$). These collineations have also been thoroughly discussed [4–9].

Apart from the conventional symmetries defined above, the idea of another type of symmetry, known as Noether symmetry (NS), was given by Emmy Noether [10]. According to Noether's theorem, if the Lagrangian of a system admits a continuous symmetry, then this symmetry corresponds to a conservation law. Consequently, if a system remains unchanged under time translation and spacial translations and rotations, this theorem yields the conservation of energy and linear and angular momenta.

The study of NS is important due to the fact that it usually provides the additional conservation laws, not given by KVs. If the Killing and Noether algebras are denoted by $K(M)$ and $N(M)$, respectively, then $K(M) \subseteq N(M)$. The NS have also a close link with homothetic vectors (HVs). In fact, if X is an HV, that is $\mathcal{L}_X g_{ab} = 2\psi g_{ab}$; ψ being a constant, then $X + 2\psi u \partial_u$ is a Noether symmetry associated with X. Here, u is the geodesics parameter of the world line of a point particle moving in a spacetime. Conversely, if $X + 2\psi u \partial_u$ is an NS, then X is an HV, provided that X is independent of u [11]. A Noether symmetry that is neither a KV, nor associated with an HV is known as a proper Noether symmetry.

NS and their comparison with Killing and homothetic vectors have been studied for some well-known spacetimes. A comparative study of KVs and NS for the conformally-flat Friedmann metric was provided in [12], and it was concluded that this metric admits proper NS. Considering some specific spherically-symmetric metrics, Bokhari et al. [13] conjectured that the NS of the Lagrangian provide additional conservation laws. The NS of some other physically-important spacetimes were examined by different authors. Camci [14] studied the NS of geodesic motion for the geodesic Lagrangian of the metric of Gödel spacetimes. Camci and Yildirim [15] worked on the NS of the geodesic Lagrangian for some classes of pp-wave spacetimes. Hickman and Yazdan [11] studied NS in Bianchi type II spacetimes. They have shown that the Noether algebra of Bianchi type II spacetimes contain Killing, as well as homothetic vectors. Ali et al. [16–18] worked on the classification of different spacetimes via NS including static plane, static spherical, and static cylindrically-symmetric spacetimes. Jamil et al. [19,20] worked on the complete classification of non-static plane and non-static spherically-symmetric spacetimes via NS. Paliathanasis et al. [21] established a relation between the Lie symmetries of the Klein–Gordon equation and conformal Killing vectors of the underlying geometry, where they also stated that the resulting Lie symmetries of the conformal algebra are also NS. Tsamparlis et al. [22] stated that for dynamical system whose equations of motion are of the form $\ddot{x}^a + \Gamma^a_{bc} \dot{x}^b \dot{x}^c + f(x^a)$; $f(x^a)$ being an arbitrary function of its argument, the computation of Lie and Noether symmetries reduces to the problem of finding the special projective collineations. Recently, the Bianchi type V and non-static plane symmetric spacetimes were completely classified via their NS [23,24].

It is important to mention here that the formulation of conservation laws may not require the Lagrangian of the system. As an example, one can see the work of Ma et al. [25], who have recently established a result giving the direct formulation of conservation laws for differential equations including the heat equation, Burgers' equation, and the Korteweg–de Vries (KdV) equation, regardless of the existence of a Lagrangian. The same authors also discussed the existence of lump solutions involving free parameters for some nonlinear partial differential equations that present Lie symmetries and that may generate associated conservation laws [26].

The Kantowski–Sachs spacetime describes a spatially-homogeneous and anisotropic universe model that admits an isometry group G_4 acting on homogeneous spacelike hypersurfaces. As the Kantowski–Sachs metric is of great interest and there is a great deal of information about this metric scattered throughout the literature, it would be useful to investigate the NS possessed by the Lagrangian of this metric and their relation with Killing and homothetic vectors.

For a physically-realistic spacetime, its energy density, represented by T_{00}, must be non-negative. This condition is usually referred to as the positive energy condition. The positive energy condition is important because if one allows both positive and negative energy regions, the empty vacuum may become unstable. In this paper, we classify the Lagrangian of the Kantowski–Sachs metric according to its NS and check the positive energy condition for the obtained models admitting different algebras of NS.

2. Determining Equations

The Kantowski–Sachs metric is given by [27]:

$$ds^2 = dt^2 - \lambda^2(t)dr^2 - v^2(t)\left(d\theta^2 + \sin^2\theta d\phi^2\right), \tag{1}$$

where the functions $\lambda(t) \neq 0$ and $v(t) \neq 0$ depend on t only. The minimum KVs admitted by the above metric are:

$$\begin{aligned} X_1 &= \partial_r, \quad X_2 = \partial_\phi, \\ X_3 &= \sin\phi\partial_\theta + \cot\theta\cos\phi\partial_\phi, \\ X_4 &= -\cos\phi\partial_\theta + \cot\theta\sin\phi\partial_\phi. \end{aligned} \tag{2}$$

One may use the EFEs with $k = 1$ to get the following non-zero components of T_{ab}.

$$\begin{aligned} T_{00} &= \frac{1}{\lambda^3 v^2}\left(v'^2\lambda^3 + 2\lambda^2 v\lambda'v' + \lambda^3\right), \\ T_{11} &= -\frac{1}{v^2}\left(2v\lambda^2 v'' + \lambda^2 v'^2 + \lambda^2\right), \\ T_{22} &= -\frac{v}{\lambda^3}\left(\lambda^3 v'' + \lambda^2 v'\lambda' + v\lambda^2\lambda''\right), \\ T_{33} &= \sin^2\theta\, T_{22}, \end{aligned} \tag{3}$$

where the primes on metric functions denote their derivatives with respect to t. The usual Lagrangian L corresponding to the Kantowski–Sachs metric (1) is given by:

$$L = \dot{t}^2 - \lambda^2(t)\dot{r}^2 - v^2(t)\left(\dot{\theta}^2 + \sin^2\theta\dot{\phi}^2\right), \tag{4}$$

where the dot represents the derivative with respect to the parameter u of the world line of a point particle moving in Kantowski–Sachs spacetime. The Lagrangian for this point particle, given in (4), represents the square of the Lagrangian for the action principle that leads to geodesics by minimizing the spacetime metric $\int ds = \sqrt{g_{ab}\frac{dx^a}{du}\frac{dx^b}{du}}du$ [28]. It may be noted that the coefficients of the quadratic terms in $\frac{dx^\mu}{du}\frac{dx^\nu}{du}$ represent the components of the metric.

A vector field $X = \eta\frac{\partial}{\partial u} + X^i\frac{\partial}{\partial x_i}$ is called an NS if the Lagrangian of a physical system remains invariant under the action of X such that the following condition holds:

$$X^{(1)}L + L(D\eta) = DF, \tag{5}$$

where:

$$X^{(1)} = X + X^i_u\frac{\partial}{\partial \dot{x}_i} \tag{6}$$

is the first prolongation of X such that $X^i_u = DX^i - \dot{x}_i D\eta$. Here, D denotes the total differential operator, given by:

$$D = \frac{\partial}{\partial u} + \dot{x}_i\frac{\partial}{\partial x_i}. \tag{7}$$

Moreover, η, X^i, and F all depend on the variables (u, x_i), where $x_i = (t(u), r(u), \theta(u), \phi(u))$. The function F is known as the gauge function. Once we find the NS and corresponding gauge function, we can use Noether's theorem to find the corresponding conservation laws in the following way:

$$I = \eta L + \left(X^i - \dot{x}_i\eta\right)\frac{\partial L}{\partial \dot{x}_i} - F. \tag{8}$$

We use the Lagrangian given in Equation (4) in the equation defining the NS, that is Equation (5), to get the following set of 19 equations.

$$F_{,u} = \eta_{,\phi} = \eta_{,\theta} = \eta_{,r} = \eta_{,t} = 0, \tag{9}$$

$$2X^0_{,t} = \eta_{,u}, \tag{10}$$

$$2\lambda' X^0 + 2\lambda X^1_{,r} = \lambda \eta_{,u}, \tag{11}$$

$$2v' X^0 + 2v X^2_{,\theta} = v \eta_{,u}, \tag{12}$$

$$X^0_{,r} - \lambda^2 X^1_{,t} = 0, \tag{13}$$

$$X^0_{,\theta} - v^2 X^2_{,t} = 0, \tag{14}$$

$$X^0_{,\phi} - v^2 \sin^2\theta X^3_{,t} = 0, \tag{15}$$

$$\lambda^2 X^1_{,\theta} + v^2 X^2_{,r} = 0, \tag{16}$$

$$\lambda^2 X^1_{,\phi} + v^2 \sin^2\theta X^3_{,r} = 0, \tag{17}$$

$$X^2_{,\phi} + \sin^2\theta X^3_{,\theta} = 0, \tag{18}$$

$$2v' X^0 + 2v \cot\theta X^2 + 2v X^3_{,\phi} = v \eta_{,u}, \tag{19}$$

$$2X^0_{,u} = F_{,t}, \tag{20}$$

$$2\lambda^2 X^1_{,u} = -F_{,r}, \tag{21}$$

$$2v^2 X^2_{,u} = -F_{,\theta}, \tag{22}$$

$$2v^2 \sin^2\theta X^3_{,u} = -F_{,\phi}. \tag{23}$$

The solution of the above system of equations would give the values of the components of the vector field generating NS and the metric functions λ and v. Consequently, the corresponding Kantowski–Sachs metrics may represent the exact solutions of EFEs. We omit writing the basic calculations and present the list of metrics, their NS, Lie algebras, and the corresponding conservation laws of all symmetries in the forthcoming sections. The bounds for the positive energy condition and the singularity of the Ricci scalar are also discussed for all the obtained metrics.

3. Five Noether Symmetries

The minimal set of NS admitted by the Lagrangian of the Kantowski–Sachs metric is:

$$\begin{aligned} X_0 &= \partial_u, \ X_1 = \partial_r, \ X_2 = \partial_\phi, \\ X_3 &= \sin\phi \partial_\theta + \cot\theta \cos\phi \partial_\phi, \\ X_4 &= -\cos\phi \partial_\theta + \cot\theta \sin\phi \partial_\phi. \end{aligned} \tag{24}$$

Out of these five NS, four are the basic KVs of the Kantowski–Sachs metric and X_0 is the symmetry corresponding to the Lagrangian. Here, X_0 represents the translation of geodesics parameter u and X_1 is the translation of spatial coordinate r. The generators X_2, X_3 and X_4 form a Lie subalgebra for the rotation group $SO(3)$. The values of the Kantowski–Sachs metric functions for which the corresponding Lagrangian admit the above minimal NS are given in the following Table 1:

Table 1. Exact form of metrics admitting five Noether symmetries (NS).

No.	$\lambda(t)$	$v(t)$
3a	$\lambda' \neq 0$	$v'' \neq 0$ and $v \neq \cosh t$
3b	$\lambda \neq \sinh t$	$v = \cosh t$
3c	$(\lambda \lambda')'' \neq 0$	$v = \text{const.} = \tilde{\zeta}$
3d	$\lambda = \sqrt{at^2 + 2bt + c};\ a \neq 0,$	$v = \text{const.} = \tilde{\zeta}$
3e	$\lambda = \sqrt{2bt + c}$	$v = \text{const.} = \tilde{\zeta}$

The corresponding Lie algebra for generators given in Equation (24) is:

$[X_3, X_2] = X_4$, $[X_2, X_4] = X_3$, $[X_4, X_3] = X_2$, $[X_i, X_j] = 0$, otherwise.

Using Equation (8), the conserved forms of these NS become:

$$\begin{aligned} I_0 &= -\dot{t}^2 + \lambda^2(t)\,\dot{r}^2 + v^2(t)\,\dot{\theta}^2 + v^2(t)\sin^2\theta\,\dot{\phi}^2, \\ I_1 &= -2\lambda^2(t)\,\dot{r}^2, \\ I_2 &= -2v^2(t)\sin^2\theta\,\dot{\phi}, \\ I_3 &= -2v^2(t)\left(\sin\phi\,\dot{\theta} + \cos\theta\cos\phi\sin\theta\,\dot{\phi}\right), \\ I_4 &= -\frac{\partial I_3}{\partial \phi}. \end{aligned} \quad (25)$$

The energy density and the Ricci scalar for Models 3a and 3b are obtained as $T_{00} = \frac{1+v'^2}{v^2} + \frac{2\lambda' v'}{v\lambda}$ and $R = -\frac{2}{v^2\lambda}\left(2v\lambda v'' + v^2\lambda'' + 2vv'\lambda' + \lambda v'^2 + \lambda\right)$. One can observe that the positive energy condition is satisfied if $\frac{\lambda'}{\lambda}$ and $\frac{v'}{v}$ have the same signs. Moreover, the Ricci scalar remains non-singular for $t \to 0$.

For the remaining three models, the energy density and the Ricci scalar are given by:

$$T_{00} = \frac{1}{\zeta^2}, \quad R = -2\left(\frac{\lambda''}{\lambda} + \frac{1}{\zeta^2}\right), \quad (26)$$

such that the energy remains positive for any arbitrary non-zero ζ and the Ricci scalar is regular for any function λ.

4. Six Noether Symmetries

The Lagrangian of the following Kantowski–Sachs metric admits six NS:

$$ds^2 = dt^2 - \gamma^2 dr^2 - v^2(t)\left(d\theta^2 + \sin^2\theta\,d\phi^2\right), \quad (27)$$

where γ is a non-zero constant, $v'' \neq 0$ and $v \neq \cosh t$. Out of these six NS, five are the same as given in Equation (24) and one extra Noether symmetry is given by:

$$X_5 = \frac{u}{\gamma^2}\partial_r, \quad F = -2r.$$

The conserved form for this symmetry is $I_5 = -2(u\dot{r} - r)$. The corresponding non-zero commutator is $[X_0, X_5] = \frac{1}{\gamma^2}X_1$.

The metric (27) is non-conformally flat, and its energy density is obtained as $T_{00} = \frac{1+v'^2}{v^2}$, which is always positive. The Ricci scalar is given by $R = -\frac{2}{v^2}\left(1 + v'^2 + 2vv''\right)$, which has no singularity at the origin.

5. Seven Noether Symmetries

There are seven metrics whose Lagrangian admit a seven-dimensional algebra of NS. Table 2 contains all these metrics along with the two extra NS, other than the minimal NS given in (24), their Lie algebra, and the corresponding conserved quantities.

Table 2. Metrics admitting seven Noether symmetries.

No.	$\lambda(t)$	$\nu(t)$	Noether Symmetries	Invariants	Lie Algebra
7(i)	$\sin(\beta t); \beta \neq 0$	$\gamma \neq 0$	$X_5 = \cosh(\beta r)\partial_t - \cot(\beta t)\sinh(\beta r)\partial_r$ $X_6 = \sinh(\beta r)\partial_t - \cot(\beta t)\cosh(\beta r)\partial_r$	$I_5 = 2\cosh(\beta r)\dot{t} + 2\cos(\beta t)\sin(\beta t)\sinh(\beta r)\dot{r}$ $I_6 = \frac{1}{\beta}\frac{\partial I_5}{\partial r}$	$[X_1, X_5] = \beta X_6, [X_1, X_6] = \beta X_5,$ $[X_6, X_5] = \beta X_1.$
7(ii)	$\cos(\beta t)$	$\gamma \neq 0$	$X_5 = \cosh(\beta r)\partial_t + \tan(\beta t)\sinh(\beta r)\partial_r$ $X_6 = \sinh(\beta r)\partial_t + \tan(\beta t)\cosh(\beta r)\partial_r$	$I_5 = 2\cosh(\beta r)\dot{t} - 2\cos(\beta t)\sin(\beta t)\sinh(\beta r)\dot{r}$ $I_6 = \frac{1}{\beta}\frac{\partial I_5}{\partial r}$	Same as in case 7(i)
7(iii)	$\sinh(\beta t); \beta \neq 0$	$\gamma \neq 0$	$X_5 = \cosh(\beta r)\partial_t - \coth(\beta t)\sinh(\beta r)\partial_r$ $X_6 = \sinh(\beta r)\partial_t - \coth(\beta t)\cosh(\beta r)\partial_r$	$I_5 = 2\cosh(\beta r)\dot{t} + 2\cosh(\beta t)\sinh(\beta r)\dot{r}$ $I_6 = \frac{1}{k}\frac{\partial I_5}{\partial r}$	Same as in case 7(i)
7(iv)	$\cosh(\beta t)$	$\gamma \neq 0$	$X_5 = \cos(\beta r)\partial_t - \tanh(\beta t)\sin(\beta r)\partial_r$ $X_6 = \sin(\beta r)\partial_t + \tanh(\beta t)\cos(\beta r)\partial_r$	$I_5 = 2\cos(\beta r)\dot{t} + 2\cosh(\beta t)\sinh(\beta t)\sin(\beta r)\dot{r}$ $I_6 = \frac{1}{\beta}\frac{\partial I_5}{\partial r}$	Same as in case 7(i)
7(v)	$e^{\beta t}$	$\gamma \neq 0$	$X_5 = \partial_t - \beta r \partial_r$ $X_6 = r\partial_t - \frac{e^{-2\beta t}}{2\beta}\partial_r - \frac{\beta r^2}{2}\partial_r$	$I_5 = 2\dot{t} + 2\beta r e^{2\beta t}\dot{r}$ $I_6 = 2r\dot{t} + \frac{\dot{r}}{\beta} + \beta r^2 e^{2\beta t}\dot{r}$	$[X_5, X_1] = \beta X_1, [X_1, X_6] = X_5,$ $[X_6, X_5] = \beta X_6$
7(vi)	$c_1 t + c_2$	$\nu = \lambda$	$X_5 = \frac{u^2}{2}\partial_s + \frac{u(c_1 t + c_2)\dot{t}}{2c_1}\partial_t; F = \frac{t(c_1 t + 2c_2)}{2c_1}$ $V_6 = u\partial_u + \frac{c_1 t + c_2}{2c_1}\partial_t$	$I_5 = \frac{u^2}{2}L + \frac{u(c_1 t + c_2)\dot{t}}{c_1} - \frac{t^2}{2} - \frac{c_2 t}{c_1}$ $I_6 = uL + \frac{(c_1 t + c_2)\dot{t}}{c_1}$	$[X_0, X_5] = X_6, [X_0, X_6] = X_0,$ $[X_6, X_5] = X_5.$
7(vii)	$(c_1 t + c_2)^{\frac{\alpha - 2\beta}{\alpha}};$ $\alpha \neq 0$	$c_1 t + c_2$	$X_5 = u\partial_u + \frac{c_1 t + c_2}{2c_1}\partial_t$ $X_6 = r\partial_r$	$I_5 = uL + \frac{(c_1 t + c_2)\dot{t}}{c_1}$ $I_6 = -2r(c_1 t + c_2)^{\frac{2(\alpha - 2\beta)}{\alpha}}\dot{r}$	$[X_0, X_5] = X_0, [X_1, X_6] = X_1,$ $[X_6, X_1] = X_5$

The metric 7(i) is non-conformally flat, whose NS other than the minimal set of NS are KVs. Here, the generators X_1, X_5 and X_6 form a Lie algebra for the rotation group $SO(3)$.

Similarly, the metric 7(ii) is non-conformally flat, and it admits seven NS, five of which are the same as given in (24), and the remaining two are KVs, represented by X_5 and X_6.

The energy density and the Ricci scalar for the models 7(i) and 7(ii) are obtained as $T_{00} = \frac{1}{\gamma^2}$ and $R = 2\left(\beta^2 - \frac{1}{\gamma^2}\right)$. The positive energy condition clearly holds, and the spacetimes remain regular at the origin, as the Ricci scalar is constant.

The non-conformally-flat metric 7(iii) admits five NS as given in (24) along with two extra KVs, which are represented by X_5 and X_6.

The Weyl tensor for the metric 7(iv) has non-zero components, and hence, it is non-conformally flat. This metric possesses seven NS, out of which five are the same as given in (24) and the extra two are KVs.

The metric 7(v) is non-conformally flat, which admits two extra NS X_5 and X_6, which are KVs. The energy density T_{00} and the Ricci scalar R for the models 7(iii), 7(iv), and 7(v) are:

$$T_{00} = \frac{1}{\gamma^2}, \quad R = -2\left(\beta^2 + \frac{1}{\gamma^2}\right), \tag{28}$$

such that the positive energy condition is satisfied and the Ricci scalar is non-singular at the origin.

There exist two extra NS other than the minimal set of five NS for the non-conformally-flat metric 7(vi). The Noether symmetry X_6 corresponds to a homothetic vector $\frac{c_1 t + c_2}{2c_1}\partial_t$, with the homothetic constant $\psi = \frac{1}{2}$. For this model, the energy density and the Ricci scalar become:

$$T_{00} = \frac{1 + 3c_1^2}{(c_1 t + c_2)^2}, \quad R = -\frac{2(3c_1^2 + 1)}{(c_1 t + c_2)^2}. \tag{29}$$

For any non-zero values of c_1 and c_2, the energy density is positive and the Ricci scalar remains non-singular at $t \to 0$.

Finally, the metric 7(vii) is non-conformally flat, which admits the minimal set of five NS along with two extra NS X_5 and X_6. Here, X_6 is a KV, and X_5 corresponds to a homothetic vector $\frac{c_1 t + c_2}{2c_1}\partial_t$, with the homothetic constant $\psi = \frac{1}{2}$. The energy density and Ricci scalar for this metric are given as:

$$T_{00} = \frac{\alpha + 3c_1^2\alpha - 4c_1^2\beta}{\alpha(c_1 t + c_2)^2}, \quad R = -\frac{2(3c_1^2\alpha^2 - 6c_1^2\alpha\beta + 4c_1^2\beta^2 + \alpha^2)}{\alpha^2(c_1 t + c_2)^2}. \tag{30}$$

For $c_2 \neq 0$, the Ricci scalar has no singularity when $t \to 0$. The energy density may vary for different values of the constants involved in T_{00}. The following two graphs in Figure 1 show positive and negative energy density for some particular choices of the values of these constants.

As far as the case in which the positive energy condition is not met, it could represent an accelerated phase of expansion that may possibly be attributed to the dark energy.

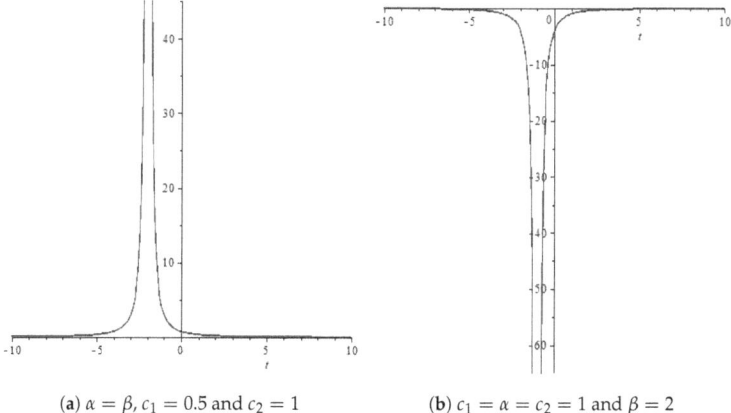

(a) $\alpha = \beta$, $c_1 = 0.5$ and $c_2 = 1$ (b) $c_1 = \alpha = c_2 = 1$ and $\beta = 2$

Figure 1. Graphs of T_{00}.

6. Eight Noether Symmetries

The following is only one Kantowski–Sachs metric whose Lagrangian admits eight NS:

$$ds^2 = dt^2 - \gamma^2 dr^2 - (\alpha t + \beta)^2 \left(d\theta^2 + \sin^2\theta d\phi^2\right), \tag{31}$$

where $\alpha \neq 0$ and $\gamma \neq 0$. The Weyl tensor for the above metric has non-vanishing components, and it is non-conformally flat. The set of eight NS for this metric contains the set of minimal five NS, given in (24), and the remaining three symmetries are:

$$X_5 = \frac{u^2}{2}\partial_u + \frac{u(\alpha t + \beta)}{2\alpha}\partial_t + \frac{ur}{2}\partial_r; \quad F = \frac{t^2}{2} + \frac{\beta t}{\alpha} - \frac{r^2\gamma^2}{2},$$

$$X_6 = \frac{u}{\gamma^2}\partial_r; \quad F = -2r,$$

$$X_7 = u\partial_u + \frac{\alpha t + \beta}{2\alpha}\partial_t + \frac{r}{2}\partial_r. \tag{32}$$

Out of the above three NS, X_7 corresponds to a homothetic vector, given by $\frac{\alpha t+\beta}{2\alpha}\partial_t + \frac{r}{2}\partial_r$. The conserved form for these symmetries and Lie algebra are listed below:

$$I_5 = \frac{u^2}{2}\left(-\dot{t}^2 + \gamma^2\dot{r}^2 + (\alpha t + \beta)^2(\dot{\theta}^2 + \sin^2\theta\dot{\phi}^2)\right) + \frac{u(\alpha t + \beta)\dot{t}}{\alpha} - ur\gamma^2\dot{r} - \frac{t^2}{2} - \frac{\beta t}{\alpha} + \frac{r^2\gamma^2}{2},$$

$$I_6 = -2u\dot{r} + 2r,$$

$$I_7 = u\left(-\dot{t}^2 + \gamma^2\dot{r}^2 + (\alpha t + \beta)^2(\dot{\theta}^2 + \sin^2\theta\dot{\phi}^2)\right) + \frac{(\alpha t + \beta)\dot{t}}{\alpha} - r\gamma^2\dot{r}. \tag{33}$$

$$[X_0, X_5] = X_7, \quad [X_0, X_7] = X_0, \quad [X_0, X_6] = \frac{1}{\gamma^2}X_1, \quad [X_1, X_5] = \frac{\gamma^2}{2}X_6,$$
$$[X_1, X_7] = \frac{1}{2}X_1, \quad [X_7, X_5] = X_7, \quad [X_7, X_6] = \frac{1}{2}X_6.$$

The energy density and the Ricci scalar for Model (31) are respectively given by $T_{00} = \frac{1+\alpha^2}{(\alpha t+\beta)^2}$ and $R = -\frac{2}{(\alpha t+\beta)^2}$. Here, the positive energy condition holds, and for $t \to 0$, the scalar R remains regular if $\beta \neq 0$.

7. Nine Noether Symmetries

Here, we present some Kantowski–Sachs metrics, whose Lagrangian admits nine NS.

(1). $$ds^2 = dt^2 - \beta^2 dr^2 - \cosh^2 t \left(d\theta^2 + \sin^2\theta d\phi^2 \right), \tag{34}$$

where $\beta \neq 0$. The Weyl tensor for the above metric vanishes, and hence, it is conformally flat. The set of NS for this metric contains the minimal set of five symmetries given in (24), and four extra NS are admitted, which are listed below:

$$\begin{aligned}
X_5 &= u\partial_r; \quad F = -2r, \\
X_6 &= \sin\theta \sin\phi \partial_t + \tanh t \cos\theta \sin\phi \partial_\theta + \csc\theta \tanh t \cos\phi \partial_\phi, \\
X_7 &= -\sin\theta \cos\phi \partial_t - \tanh t \cos\theta \cos\phi \partial_\theta + \csc\theta \tanh t \sin\phi \partial_\phi, \\
X_8 &= -\cos\theta \partial_t + \tanh t \sin\theta \partial_\theta.
\end{aligned} \tag{35}$$

Here, X_6, X_7, and X_8 are KVs, while X_5 is a non-trivial Noether symmetry. The Lie algebra in this case gives:

$$[X_0, X_5] = X_1, \quad [X_6, X_2] = [X_4, X_8] = X_7, \quad [X_2, X_7] = [X_3, X_8] = X_6,$$
$$[X_6, X_3] = [X_7, X_4] = X_8, \quad [X_6, X_7] = X_2, \quad [X_6, X_8] = X_3, \quad [X_7, X_8] = X_4,$$

and the conservation laws are:

$$\begin{aligned}
I_5 &= -2(u\beta^2 \dot{r} - r), \\
I_6 &= 2\sin\theta \sin\phi \dot{t} - 2\sinh t \cosh t (\cos\theta \sin\phi \dot\theta + \sin\theta \cos\phi \dot\phi), \\
I_7 &= -2\sin\theta \cos\phi \dot{t} + 2\sinh t \cosh t (\cos\theta \cos\phi \dot\theta - \sin\theta \sin\phi \dot\phi), \\
I_8 &= -2\cos\theta \dot{t} - 2\sinh t \cosh t \sin\theta \dot\theta.
\end{aligned} \tag{36}$$

The energy density and Ricci scalar for the metric (34) are respectively given by $T_{00} = 1$ and $R = -6$, such that the positive energy condition is satisfied and R is regular at the origin.

(2). $$ds^2 = dt^2 - \alpha^2 dr^2 - \beta^2 (d\theta^2 + \sin^2\theta d\phi^2), \tag{37}$$

where $\alpha \neq 0$ and $\beta \neq 0$. For the above metric, the Weyl tensor has non-zero components. The extra four NS other than the minimal set of NS, given in (24), for this metric are:

$$\begin{aligned}
X_5 &= \partial_t, \\
X_6 &= -u\partial_t; \quad F = -2t, \\
X_7 &= \frac{u}{\alpha^2} \partial_r; \quad F = -2r, \\
X_8 &= r\partial_t + \frac{t}{\alpha^2} \partial_r.
\end{aligned} \tag{38}$$

Here, X_5 and X_8 are KVs, while X_6 and X_7 are non-trivial NS. The Lie algebra and the conserved quantities are given as:

$$[X_6, X_0] = [X_8, X_1] = X_5, \quad [X_0, X_7] = [X_5, X_8] = \tfrac{1}{\alpha^2} X_1,$$
$$[X_8, X_6] = X_7, \quad [X_8, X_7] = \tfrac{1}{\alpha^2} X_6.$$

$$I_5 = 2\dot{t}, \quad I_6 = -2u\dot{t} + 2t, \quad I_7 = -2u\dot{r} + 2r,$$
$$I_8 = 2r\dot{t} - 2t\dot{r}. \tag{39}$$

The corresponding T_{00} and R for the above metric are given by $T_{00} = \frac{1}{\beta^2}$ and $R = -\frac{2}{\beta^2}$. Clearly, the energy density is positive for any value of β, and the Ricci scalar is regular at the origin.

(3).
$$ds^2 = dt^2 - (\alpha t + \beta)^2 dr^2 - \xi^2(d\theta^2 + \sin^2\theta d\phi^2), \tag{40}$$

where $\xi \neq 0$. The above metric is non-conformally flat, and it admits the following four NS along with five minimal symmetries, given in Equation (24).

$$\begin{aligned}
X_5 &= -u\cosh(\alpha r)\partial_t + \frac{u}{\alpha t + \beta}\sinh(\alpha r)\partial_r; \quad F = -\frac{2\beta}{\alpha}\cosh(\alpha r), \\
X_6 &= -u\sinh(\alpha r)\partial_t + \frac{u}{\alpha t + \beta}\cosh(\alpha r)\partial_r; \quad F = -\frac{2\beta}{\alpha}\sinh(\alpha r), \\
X_7 &= \cosh(\alpha r)\partial_t - \frac{\sinh(\alpha r)}{\alpha t + \beta}\partial_r, \\
X_8 &= \sinh(\alpha r)\partial_t - \frac{\cosh(\alpha r)}{\alpha t + \beta}\partial_r.
\end{aligned} \tag{41}$$

From the above, one can see that X_5 and X_6 are non-trivial NS, while X_7 and X_8 are KVs. In this case, the Lie algebra becomes:

$$[X_5, X_0] = X_7, \quad [X_6, X_0] = X_8, \quad [X_1, X_5] = \alpha X_6,$$
$$[X_1, X_6] = \alpha X_5, \quad [X_1, X_7] = \alpha X_8, \quad [X_1, X_8] = \alpha X_7.$$

The conserved forms of the symmetries are given as follows:

$$\begin{aligned}
I_5 &= -2u\left(\cosh(\alpha r)\dot{t} + (\alpha t + \beta)\sinh(\alpha r)\dot{r}\right) + \frac{2\beta}{\alpha}\cosh(\alpha r), \\
I_6 &= \frac{1}{\alpha}\frac{\partial I_5}{\partial r}, \\
I_7 &= 2\left(\cosh(\alpha r)\dot{t} + (\alpha t + \beta)\sinh(\alpha r)\dot{r}\right), \\
I_8 &= \frac{1}{\alpha}\frac{\partial I_7}{\partial r}.
\end{aligned} \tag{42}$$

For the above model, we have $T_{00} = \frac{1}{\xi^2}$ and $R = -\frac{2}{\xi^2}$ such that the energy density is positive and the Ricci scalar has no singularity at the origin.

8. Eleven Noether Symmetries

For $\lambda = \sinh t$ and $v = \cosh t$, we have the following Kantowski–Sachs metric, whose Lagrangian admits eleven NS:

$$ds^2 = dt^2 - \sinh^2 t\, dr^2 - \cosh^2 t(d\theta^2 + \sin^2\theta d\phi^2). \tag{43}$$

This metric has a zero Weyl tensor, and hence, it is conformally flat. The energy density and the Ricci scalar for the above model are obtained as $T_{00} = 3$ and $R = -12$. Here, the positive energy condition is clearly satisfied, and R is non-singular at the origin. The extra six NS (KVs) for the above metric are as follows:

$$\begin{aligned}
X_5 &= \sin\theta\sin\phi(\sinh r\,\partial_t - \coth t\cosh r\,\partial_r) + \tanh t\sinh r(\cos\theta\sin\phi\,\partial_\theta + \csc\theta\cos\phi\,\partial_\phi), \\
X_6 &= \text{same as } X_5 \text{ with } \sinh r \leftrightarrow \cosh r \\
X_7 &= -\sin\theta\cos\phi(\sinh r\partial_t - \coth t\cosh r\partial_r) - \tanh t\sinh r(\cos\theta\cos\phi\partial_\theta - \csc\theta\sin\phi\partial_\phi), \\
X_8 &= \text{same as } X_7 \text{ with } \sinh r \leftrightarrow \cosh r \\
X_9 &= -\cos\theta(\sinh r\partial_t - \coth t\cosh r\,\partial_r) + \tanh t\sinh r\sin\theta\partial_\theta, \\
X_{10} &= \text{same as } X_9 \text{ with } \sinh r \leftrightarrow \cosh r
\end{aligned} \tag{44}$$

One may easily simplify Equation (8) to find the expressions for the corresponding conservation laws for the above generators. The Lie algebra in this case is obtained as:

$[X_1, X_5] = X_6,$ $[X_1, X_6] = X_5,$ $[X_1, X_7] = X_8,$ $[X_1, X_8] = X_7,$ $[X_1, X_9] = X_{10},$
$[X_1, X_{10}] = X_9,$ $[X_5, X_2] = X_7,$ $[X_6, X_2] = X_8,$ $[X_2, X_7] = X_5,$ $[X_2, X_8] = X_6,$
$[X_5, X_3] = X_9,$ $[X_6, X_3] = X_{10},$ $[X_3, X_9] = X_5,$ $[X_3, X_{10}] = X_6,$ $[X_7, X_4] = X_9,$
$[X_8, X_4] = X_{10},$ $[X_4, X_9] = X_7,$ $[X_4, X_{10}] = X_8,$ $[X_5, X_6] = X_1,$ $[X_7, X_5] = X_2,$
$[X_9, X_5] = X_3,$ $[X_6, X_8] = X_2,$ $[X_6, X_{10}] = X_3,$ $[X_7, X_8] = X_1,$ $[X_9, X_7] = X_4,$
$[X_8, X_{10}] = X_4,$ $[X_9, X_{10}] = X_1,$

9. Summary and Discussion

In this paper, we have classified the Lagrangian of the Kantowski–Sachs metric via its NS. The set of determining equations is obtained and then integrated in several cases. It is observed that the Kantowski–Sachs metric admits a 5-, 6-, 7-, 8-, 9-, or 11-dimensional Lie algebra of NS for different values of the metric functions. The number of non-trivial NS for this metric is shown to be one, two, or three, while the number of KVs is found to be 4, 5, 6, 7, or 10.

We have found five different metrics, each admitting the minimal set of five NS, out of which four are the minimum KVs of the Kantowski–Sachs metric and one is a non-trivial Noether symmetry, which is ∂_u. The gauge function is trivial here.

There is only one metric (27) that admits six NS. This set of six NS contains the minimal set of five NS along with one extra Noether symmetry with the gauge function $F = -2r$.

In the case of seven-dimensional Noether algebra, we have obtained seven different metrics. For the first five metrics, 7(i)–7(v), we have six KVs and one Noether symmetry ∂_u, with a trivial gauge function. For metric 7(vi), we have the minimal set of NS along with two extra NS, given by X_5 and X_6. The gauge function corresponding to X_5 is found to be $F = \frac{t(c_1 t + 2c_2)}{2c_1}$. It can bee seen that the Noether symmetry X_6 for this metric corresponds to an HV $\frac{c_1 t + c_2}{2c_1} \partial_t$, with homothetic constant $\psi = \frac{1}{2}$. Similar results are obtained for the metric 7(vii). Here, the number of KVs is five.

There exists only one metric (31) possessing eight NS, of which five are the minimal NS for the Kantowski–Sachs metric, and three extra NS are obtained, which are presented in (32). One of these three NS, denoted as X_7, corresponds to an HV $\frac{\alpha t + \beta}{2\alpha} \partial_t + \frac{r}{2} \partial_r$. The number of KVs for this metric is only four.

In the case of the nine-dimensional Lie algebra of NS, we have found three metrics (34), (37), and (40). For the metric (34), we have the minimal set of five NS along with four extra symmetries, of which three are KVs and one is a non-trivial Noether symmetry $u \partial_r$ with the gauge function $F = -2r$. The number of KVs in this case is seven. For the remaining two metrics (37) and (40), we have six KVs along with three non-trivial NS.

Finally, we have only one metric (43) where the dimension of the algebra of NS is 11. Out of eleven NS, ten are the KVs, and there is only one non-trivial Noether symmetry, given by ∂_u.

For almost all the obtained metrics, it is observed that the positive energy condition is satisfied, and the corresponding Ricci scalar has no singularity at the origin.

Author Contributions: This work is an original research work prepared by S.S.A. under the joint mentorship of T.H. and A.H.B.

Funding: This research received no external funding.

Acknowledgments: We are thankful to the referees for their invaluable suggestions due to which the manuscript was significantly improved. Sumaira Saleem Akhtar also acknowledges the Higher Education Commission of Pakistan for granting the Indigenous fellowship for her Ph.D.

Conflicts of Interest: The authors declare no conflict of interest.

References

1. Stephani, H.; Kramer, D.; Maccallum, M.; Hoenselaers, C.; Herlt, E. *Exact Solutions of Einstein's Field Equations*, 2nd ed.; Cambridge University Press: Cambridge, UK, 2003.
2. Meisner, C.W.; Thorne, K.S.; Wheeler, J.A. *Gravitation*; Benjamin: New York, NY, USA, 1973.

3. Petrov, A.Z. *Einstein Spaces*; Pergamon: Oxford, UK, 1969.
4. Bokhari, A.H.; Kashif, A.R.; Qadir, A. Classification of curvature collineations of plane symmetric static spacetimes. *J. Math. Phys.* **2000**, *41*, 2167–2172. [CrossRef]
5. Hall, G.S.; da Costa, J. Curvature collineations in general relativity. I. *J. Math. Phys.* **1991**, *32*, 2848–2853. [CrossRef]
6. Hall, G.S.; da Costa, J. Curvature collineations in general relativity. II. *J. Math. Phys.* **1991**, *32*, 2854–2862. [CrossRef]
7. Hussain, T.; Akhtar, S.S.; Khan, S. Ricci inheritance collineation in Bianchi type I spacetimes. *Eur. Phys. J. Plus* **2015**, *130*, 44. [CrossRef]
8. Hussain, T.; Akhtar, S.S.; Bokhari, A.H.; Khan, S. Ricci inheritance collineations in Bianchi type II spacetime. *Mod. Phys. Lett. A* **2016**, *31*, 1650102. [CrossRef]
9. Akhtar, S.S.; Hussain, T.; Bokhari, A.H.; Khan, F. Conformal collineations of the Ricci and energy–momentum tensors in static plane symmetric space–times. *Theor. Math. Phys.* **2018**, *195*, 595–601. [CrossRef]
10. Noether, E. Invariant variations problems. *Nachr. Konig. Gissell. Wissen. Got-tingen Math.-Phys. Kl.* **1918**, *2*, 235–257; English translation in *Transp. Theory Stat. Phys.* **1971**, *1*, 186–207. [CrossRef]
11. Hickman, M.; Yazdan, S. Noether symmetries of Bianchi type II spacetimes. *Gen. Relativ. Gravit.* **2017**, *49*, 65.
12. Bokhari, A.H.; Kara, A.H. Noether versus Killing symmetry of conformally flat Friedmann metric. *Gen. Relativ. Gravit.* **2007**, *39*, 2053–2059. [CrossRef]
13. Bokhari, A.H.; Kara, A.H.; Kashif, A.R.; Zaman, F.D. Noether symmetries versus Killing vectors and isometries of spacetimes. *Int. J. Theor. Phys.* **2006**, *45*, 1029–1039. [CrossRef]
14. Camci, U. Symmetries of geodesic motion in Gödel-type spacetimes. *J. Cosmol. Astropart. Phys.* **2014**, *2014*. [CrossRef]
15. Camci, U.; Yildirim, A. Noether gauge symmetry classes for pp-wave spacetimes. *Int. J. Geom. Meth. Mod. Phys.* **2015**, *12*, 1550120. [CrossRef]
16. Ali, F.; Feroze, T. Classification of Plane Symmetric Static Space-Times According to Their Noether Symmetries. *Int. J. Theor. Phys.* **2013**, *52*, 3329–3342. [CrossRef]
17. Ali, F.; Feroze, T.; Ali, S. Complete classification of spherically symmetric static spacetimes via Noether symmetries. *Theor. Math. Phys.* **2015**, *184*, 973–985. [CrossRef]
18. Ali, F.; Feroze, T. Complete classification of cylindrically symmetric static spacetimes and the corresponding conservation laws. *Sigma Math.* **2016**, *4*, 50. [CrossRef]
19. Jamil, B.; Feroze, T. Conservation laws corresponding to the Noether symmetries of the geodetic Lagrangian in spherically symmetric spacetimes. *Int. J. Mod. Phys. D* **2017**, *26*. [CrossRef]
20. Jamil, B.; Feroze, T.; Vargas, A. Geometrical/Physical interpretation of the conserved quantities corresponding to Noether symmetries of plane symmetric spacetimes. *Adv. Math. Phys.* **2017**, *2017*, 4384093. [CrossRef]
21. Paliathanasis, A.; Tsamparlis, M.; Mustafa, M.T. Symmetry analysis of the Klein–Gordon equation in Bianchi I spacetimes. *Int. J. Geom. Meth. Mod. Phys.* **2015**, *12*, 1550033. [CrossRef]
22. Tsamparlis, M.; Paliathanasis, A. The geometric nature of Lie and Noether symmetries. *Gen. Relativ. Gravit.* **2011**, *43*, 1861–1881. [CrossRef]
23. Ali, S.; Hussain, I. A study of positive energy condition in Bianchi V spacetimes via Noether symmetries. *Eur. Phys. J. C* **2016**, *76*. [CrossRef]
24. Usamah, S.A.; Bokhari, A.H.; Kara, A.H.; Shabbir, G. Classification of Variational Conservation Laws of General Plane Symmetric Spacetimes. *Commun. Theor. Phys.* **2017**, *68*, 335–341.
25. Ma, W.X. Conservation laws by symmetries and adjoint symmetries. *Discret. Contin. Dyn. Syst. Ser. S* **2018**, *11*, 707–721. [CrossRef]
26. Ma, W.X.; Zhou, Y. Lump solutions to nonlinear partial differential equations via Hirota bilinear forms. *J. Diff. Equ.* **2018**, *264*, 2633–2659. [CrossRef]
27. Kantowski, R.; Sachs, R.K. Some Spatially Homogeneous Anisotropic Relativistic Cosmological Models. *J. Math. Phys.* **1966**, *7*, 443–446. [CrossRef]
28. Hand, L.N.; Finch, J.D. *Analytical Mechanics*; Cambridge University Press: Cambridge, UK, 2008.

© 2018 by the authors. Licensee MDPI, Basel, Switzerland. This article is an open access article distributed under the terms and conditions of the Creative Commons Attribution (CC BY) license (http://creativecommons.org/licenses/by/4.0/).

Article

$F(R, G)$ Cosmology through Noether Symmetry Approach

Ugur Camci

Siteler Mahallesi, 1307 Sokak, Ahmet Kartal Konutlari, A-1 Blok, No. 7/2, 07070 Konyaalti, Antalya, Turkey; ugurcamci@gmail.com

Received: 2 November 2018; Accepted: 3 December 2018; Published: 5 December 2018

Abstract: The $F(R, G)$ theory of gravity, where R is the Ricci scalar and G is the Gauss-Bonnet invariant, is studied in the context of existence the Noether symmetries. The Noether symmetries of the point-like Lagrangian of $F(R, G)$ gravity for the spatially flat Friedmann-Lemaitre-Robertson-Walker cosmological model is investigated. With the help of several explicit forms of the $F(R, G)$ function it is shown how the construction of a cosmological solution is carried out via the classical Noether symmetry approach that includes a functional boundary term. After choosing the form of the $F(R, G)$ function such as the case $(i): F(R, G) = f_0 R^n + g_0 G^m$ and the case $(ii): F(R, G) = f_0 R^n G^m$, where n and m are real numbers, we explicitly compute the Noether symmetries in the vacuum and the non-vacuum cases if symmetries exist. The first integrals for the obtained Noether symmetries allow to find out exact solutions for the cosmological scale factor in the cases (i) and (ii). We find several new specific cosmological scale factors in the presence of the first integrals. It is shown that the existence of the Noether symmetries with a functional boundary term is a criterion to select some suitable forms of $F(R, G)$. In the non-vacuum case, we also obtain some extra Noether symmetries admitting the equation of state parameters $w \equiv p/\rho$ such as $w = -1, -2/3, 0, 1$ etc.

Keywords: Noether symmetry approach; FLRW spacetime; action integral; variational principle; first integral; modified theories of gravity; Gauss-Bonnet cosmology

1. Introduction

Recent observational data indicate that the current expansion of the universe is accelerating [1-8], not only expanding. Then this acceleration is explained by the existence of a dark energy, which could result from a cosmological constant Λ as the simplest candidate with the equation of state parameter $w_\Lambda = -1$, or may also be explained in the context of modified gravity models. The nature and origin of the dark energy has not been persuasively explained yet. In addition to the cosmological constant, there are different kinds of candidates for dark energy such as quintessence or phantom in the literature, and it is not even clear what type of candidates to the dark energy occur in the present universe. Therefore, there have been a number of attempts [9-15] to modify gravity to explain the origin of dark energy.

A possible modification of the standard general relativistic gravitational Lagrangian includes a wider number of curvature invariants $R, R_{ij}R^{ij}$ and $R_{ijkl}R^{ijkl}$ among others. In the so-called Gauss-Bonnet (GB) gravity theories the gravitational Lagrangian consists of a $F(R, G)$ function, where the GB invariant G is defined as $G = R^2 - 4R_{ij}R^{ij} + R_{ijkl}R^{ijkl}$. Considering the GB invariant G in dynamical equations one can recover all the curvature budget coming from the Riemann tensor. Due to of the fact that the GB invariant comes out from defining quantum fields in curved spacetimes, it should be important to take it in the context of the extended theories of gravity. It is shown in [13] that the quintessence paradigm can be recovered in the framework of $F(R, G)$ theories of gravity. The $F(R, G)$ gravity theories are generalizations $f(R)$ and $f(G)$ theory of gravities which are offered by higher order gravities, and use combinations of higher order curvature invariants constructed from the Ricci and Gauss-Bonnet scalars. In [14], some classes of $F(R,G)$ gravity have been studied with respect to the successful realization of the dark energy and of the inflationary era. We refer to readers the latest review [15] on developments of modified gravity in cosmology, emphasizing on inflation, bouncing cosmology and late-time acceleration era.

If a Lagrangian $\mathcal{L}$ for a given dynamical system admits any symmetry, this property should strongly be related with Noether symmetries that describe physical features of differential equations possessing a Lagrangian $\mathcal{L}$ in terms of first integrals admitted by them [16,17]. This can actually be seen in two ways. Firstly, one can consider a *strict Noether symmetry approach* [18–21] which yields $\pounds_X \mathcal{L} = 0$, where $\pounds_X$ is the Lie derivative operator along **X**. On the other side, one could use the *classical Noether symmetry approach* with a functional term [22–25] which is a generalization of the strict Noether symmetry approach in the sense that the Noether symmetry equation includes a divergence of a functional boundary term. The classical Noether symmetry approach was originally established by Emmy Noether [26] and it gives a connection between a Noether symmetry and the existence of a first integral expressed in a simple form. Not only the classical Noether symmetries but also the strict ones are useful in a variety of problems arising from physics and applied mathematics. Both types of symmetries lead to the first integrals. Which type of symmetry works, i.e., gives any conserved quantity, in the first instance this is what is important. The classical Noether symmetries are directly related with the conserved quantities (first integrals) or conservation laws [17]. The strict Noether symmetry approach represents how Noether's theorem and cyclic variables are related. It is known that the conserved quantities are also related to the existence of cyclic variables into the dynamics by the strict Noether symmetry. However, it is usually required a clever choice of cyclic variables because of that the equations for the change of coordinates have not a unique solution which is also not well defined on the whole space, and thus it is not unique to find those of the cyclic variables (see References [27] for details). Furthermore, we refer to the interested readers the recent review on symmetries in differential equations [28].

The cosmological principle assume that the universe is homogeneous and isotropic in large scale structure and the geometrical model that satisfies these properties is Friedmann-Lemaitre-Robertson-Walker (FLRW) spacetime. In [19], it has been discussed the strict Noether symmetry approach for spatially flat FLRW spacetime in GB cosmology, where it was pointed out that the existence of Noether symmetries is capable of selecting suitable $F(R,G)$ models to integrate dynamics by the identification of suitable cyclic variables. After this work, the classical Noether symmetries of flat FLRW spacetime have been computed by [25], where the authors were used Noether symmetries as a geometric criterion to select the form of $F(R,G)$ function. Due to the richness of the classical Noether symmetry approach, we deduced throughout this study that it is better to use the classical Noether symmetry approach to find Noether symmetries in $F(R,G)$ gravity as in [25], rather that the approach used in [19]. If there exists any Noether symmetry with a selection of physically interesting forms of $F(R,G)$ function, then this allows us to write out the constants of motion which reduce dynamics. Furthermore, the reduced dynamics results exactly solvable cosmological model by a straightforward way. In fact, choosing an appropriate $F(R,G)$ Lagrangian, it is possible to find out conserved Noether currents which will be useful to solve dynamics. This approach is very powerful due to the fact that it allows us to find a closed system of equations, where we do not need to impose the particular form of $F(R,G)$ which is selected by the classical Noether symmetry itself. To this aim, it is possible to consider flat FLRW background metric and demonstrate that it is possible to find exact solutions via the Noether Symmetry Approach. In this study we again underline the generality of Noether's Theorem in its original form by considering the standard cosmological model.

This paper is organized as follows. In the following section, we will present an analysis of the classical Noether symmetry approach including a boundary function for the point-like $F(R,G)$ Lagrangian according to the spatially flat FLRW background. In Section 3, we will apply the classical Noether theorem to the $F(R,G)$ Lagrangian obtained in Section 2 for the flat FLRW model. In Section 4, we classify the Noether symmetries with respect to some specific forms of $F(R, G)$, and search the cosmological solutions of $F(R, G)$ gravity by considering both the vacuum and the non-vacuum cases. Finally, in Section 5, we will provide a summary of the main results obtained in the paper.

2. $F(R, G)$ Gravity

In this section we briefly present the general formalism of $F(R, G)$ gravity. The action for $F(R, G)$ gravity is given by

$$S = \int d^4x \sqrt{-g} \left[\frac{1}{2\kappa^2} F(R, G) + \mathcal{L}_m \right], \quad (1)$$

where $\kappa^2 = 8\pi G_N$, G_N is the Newton constant and $\mathcal{L}_m$ represents the matter Lagrangian. Variation of the action (1) with respect to the metric tensor g_{ij} we obtain the modified field equations

$$\begin{aligned}
F_R G_{ij} &= \kappa^2 T_{ij}^m + \frac{1}{2} g_{ij}(F - R F_R) + \nabla_i \nabla_j F_R - g_{ij} \Box F_R \\
&+ F_G \left(-2 R R_{ij} + 4 R_{ik} R_j^k - 2 R_i^{klm} R_{jklm} + 4 g^{kl} g^{mn} R_{ikjm} R_{ln} \right) \\
&+ 2 \left(\nabla_i \nabla_j F_G \right) R - 2 g_{ij} \left(\Box F_G \right) R + 4 \left(\Box F_G \right) R_{ij} - 4 \left(\nabla_k \nabla_i F_G \right) R_j^k - 4 \left(\nabla_k \nabla_j F_G \right) R_i^k \\
&+ 4 g_{ij} \left(\nabla_k \nabla_l F_G \right) R^{kl} - 4 \left(\nabla_l \nabla_n F_G \right) g^{kl} g^{mn} R_{ikjm},
\end{aligned} \quad (2)$$

where we have defined the following expressions

$$F_R \equiv \frac{\partial F(R, G)}{\partial R}, \quad F_G \equiv \frac{\partial F(R, G)}{\partial G}. \quad (3)$$

In the above field equations, ∇_i is the covariant derivative operator associate with g_{ij}, $\Box \equiv g^{ij} \nabla_i \nabla_j$ is the covariant d'Alembertian operator, and T_{ij}^m describes the ordinary matter. It is clear from the field Equation (2) that the form of $F(R, G)$ determine the dynamical behaviour of the theory.

In this study, we consider the spatially flat FLRW metric

$$ds^2 = -dt^2 + a(t)^2 \left(dx^2 + dy^2 + dz^2 \right) \quad (4)$$

where $a(t)$ is the scale factor of the Universe. Then, the Hubble parameter H is usually defined by $H \equiv \dot{a}/a$, and R and G become

$$R = 6 \left(\frac{\ddot{a}}{a} + \frac{\dot{a}^2}{a^2} \right) = 6(\dot{H} + 2H^2), \quad G = 24 \frac{\dot{a}^2 \ddot{a}}{a^3} = 24 H^2 \left(\dot{H} + H^2 \right), \quad (5)$$

where the overdot denotes a derivative with respect to the time coordinate, t. For a perfect fluid matter with comoving observer $u_i = \delta_i^0$, the energy momentum tensor is $T_{ij} = (\rho + p) u_i u_j + p g_{ij}$, where ρ is the energy density and p is the isotropic pressure measured by the observer u_i. Let us assume that the matter fluid will be given under the form of a perfect fluid with the equation of state $p = w\rho$ satisfying the standard continuity equation $\dot{\rho} + 3(1 + w) \rho \dot{a}/a = 0$ which yields a solution $\rho = \rho_{m0} a^{-3(1+w)}$, where ρ_{m0} is the energy density of the present universe, and w is a constant parameter. Thus, in the flat FLRW background with a perfect fluid matter, the field Equation (2) for the $F(R, G)$ gravity are given by

$$3 F_R \frac{\dot{a}^2}{a^2} = \kappa^2 \rho + \frac{1}{2}(R F_R + G F_G - F) - 3 \dot{F}_R \frac{\dot{a}}{a} - 12 \dot{F}_G \frac{\dot{a}^3}{a^3}, \quad (6)$$

$$F_R \left(\frac{2\ddot{a}}{a} + \frac{\dot{a}^2}{a^2} \right) = -\kappa^2 p + \frac{1}{2}(R F_R + G F_G - F) - 2 \dot{F}_R \frac{\dot{a}}{a} - \ddot{F}_R - 4 \frac{\dot{a}}{a} \left(\frac{\dot{a}}{a} \ddot{F}_G + \frac{2\ddot{a}}{a} \dot{F}_G \right). \quad (7)$$

In terms of the Hubble parameter H, the gravitational field Equations (6) and (7) for $F(R, G)$ gravity have the following form

$$H^2 = \frac{\kappa^2}{3}\rho_{eff}, \qquad 2\dot{H} + 3H^2 = -\kappa^2 p_{eff}, \qquad (8)$$

where ρ_{eff} and p_{eff} are respectively the effective energy density and pressure of the universe, which are defined as

$$\rho_{eff} \equiv \frac{1}{F_R}\left\{\rho + \frac{1}{2\kappa^2}\left[RF_R + GF_G - F - 6H\dot{F}_R - 24H^3\dot{F}_G\right]\right\}, \qquad (9)$$

$$p_{eff} \equiv \frac{1}{F_R}\left\{p + \frac{1}{2\kappa^2}\left[F - RF_R - GF_G + 4H\dot{F}_R + 2\ddot{F}_R + 16H(\dot{H} + H^2)\dot{F}_G + 8H^2\ddot{F}_G\right]\right\}. \qquad (10)$$

Here we observe from (8) that $\rho_{eff} + p_{eff} = -\frac{2}{\kappa^2}\dot{H}$.

3. Noether Symmetry Approach

Recently the strict Noether symmetries of GB cosmology for the flat FLRW spacetime have been calculated, and choosing some functional form of the $F(R, G)$, the Noether symmetries related to these functional forms have been achieved [19]. Afterwards, the classical Noether symmetries have also been calculated by [25]. Both of these studies were performed in the vacuum case. In this work, after reviewing the vacuum case, we aim to generalize these studies to the non-vacuum case using the classical Noether symmetry approach described below.

The Noether symmetry generator for any point-like Lagrangian $\mathcal{L}$ is

$$\mathbf{X} = \xi(t, a, R, G)\frac{\partial}{\partial t} + \eta^1(t, a, R, G)\frac{\partial}{\partial a} + \eta^2(t, a, R, G)\frac{\partial}{\partial R} + \eta^3(t, a, R, G)\frac{\partial}{\partial G}, \qquad (11)$$

if there exists a function $K(t, a, R, G)$ and the Noether symmetry condition

$$\mathbf{X}^{[1]}\mathcal{L} + \mathcal{L}(D_t\xi) = D_t K \qquad (12)$$

is satisfied, where $D_t = \frac{\partial}{\partial t} + \dot{q}^i\frac{\partial}{\partial q^i}$ is the total derivative operator and $\mathbf{X}^{[1]}$ is the first prolongation of Noether symmetry generator $\mathbf{X}$, i.e.

$$\mathbf{X}^{[1]} = \mathbf{X} + \dot{\eta}^i\left(t, q^i, \dot{q}^i\right)\frac{\partial}{\partial \dot{q}^i} \qquad (13)$$

where $\dot{\eta}^i(t, q^k, \dot{q}^k) = D_t\eta^i - \dot{q}^i D_t\xi$, $q^i = \{a, R, G\}$ are the generalized coordinates in the three-dimensional configuration space $Q \equiv \{q^i, i = 1, 2, 3\}$ of the Lagrangian, whose tangent space is $TQ \equiv \{q^i, \dot{q}^i\}$. The energy functional $E_\mathcal{L}$ or the Hamiltonian of the Lagrangian $\mathcal{L}$ is defined by

$$E_\mathcal{L} = \dot{q}^i\frac{\partial \mathcal{L}}{\partial \dot{q}^i} - \mathcal{L}. \qquad (14)$$

Using above definition of energy functional, the corresponding Noether flow I, which is a constant called the first integral of motion, has the expression

$$I = -\xi E_\mathcal{L} + \eta^i\frac{\partial \mathcal{L}}{\partial \dot{q}^i} - K, \qquad (15)$$

which is a *conserved quantity*. The Noether flow (15) satisfies the conservation relation $D_t I = 0$.

It is obviously seen from a general point of view that R and G are functions of $a, \dot{a}$ and $\ddot{a}$, which yields non-canonical dynamics. The Lagrange multipliers plays a main role so as to get a canonical point-like Lagrangian [29]. Using this key future in [19], it has been accomplished that the point-like Lagrangian for $F(R,G)$ gravity becomes canonical with suitable Lagrange multipliers, where both R and G behave like effective scalar fields. We left the details for finding a canonical point-like Lagrangian by Lagrange multipliers method to the Reference [19]. For the spatially flat FLRW spacetime (4), the Lagrangian for the action of $F(R,G)$ gravity (1) has the form

$$\mathcal{L} = -6F_R a \dot{a}^2 - 6a^2 \dot{a} \dot{F}_R - 8 \dot{F}_G \dot{a}^3 + a^3 (F - RF_R - GF_G) - 2\kappa^2 \rho_{m0} a^{-3w}, \quad (16)$$

where $\dot{F}_R = F_{RR} \dot{R} + F_{RG} \dot{G}$ and $\dot{F}_G = F_{GR} \dot{R} + F_{GG} \dot{G}$. By variation of the above Lagrangian with respect to the configuration space variables a, R and G, we find respectively that

$$F_R \left(\frac{2\ddot{a}}{a} + \frac{\dot{a}^2}{a^2} \right) + \frac{2\dot{a}}{a} \dot{F}_R + \ddot{F}_R - \frac{4\dot{a}}{a} \left(\frac{2\ddot{a}}{a} \dot{F}_G + \frac{\dot{a}}{a} \ddot{F}_G \right) + \frac{1}{2}(RF_R + GF_G - F) = \kappa^2 p, \quad (17)$$

$$6F_{RR} \left(\frac{\ddot{a}}{a} + \frac{\dot{a}^2}{a^2} - \frac{R}{6} \right) - F_{GR} \left(24 \frac{\dot{a}^2 \ddot{a}}{a^3} - G \right) = 0, \quad (18)$$

$$6F_{GR} \left(\frac{\ddot{a}}{a} + \frac{\dot{a}^2}{a^2} - \frac{R}{6} \right) - F_{GG} \left(24 \frac{\dot{a}^2 \ddot{a}}{a^3} - G \right) = 0, \quad (19)$$

in which the Equation (17) is equivalent to the field Equation (7). Then we note that R and G coincides with the definitions of the Ricci scalar and Gauss-Bonnet invariant given by (5), respectively. Now, we calculate the energy functional $E_{\mathcal{L}}$ for the Lagrangian density (16) which has the form

$$E_{\mathcal{L}} = 2a^3 \left[3F_R \frac{\dot{a}^2}{a^2} + 3 \frac{\dot{a}}{a} \dot{F}_R + 12 \frac{\dot{a}^3}{a^3} \dot{F}_G - \frac{1}{2}(RF_R + GF_G - F) - \kappa^2 \rho \right]. \quad (20)$$

It is explicitly seen that the energy function $E_{\mathcal{L}}$ vanishes due to the (0,0)-field Equation (6).

Let us consider the Noether symmetry conditions (12) for the point-like Lagrangian (16) to seek the dependent variables $\xi, \eta^1, \eta^2, \eta^3$ which will be solved in order that the Lagrangian (16) would admit any Noether symmetry (11). For the flat FLRW spacetime (4), the Noether symmetry conditions (12) yield 27 partial differential equations as the following

$$F_{GR}\xi_{,a} = 0, \quad F_{GG}\xi_{,a} = 0, \quad F_{GR}\xi_{,R} = 0, \quad F_{RR}\xi_{,R} = 0, \quad F_{GR}\xi_{,G} = 0, \quad F_{GG}\xi_{,G} = 0,$$

$$F_{RR}\eta^1_{,R} = 0, \quad F_{GR}\eta^1_{,R} = 0, \quad F_{GR}\eta^1_{,G} = 0, \quad F_{GG}\eta^1_{,G} = 0, \quad F_{GG}\xi_{,R} + F_{GR}\xi_{,G} = 0,$$

$$F_{GR}\xi_{,R} + F_{RR}\xi_{,G} = 0, \quad F_{GR}\eta^1_{,R} + F_{RR}\eta^1_{,G} = 0, \quad F_{GG}\eta^1_{,R} + F_{GR}\eta^1_{,G} = 0, \quad F_{GR}\eta^2_{,a} + F_{GG}\eta^3_{,a} = 0,$$

$$6a \left(2F_R \eta^1_{,t} + aF_{RR}\eta^2_{,t} + aF_{GR}\eta^3_{,t} \right) + V\xi_{,a} + K_{,a} = 0, \quad 6a^2 F_{RR}\eta^1_{,t} + V\xi_{,R} + K_{,R} = 0,$$

$$6a^2 F_{GR}\eta^1_{,t} + V\xi_{,G} + K_{,G} = 0, \quad 4 \left(F_{GR}\eta^2_{,t} + F_{GG}\eta^3_{,t} \right) - 3aF_R\xi_{,a} = 0,$$

$$4F_{GR}\eta^1_{,t} - a \left(F_R \xi_{,R} + aF_{RR}\xi_{,a} \right) = 0, \quad 4F_{GG}\eta^1_{,t} - a \left(F_R \xi_{,G} + aF_{GR}\xi_{,a} \right) = 0,$$

$$F_{GRR}\eta^2 + F_{GGR}\eta^3 + F_{GG}\eta^3_{,R} + F_{GR} \left(3\eta^1_{,a} + \eta^2_{,R} - 3\xi_{,t} \right) = 0, \quad (21)$$

$$F_{GGR}\eta^2 + F_{GGG}\eta^3 + F_{GR}\eta^2_{,G} + F_{GG} \left(3\eta^1_{,a} + \eta^3_{,G} - 3\xi_{,t} \right) = 0,$$

$$F_R \left(\frac{\eta^1}{a} + 2\eta^1_{,a} - \xi_{,t} \right) + F_{RR}\eta^2 + F_{GR}\eta^3 + a \left(F_{RR}\eta^2_{,a} + F_{GR}\eta^3_{,a} \right) = 0,$$

$$F_{RR} \left(2\frac{\eta^1}{a} + \eta^1_{,a} + \eta^2_{,R} - \xi_{,t} \right) + \frac{2}{a} F_R \eta^1_{,R} + F_{RRR}\eta^2 + F_{RRG}\eta^3 + F_{GR}\eta^3_{,R} = 0,$$

$$F_{GR} \left(2\frac{\eta^1}{a} + \eta^1_{,a} + \eta^3_{,G} - \xi_{,t} \right) + \frac{2}{a} F_R \eta^1_{,G} + F_{RRG}\eta^2 + F_{GGR}\eta^3 + F_{RR}\eta^2_{,G} = 0,$$

$$V_{,a}\eta^1 + V_{,R}\eta^2 + V_{,G}\eta^3 + V\xi_{,t} + K_{,t} = 0,$$

where V is defined as $V(a, R, G) = a^3(RF_R + GF_G - F) + 2\kappa^2 \rho_{m0} a^{-3w}$, which can be considered as an effective potential for the $F(R,G)$ gravity. Here, R and G act as two different scalar fields whose regimes can lead different phases of the cosmological evolution.

We note here that $\eta^1 = 0, \eta^2 = 0, \eta^3 = 0, K = const.$ and $\xi = const.$ are trivial solutions for the Noether symmetry Equation (21). This result implies that any form of $F(R,G)$ function admits the trivial Noether symmetry $X_1 = \partial/\partial t$, i.e., energy conservation, whose Noether first integral or the Hamiltonian of the system vanishes, $I = -E_\mathcal{L} = 0$. In the following section, we consider the form of $F(R,G)$ to find the corresponding Noether symmetries and solutions to the corresponding first integrals for each of the vacuum an the non-vacuum cases.

4. Noether Symmetries and Cosmological Solutions

Using the symmetry condition (12) to the point-like Lagrangian (16), which will fix the form of $F(R,G)$, several different cases were classified in [25] according to whether the derivative F_{RG} vanishes or not. If $F_{RG} = 0$, it means $F(R,G) = f(R) + g(G)$, which is considered as the **case (i)** below, taking $f(R) = f_0 R^n$ and $g(G) = g_0 G^m$. Otherwise, if $F_{RG} \neq 0$, we will take the form of $F(R,G)$ function as the **case (ii)**, i.e., $F(R,G) = f_0 R^n G^m$, where n and m are real numbers.

4.1. Vacuum Case

In this case, we assume the vacuum where $L_m = 0$, i.e., $\rho_{m0} = 0$.

Case (i): $F(R,G) = f(R) + g(G)$. For this case, we choose the functional forms $f(R) = f_0 R^n$ and $g(G) = g_0 G^m$. Then we examine the following subcases where the powers n and m are fixed to some values, which are compatible with the Noether symmetries.

- $n = m = 1$: Then, the Noether symmetry Equation (21) imply that

$$\xi = c_1 + c_2 t + c_3 \frac{t^2}{2}, \quad \eta^1 = \frac{a}{3}(c_2 + c_3 t) + \frac{c_4 t + c_5}{\sqrt{a}}, \quad \eta^2, \eta^3 \text{ arbitrary,} \quad (22)$$

$$K = -\frac{4}{3} f_0 c_3 a^3 - 8 f_0 c_4 a^{\frac{3}{2}}. \quad (23)$$

This solution to Equation (21) was given in [25] by (55) together with non-trivial function (23). Thus, the Noether symmetry generators from the solution (22) together with (23) take the following forms:

$$X_1 = \frac{\partial}{\partial t}, \quad X_2 = \frac{1}{\sqrt{a}} \frac{\partial}{\partial a}, \quad X_3 = 3t \frac{\partial}{\partial t} + a \frac{\partial}{\partial a} \quad \text{with } K = 0, \quad (24)$$

$$X_4 = \frac{3t^2}{2} \frac{\partial}{\partial t} + ta \frac{\partial}{\partial a} \quad \text{with } K = -4 f_0 a^3; \quad X_5 = \frac{t}{\sqrt{a}} \frac{\partial}{\partial a} \quad \text{with } K = -8 f_0 a^{\frac{3}{2}}, \quad (25)$$

which give the non-vanishing commutators

$$[X_1, X_3] = 3X_1, \quad [X_1, X_4] = X_3, \quad [X_1, X_5] = X_2, \quad (26)$$

$$[X_2, X_3] = \frac{3}{2} X_2, \quad [X_2, X_4] = X_5, \quad [X_3, X_4] = 3X_4, \quad [X_3, X_5] = \frac{3}{2} X_5. \quad (27)$$

The first integrals of the above vector fields are the Hamiltonian, $I_1 = -E_\mathcal{L} = 0$, and the quantities

$$I_2 = -12 f_0 \sqrt{a} \dot{a}, \quad I_3 = -12 a^2 \dot{a}, \quad I_4 = -4 f_0 a^2 (3t\dot{a} - a), \quad I_5 = -4 f_0 \sqrt{a} (3t\dot{a} - 2a). \quad (28)$$

Here we note that it is only found one Noether symmetry in Reference [19] which is X_2 given in (24), and the remaining ones are not appeared in this reference. It follows from $E_\mathcal{L} = 0$ that

$\dot{a} = 0$, that is, $a(t) = a_0 = constant$, which is the Minkowski spacetime recovered in vacuum and so $I_2 = I_3 = 0$, $I_4 = 4f_0 a_0^3$ and $I_5 = 8f_0 a_0^{3/2}$ by (28).

- n arbitrary (with $n \neq 0, 1, \frac{3}{2}, \frac{7}{8}$), $m = 1$: For this case, it follows from (21) that there are *two* Noether symmetries,

$$X_1 = \frac{\partial}{\partial t}, \quad X_2 = 3t\frac{\partial}{\partial t} + (2n-1)a\frac{\partial}{\partial a} - 6R\frac{\partial}{\partial R}, \tag{29}$$

which gives the non-vanishing Lie algebra $[X_1, X_2] = 3X_1$. The first integrals are $I_1 = -E_{\mathcal{L}} = 0$, that means

$$\frac{\dot{a}^2}{a^2} + (n-1)\frac{\dot{a}\dot{R}}{aR} - \frac{(n-1)}{6n}R = 0, \tag{30}$$

by using (20), and

$$I_2 = 6f_0 n a^3 R^{n-1} \left[2(n-2)\frac{\dot{a}}{a} - (n-1)(2n-1)\frac{\dot{R}}{R} \right], \tag{31}$$

for X_1 and X_2, respectively. Then, solving the first integral (31) in terms of a, one gets

$$a(t) = R^{-\frac{(n-1)(2n-1)}{2(n-2)}} \left[a_0 + \frac{I_2}{4 f_0 n (n-2)} \int R^{\frac{(n-1)(8n-7)}{2(2-n)}} dt \right]^{\frac{1}{3}}, \tag{32}$$

where a_0 is an integration constant, and $n \neq 2$. Substituting R given in (5) to the Equation (30), it follows from the integration of resulting equation with respect to t that

$$a^{-\frac{1}{n-1}} \dot{a} = a_1 R^n, \tag{33}$$

which is a constraint equation for a, and it gives

$$\frac{\dot{a}}{a} = a_1 R^2, \tag{34}$$

for $n = 2$, where a_1 is a constant of integration. Thus, the curvature scalar R given by (5) together with the relation (34) becomes

$$\dot{R} + a_1 R^3 = \frac{1}{12 a_1}, \tag{35}$$

which is Abel's differential equation of first kind, and has the following solution

$$R(t) = 4a_1^2(a_1 t - a_2) \left[1 + \frac{4a_1^2(a_1 t - a_2)}{\Delta(t)} \right] + \Delta(t), \tag{36}$$

where a_2 is an integration constant, and $\Delta(t)$ is defined as

$$\Delta(t) = a_1^{2/3} \left[64 a_1^4 (a_1 t - a_2)^3 - 3 + 3\sqrt{2}\sqrt{3 - 64 a_1^2 (a_1 t - a_2)^3} \right]^{\frac{1}{3}}.$$

The first integral (31) for $n = 2$ yields $I_2 = -36 f_0 a^3 \dot{R}$, and then the Equation (35) gives rise to the scale factor as

$$a(t) = \left[\frac{a_1 I_2}{3 f_0 (12 a_1^2 R(t)^3 - 1)} \right]^{\frac{1}{3}}. \tag{37}$$

- $n = \frac{3}{2}, m = 1$: This case admits extra Noether symmetries as pointed out in Reference [30]. The existence of the extra Noether symmetries put even further first integrals which raise the possibility to find an exact solution. The Noether symmetries obtained from (21) are X_1 and

$$X_2 = \frac{1}{a}\frac{\partial}{\partial a} - \frac{2R}{a^2}\frac{\partial}{\partial R}, \quad X_3 = 3t\frac{\partial}{\partial t} + 2a\frac{\partial}{\partial a} - 6R\frac{\partial}{\partial R}, \quad X_4 = tX_2 \quad \text{with } K = -9f_0 a\sqrt{R}, \tag{38}$$

with the non-vanishing Lie brackets

$$[X_1, X_3] = 3X_1, \quad [X_1, X_4] = X_2, \quad [X_2, X_3] = 4X_2, \quad [X_3, X_4] = -X_4. \tag{39}$$

The corresponding Noether constants are $I_1 = -E_{\mathcal{L}} = 0$, which give

$$\frac{\dot{a}^2}{a^2} + \frac{\dot{a}\dot{R}}{2aR} - \frac{R}{18} = 0, \tag{40}$$

and

$$I_2 = -9f_0 a\sqrt{R}\left(\frac{\dot{a}}{a} + \frac{\dot{R}}{2R}\right), \quad I_3 = -9f_0 a^3\sqrt{R}\left(\frac{\dot{a}}{a} + \frac{\dot{R}}{R}\right), \quad I_4 = I_2 t + 9f_0 a\sqrt{R}. \tag{41}$$

Using above first integrals, we find the scale factor and the Ricci scalar as follows:

$$a(t) = \frac{1}{6\tilde{I}_2}\sqrt{(\tilde{I}_2 t - \tilde{I}_4)^4 + 18\tilde{I}_2\tilde{I}_3}, \quad R(t) = \frac{36\tilde{I}_2^2(\tilde{I}_2 t - \tilde{I}_4)^2}{(\tilde{I}_2 t - \tilde{I}_4)^4 + 18\tilde{I}_2\tilde{I}_3}, \tag{42}$$

where it is defined $\tilde{I}_2 = -9f_0 I_2$, $\tilde{I}_3 = -9f_0 I_3$ and $\tilde{I}_4 = -9f_0 I_4$.

- $n = \frac{7}{8}$, $m = 1$: In addition to X_1, this case includes extra *two* Noether symmetries [30]

$$X_2 = 4t\frac{\partial}{\partial t} + a\frac{\partial}{\partial a} - 8R\frac{\partial}{\partial R}, \quad X_3 = 2t^2\frac{\partial}{\partial t} + ta\frac{\partial}{\partial a} - 8tR\frac{\partial}{\partial R} \quad \text{with } K = -\frac{21}{4}f_0 a^3 R^{-\frac{1}{8}}. \tag{43}$$

Then the non-zero Lie brackets are

$$[X_1, X_2] = 4X_1, \quad [X_1, X_3] = X_2, \quad [X_2, X_3] = 4X_3. \tag{44}$$

Thus the first integrals of this case are $I_1 = -E_{\mathcal{L}} = 0$, which yield

$$\frac{\dot{a}^2}{a^2} - \frac{\dot{a}\dot{R}}{8aR} + \frac{R}{42} = 0, \tag{45}$$

and

$$I_2 = \frac{21}{4}f_0 R^{-\frac{1}{8}}\left(-3a^2\dot{a} + a^3\frac{\dot{R}}{8R}\right), \quad I_3 = I_2 t + \frac{21}{4}f_0 a^3 R^{-\frac{1}{8}}. \tag{46}$$

Substituting the Ricci scalar R given in (5) to the Equation (45), and integrating the resulting equation, one gets

$$a^8 \dot{a} = a_0 R^{\frac{7}{8}}, \tag{47}$$

where a_0 is a constant of integration. Defining $\tilde{I}_2 = -4I_2/(21f_0)$ and $\tilde{I}_3 = -4I_3/(21f_0)$, the first integrals (46) become

$$\tilde{I}_2 = \left(a^3 R^{-\frac{1}{8}}\right)', \quad \tilde{I}_3 = \tilde{I}_2 t - a^3 R^{-\frac{1}{8}}, \tag{48}$$

which give

$$a(t) = \left[R^{\frac{1}{8}}(\tilde{I}_2 t - \tilde{I}_3)\right]^{\frac{1}{3}}. \tag{49}$$

Putting the latter form of scale factor into (47), after integration for R, one finds

$$R(t) = \frac{(\tilde{I}_2 t - \tilde{I}_3)^4}{\left[\frac{2a_0}{\tilde{I}_2} + R_0(\tilde{I}_2 t - \tilde{I}_3)^6\right]^2}, \tag{50}$$

then the scale factor becomes

$$a(t) = \sqrt{\bar{I}_2 t - \bar{I}_3} \left[\frac{2a_0}{\bar{I}_2} + R_0(\bar{I}_2 t - \bar{I}_3)^6 \right]^{-\frac{1}{12}},\qquad(51)$$

where R_0 is an integration constant.

- $n = \frac{1}{2}$, $m = \frac{1}{4}$: Here there exist *two* Noether symmetries,

$$\mathbf{X}_1 = \frac{\partial}{\partial t},\qquad \mathbf{X}_2 = t\frac{\partial}{\partial t} - 2R\frac{\partial}{\partial R} - 4G\frac{\partial}{\partial G},\qquad(52)$$

with the non-vanishing Lie algebra $[\mathbf{X}_1, \mathbf{X}_2] = \mathbf{X}_1$. Then the first integrals related with these Noether symmetries are $I_1 = -E_{\mathcal{L}} = 0$, which yield

$$\frac{\dot{a}^2}{a^2} - \frac{\dot{a}\dot{R}}{2aR} + \frac{R}{6} + \frac{g_0}{4f_0}\sqrt{R}G^{\frac{1}{4}}\left(1 - \frac{6\dot{a}^3\dot{G}}{a^3G^2}\right) = 0,\qquad(53)$$

and

$$I_2 = -6\left(\frac{f_0}{2}R^{-\frac{1}{2}}a^2\dot{a} + g_0 G^{-\frac{3}{4}}\dot{a}^3\right).\qquad(54)$$

The Noether symmetries (52) have also been obtained in [25] with the symmetry vector (41). In order to determine the invariant functions of the Noether symmetry $\mathbf{X}_2$ given in (52), after solving the Lagrange system [28]

$$\frac{dt}{t} = \frac{dR}{-2R} = \frac{dG}{-4G},\qquad(55)$$

one find the solutions for $R(t)$ and $G(t)$ as

$$R(t) = \frac{R_0}{t^2},\qquad G(t) = \frac{G_0}{t^4}.\qquad(56)$$

Here we get a power-law solution $a(t) = a_0 t^2$, where the Equations (53) and (54) yield

$$g_0 = -\frac{4f_0 G^{\frac{3}{4}}(R_0 + 36)}{6\sqrt{R_0}(G_0 + 192)},\qquad I_2 = \frac{6f_0 a_0^3}{\sqrt{R_0}}\left[\frac{16(R_0 + 36)}{3(G_0 + 192)} - 1\right].\qquad(57)$$

For the obtained R and G in (56), if we take into account the definitions of R and G given by (5), then we get the values of constants as $R_0 = 36$ and $G_0 = 192$. Thus, the relation (57) becomes $g_0 = -(4/3)^{1/4} f_0$ and $I_2 = 0$.

- $n = 1$, $m = \frac{1}{2}$: In this case, there are also *two* Noether symmetries

$$\mathbf{X}_1 = \frac{\partial}{\partial t},\qquad \mathbf{X}_2 = 3t\frac{\partial}{\partial t} + a\frac{\partial}{\partial a} - 12G\frac{\partial}{\partial G},\qquad(58)$$

which give rise to the first integrals $I_1 = -E_{\mathcal{L}} = 0$, which can be written by using (20) as follows

$$\frac{\dot{a}^2}{a^2} + \frac{g_0}{f_0}\sqrt{G}\left(\frac{1}{12} - \frac{\dot{a}^3 \dot{G}}{a^3 G^2}\right) = 0,\qquad(59)$$

and

$$I_2 = 6\left[-2f_0 a^2 \dot{a} + \frac{g_0}{\sqrt{G}} a\dot{a}^2\left(\frac{\dot{G}}{G} - 4\frac{\dot{a}}{a}\right)\right].\qquad(60)$$

Here we have to point out that the Noether symmetries (58) are of the form (41) in [25]. By solving the Lagrange system for $\mathbf{X}_2$

$$\frac{dt}{3t} = \frac{da}{a} = \frac{dG}{-12G},\qquad(61)$$

the invariant functions can be obtained as

$$a(t) = a_0 t^{\frac{1}{3}}, \qquad G(t) = \frac{G_0}{t^4}. \tag{62}$$

Then, substituting these into the Equations (59) and (60), we can find the constraint relations

$$g_0 = -\frac{12 f_0 \sqrt{|G_0|}}{9 G_0 + 16}, \qquad I_2 = 4 f_0 a_0^3 \left[\frac{32}{3(9 G_0 + 16)} - 1 \right]. \tag{63}$$

Here the definition of G in terms of $a(t)$ by (5) gives rise to the value $G_0 = -16/27$, which means $g_0 = -\sqrt{3} f_0/2$ and $I_2 = 0$ after substituting G_0 into (63).

Case (ii): $F(R,G) = f_0 f(R) g(G)$. Here we will consider the functional forms $f(R) = R^n$ and $g(G) = G^m$. These types of functional forms are appeared in some references such as [9,19,25,31,32].

- n, m arbitrary: This theory admits the following Noether symmetries

$$X_1 = \frac{\partial}{\partial t}, \qquad X_2 = 3t \frac{\partial}{\partial t} + (4m + 2n - 1) a \frac{\partial}{\partial a} - 6R \frac{\partial}{\partial R} - 12 G \frac{\partial}{\partial G}, \tag{64}$$

and the corresponding first integrals are

$$\frac{\dot{a}^2}{a^2} + \frac{\dot{a}}{a}\left[(n-1)\frac{\dot{R}}{R} + m\frac{\dot{G}}{G}\right] + \frac{12mR \dot{a}^3}{G a^3}\left[\frac{\dot{R}}{R} + \frac{(m-1)}{n}\frac{\dot{G}}{G}\right] - (n+m-1)\frac{R}{6n} = 0, \tag{65}$$

$$I_2 = 6 f_0 R^n G^m a^3 \left\{ 2(2m + n - 2) \frac{\dot{a}}{a} \left(\frac{n}{R} + \frac{4m\dot{a}^2}{G a^2} \right) \right.$$
$$\left. - (4m + 2n - 1) \left(\frac{n}{R} \left[(n-1)\frac{\dot{R}}{R} + m\frac{\dot{G}}{G} \right] + \frac{4m\dot{a}^2}{G a^2} \left[n\frac{\dot{R}}{R} + (m-1)\frac{\dot{G}}{G} \right] \right) \right\}. \tag{66}$$

These are very general statements and one can find any solution choosing the arbitrary powers n and m. The invariant functions for the vector field X_2 can be determined by solving the associated Lagrange system

$$\frac{dt}{3t} = \frac{da}{(4m + 2n - 1)a} = \frac{dR}{-6R} = \frac{dG}{-12G}, \tag{67}$$

which yields

$$a(t) = a_0 t^{\frac{4m+2n-1}{3}}, \qquad R(t) = \frac{R_0}{t^2}, \qquad G(t) = \frac{G_0}{t^4}. \tag{68}$$

Now one can find the constants R_0 and G_0 in terms of powers of $a(t)$ as $R_0 = 2(4m + 2n - 1)(8m + 4n - 5)/3$ and $G_0 = 16(4m + 2n - 1)^3(2m + n - 2)/27$ by considering (5). Thus, using the obtained quantities by (68) in (65) and (66), we find the constraints

$$(10m + 2n - 1)(4m + 2n - 1)(8m + 4n - 5) = 0, \qquad I_2 = 6(2m+n) f_0 a_0^3 R_0^n G_0^m. \tag{69}$$

- $m = 1 - n$: This case is considered in the reference [19] as a simplest non-trivial case with the selection of $n = 2$. In general, the solution of Noether symmetry equations (21) becomes

$$\xi = c_1 + c_2 t, \quad \eta^1 = c_2(3 - 2n)\frac{a}{3}, \quad \eta^2 = \eta^2(t, a, R, G), \quad \eta^3 = \frac{G}{R}(-2 c_2 R + \eta^2), \quad K = c_3, \tag{70}$$

where c_i's ($i = 1, 2, 3$) are constant parameters, and η^2 is an arbitrary function of t, a, R and G. This arbitrariness means that there are infinitely many Noether symmetries and it gives us to decide a selection of consistent solution for the scale factor a. Therefore, we choose $\eta^2 = -2 c_2 R$ to get a consistent power-law solution for the scale factor a, using the associated Lagrange system. It has to be mentioned here that this type of selection is not necessary for non power-law solutions. We proceed considering $\eta^2 = -2c_2R$ at (70), which yields that there are *two* Noether symmetries

$$X_1 = \frac{\partial}{\partial t}, \quad X_2 = 3t\frac{\partial}{\partial t} + (3-2n)a\frac{\partial}{\partial a} - 6R\frac{\partial}{\partial R} - 12G\frac{\partial}{\partial G}. \tag{71}$$

The first integrals of the above vector fields are

$$\frac{\dot{a}}{a} + (n-1)\left(1 - \frac{4R\dot{a}^2}{Ga^2}\right)\left(\frac{\dot{R}}{R} - \frac{\dot{G}}{G}\right) = 0, \tag{72}$$

and

$$I_2 = -6f_0n\left(\frac{R}{G}\right)^{n-1}a^2\dot{a}\left\{4n - 3 + 24(n-1)\frac{R\dot{a}^2}{Ga^2}\right\}. \tag{73}$$

By choosing the variable $\zeta = \frac{R}{G}$, the first integrals (72) and (73) take the form

$$\frac{\dot{a}}{a} + (n-1)\left(\frac{1}{\zeta} - 4\frac{\dot{a}^2}{a^2}\right)\dot{\zeta} = 0, \quad 6f_0n\zeta^{n-1}a^2\dot{a}\left(4n - 3 + 24(n-1)\zeta\frac{\dot{a}^2}{a^2}\right) + I_2 = 0. \tag{74}$$

For the selection of $n = 2$, it is seen that the first equation of (74) is similar to the Equation (38) of the Reference [19]. After solving the associated Lagrange system for the vector field X_2 given in (71), we have

$$a(t) = a_0 t^{\frac{3-2n}{3}}, \quad R(t) = \frac{R_0}{t^2}, \quad G(t) = \frac{G_0}{t^4}, \tag{75}$$

Using the definitions of R and G in (5), the constants R_0 and G_0 are found as $R_0 = (16n^2 - 36n + 18)/3$ and $G_0 = 16n(2n-3)^3/27$.

As a simple selection for the component η^2, we choose $\eta^2 = 0$ in (70). Then there are again *two* Noether symmetries

$$X_1 = \frac{\partial}{\partial t}, \quad X_2 = 3t\frac{\partial}{\partial t} + (3-2n)a\frac{\partial}{\partial a} - 6G\frac{\partial}{\partial G}. \tag{76}$$

The Noether constants for these vector fields are $I_1 = -E_\mathcal{L} = 0$, which yield the same relation with (72), and

$$I_2 = 6f_0na^3\left(\frac{R}{G}\right)^{n-1}\left\{(2n-3)\left[\frac{2\dot{a}}{a} + (n-1)\left(1 - \frac{4R\dot{a}^2}{Ga^2}\right)\left(\frac{\dot{R}}{R} - \frac{\dot{G}}{G}\right)\right]\right.$$
$$\left. + 2(1-n)\frac{\dot{a}}{a}\left(3 - \frac{4R\dot{a}^2}{Ga^2}\right)\right\}, \tag{77}$$

which becomes

$$I_2 = 6f_0n\left(\frac{R}{G}\right)^{n-1}a^2\dot{a}\left\{2n - 3 + 2(1-n)\left(3 - \frac{4R\dot{a}^2}{Ga^2}\right)\right\}, \tag{78}$$

by using (72). It is easily seen that the Noether symmetry X_2 in (76) does not have a consistent solution for a power-law form of the scale factor a. The reason of this inconsistency is follows from analysing of the associated Lagrange system for X_2 in such a way that it gives the scale factor $a(t)$ as in (75), but $G(t) = G_0 t^{-2}$ which contradicts the form of $G(t) \sim t^{-4}$ from the definition (5).

4.2. Non-Vacuum Case

In this section, we assume that the matter has a constant equation of state (EoS) parameter $w \equiv p/\rho$ with the perfect fluid matter. We mention that Equations (9) and (10) imply that the contribution of the $F(R,G)$ gravity can formally be included in the effective energy density and pressure of the universe. For the GR with $F(R,G) = R$, $\rho_{eff} = \rho$ and $p_{eff} = p$, and so the Equations (9) and (10) are the FLRW equations.

Case (i): $F(R,G) = f(R) + g(G)$.

For this case, we again choose $f(R) = f_0 R^n$, $g(G) = g_0 G^m$, and determine the Noether symmetries in the presence of matter.

- $n = m = 1$: This gives the usual GR theory. For some value of the constant EoS parameter, we would like to give the Noether symmetries in the following. First of all, for $w = -1$ (the cosmological constant), the present value of the energy density becomes $\rho_{m0} = 4 f_0/(3\kappa^2 \alpha^2)$, and there exist *five* Noether symmetries

$$X_1 = \frac{\partial}{\partial t}, \quad X_2 = \frac{e^{\frac{t}{\alpha}}}{\sqrt{a}} \frac{\partial}{\partial a} \quad \text{with } K = -\frac{8 f_0}{\alpha} a^{\frac{3}{2}} e^{\frac{t}{\alpha}}, \quad X_3 = \frac{e^{-\frac{t}{\alpha}}}{\sqrt{a}} \frac{\partial}{\partial a} \quad \text{with } K = \frac{8 f_0}{\alpha} a^{\frac{3}{2}} e^{-\frac{t}{\alpha}},$$

$$X_4 = e^{\frac{2t}{\alpha}} \frac{\partial}{\partial t} + \frac{2}{3\alpha} e^{\frac{2t}{\alpha}} a \frac{\partial}{\partial a} \quad \text{with } K = -\frac{16 f_0}{3\alpha^2} a^3 e^{\frac{2t}{\alpha}}, \tag{79}$$

$$X_5 = e^{-\frac{2t}{\alpha}} \frac{\partial}{\partial t} - \frac{2}{3\alpha} e^{-\frac{2t}{\alpha}} a \frac{\partial}{\partial a} \quad \text{with } K = -\frac{16 f_0}{3\alpha^2} a^3 e^{-\frac{2t}{\alpha}},$$

with the non-vanishing commutators

$$[X_1, X_2] = \frac{1}{\alpha} X_2, \quad [X_1, X_3] = -\frac{1}{\alpha} X_3, \quad [X_1, X_4] = \frac{2}{\alpha} X_4,$$

$$[X_1, X_5] = -\frac{2}{\alpha} X_5, \quad [X_2, X_5] = -\frac{2}{\alpha} X_3, \quad [X_3, X_4] = \frac{2}{\alpha} X_2, \quad [X_4, X_5] = -\frac{4}{\alpha} X_1, \tag{80}$$

where α is a constant. Then the first integrals are $I_1 = -E_{\mathcal{L}} = 0$, that gives $\kappa^2 \rho_{m0} = 3 f_0 \dot{a}^2/a^2$, and the quantities

$$\bar{I}_2 = e^{\frac{t}{\alpha}} \sqrt{a} \left(-3\dot{a} + \frac{2}{\alpha} a \right), \quad \bar{I}_3 = -e^{-\frac{t}{\alpha}} \sqrt{a} \left(3\dot{a} + \frac{2}{\alpha} a \right),$$

$$\bar{I}_4 = \frac{2}{3\alpha} e^{\frac{2t}{\alpha}} \left(-3a^2 \dot{a} + \frac{2}{\alpha} a^3 \right), \quad \bar{I}_5 = -\frac{2}{3\alpha} e^{-\frac{2t}{\alpha}} \left(3a^2 \dot{a} + \frac{2}{\alpha} a^3 \right), \tag{81}$$

where we have defined $I_2 = 4 f_0 \bar{I}_2$, $I_3 = 4 f_0 \bar{I}_3$, $I_4 = 4 f_0 \bar{I}_4$ and $I_5 = 4 f_0 \bar{I}_5$. After solving these first integrals for a, we find that the Noether constants become $\bar{I}_3 = 0$, $\bar{I}_5 = 0$, $\bar{I}_4 = \bar{I}_2^2/6$, and the scale factor is

$$a(t) = a_0 \exp\left(-\frac{2t}{3\alpha}\right), \tag{82}$$

where $a_0 = (\alpha/4)^{2/3}$. This is the well-known de Sitter solution.

In the case of $w = -1/2$, we also find *five* Noether symmetries

$$X_1 = \frac{\partial}{\partial t}, \quad X_2 = \frac{1}{\sqrt{a}} \frac{\partial}{\partial a} \quad \text{with } K = -48 f_0 t, \quad X_3 = \frac{t}{\sqrt{a}} \frac{\partial}{\partial a} \quad \text{with } K = -8 f_0 a^{\frac{3}{2}} - 24 f_0 t^2,$$

$$X_4 = t \frac{\partial}{\partial t} + \left(\frac{a}{3} + \frac{3 t^2}{\sqrt{a}} \right) \frac{\partial}{\partial a} \quad \text{with } K = -48 f_0 t a^{\frac{3}{2}} - 48 f_0 t^3, \tag{83}$$

$$X_5 = \frac{t^2}{2} \frac{\partial}{\partial t} + t \left(\frac{a}{3} + \frac{t^2}{\sqrt{a}} \right) \frac{\partial}{\partial a} \quad \text{with } K = -24 f_0 t^2 a^{\frac{3}{2}} - 4 f_0 a^3 - 12 f_0 t^4.$$

Thus the non-vanishing Lie brackets of the above vector fields are

$$[X_1, X_3] = X_2, \quad [X_1, X_4] = X_1 + 6X_3, \quad [X_1, X_5] = X_4,$$
$$[X_2, X_4] = \frac{1}{2}X_2, \quad [X_2, X_5] = \frac{1}{2}X_3, \quad [X_3, X_4] = -\frac{1}{2}X_3, \quad [X_4, X_5] = X_5. \tag{84}$$

Under the change of the Noether constants $I_2 \to 12f_0 I_2$, $I_3 \to 12f_0 I_3$, $I_4 \to 12f_0 I_4$, $I_5 \to 12f_0 I_5$ for the Noether symmetries (83), the first integrals for $X_1, ..., X_5$ become

$$3f_0 \frac{\dot{a}^2}{\sqrt{a}} = \kappa^2 \rho_{m0},$$

$$I_2 = -\sqrt{a}\dot{a} + 4t, \quad I_3 = -t\sqrt{a}\dot{a} + \frac{2}{3}a^{\frac{3}{2}} + 2t^2,$$

$$I_4 = \frac{a^2}{3}\dot{a} - 3t^2\sqrt{a}\dot{a} + 4ta^{\frac{3}{2}} + 4t^3, \quad I_5 = -\frac{t}{3}a^2\dot{a} - t^3\sqrt{a}\dot{a} + 2t^2 a^{\frac{3}{2}} + \frac{a^3}{9} + t^4, \tag{85}$$

Taking into account these first integrals, we find that

$$a(t) = a_0 (4t - I_2)^{\frac{4}{3}}, \tag{86}$$

$$\rho_{m0} = \frac{16 f_0}{\kappa^2}, \quad I_3 = \frac{I_2^2}{8}, \quad I_4 = \frac{I_2^3}{16}, \quad I_5 = \frac{I_2^4}{4}, \tag{87}$$

where $a_0 = (3/16)^{2/3}$.

For $w = 0$ (the dust), the dynamical system admits the following *five* Noether symmetries

$$X_1 = \frac{\partial}{\partial t}, \quad X_2 = \frac{1}{\sqrt{a}}\frac{\partial}{\partial a}, \quad X_3 = \frac{t}{\sqrt{a}}\frac{\partial}{\partial a} \quad \text{with } K = -8f_0 a^{\frac{3}{2}},$$

$$X_4 = t\frac{\partial}{\partial t} + \frac{a}{3}\frac{\partial}{\partial a} \quad \text{with } K = -\kappa^2 \rho_{m0} t, \tag{88}$$

$$X_5 = \frac{t^2}{2}\frac{\partial}{\partial t} + \frac{ta}{3}\frac{\partial}{\partial a} \quad \text{with } K = -\frac{4f_0}{3}a^3 - \kappa^2 \rho_{m0} t^2,$$

and then the non-vanishing commutators are

$$[X_1, X_3] = X_2, \quad [X_1, X_4] = X_1, \quad [X_1, X_5] = X_4,$$
$$[X_2, X_4] = \frac{1}{2}X_2, \quad [X_2, X_5] = \frac{1}{2}X_3, \quad [X_3, X_4] = -\frac{1}{2}X_3, \quad [X_4, X_5] = X_5. \tag{89}$$

The corresponding first integrals of the Noether symmetries (88) are

$$I_1 = -E_\mathcal{L} = 0, \quad I_2 = -12f_0 \sqrt{a}\dot{a}, \quad I_3 = I_2 t + 8f_0 a^{\frac{3}{2}}, \tag{90}$$

$$I_4 = -4f_0 a^2 \dot{a} + 2\kappa^2 \rho_{m0} t, \quad I_5 = 4f_0\left(-ta^2\dot{a} + \frac{1}{3}a^3\right) + \kappa^2 \rho_{m0} t^2. \tag{91}$$

Using the above first integrals one can find the scale factor and the constraints on Noether constants as follows

$$a(t) = a_0 (I_3 - I_2 t)^{\frac{2}{3}}, \tag{92}$$

$$\rho_{m0} = \frac{I_2^2}{48 f_0 \kappa^2}, \quad I_4 = \frac{I_2 I_3}{24 f_0}, \quad I_5 = \frac{I_3^2}{48 f_0}, \tag{93}$$

where $a_0 = (8f_0)^{-2/3}$.

Finally, for $w = 1$ (stiff matter), we find *three* Noether symmetries

$$X_1 = \frac{\partial}{\partial t}, \quad X_2 = 3t\frac{\partial}{\partial t} + a\frac{\partial}{\partial a}, \quad X_3 = \frac{3}{2}t^2\frac{\partial}{\partial t} + ta\frac{\partial}{\partial a} \quad \text{with } K = -4f_0a^3, \quad (94)$$

which yields the non-vanishing Lie algebra: $[X_1, X_2] = 3X_1$, $[X_1, X_3] = X_2$, $[X_2, X_3] = 3X_3$. The Noether constants for X_1, X_2 and X_3 are

$$I_1 = 0 \Rightarrow 3f_0 a^4 \dot{a}^2 = \kappa^2 \rho_{m0}, \quad I_2 = -12 f_0 a^2 \dot{a}, \quad I_3 = I_2 t + 4 f_0 a^3, \quad (95)$$

having the solution

$$a(t) = a_0 \left(I_3 - I_2 t\right)^{\frac{1}{3}}, \quad \rho_{m0} = \frac{I_2^2}{48 f_0 \kappa^2}, \quad (96)$$

where $a_0 = (4f_0)^{-1/3}$.

- n arbitrary (with $n \neq 0, \frac{3}{2}, \frac{7}{8}$), $m = 1$: In this case we have the same Noether symmetries X_1, X_2 given by (29) in the vacuum case. For this case we are led to the constant EoS parameter w as

$$w = \frac{1}{2n - 1}. \quad (97)$$

Using this EoS parameter, the first integral for X_1 gives

$$\frac{\dot{a}^2}{a^2} + (n-1)\frac{\dot{a}\dot{R}}{aR} - \frac{(n-1)}{6n}R = \frac{\kappa^2 \rho_{m0}}{3f_0 n} a^{-\frac{6n}{2n-1}} R^{1-n}. \quad (98)$$

The scale factor for this case has the same form with (32), which is not a power-law form, and the Equations (31) and (98) are the constraint equations to be considered. It is interesting to see from (97) that $n = 0$ if $w = -1$ (the cosmological constant) which is excluded in this case, $n = 1$ if $w = 1$ (the stiff matter), and $n = 2$ if $w = 1/3$ (the relativistic matter), etc. Therefore, this case includes some important values of the EoS parameter.

This model admits power-law solution of the form $a(t) = a_0 t^{(2n-1)/3}$, and the Ricci scalar and the GB invariant become $R(t) = R_0 t^{-2}$ and $G(t) = G_0 t^{-4}$, where the constants R_0 and G_0 follow from (5) as $R_0 = 2(2n-1)(4n-5)/3$ and $G_0 = 16(n-2)(2n-1)^3/27$. Meanwhile, the constraint relations (31) and (98) for this power-law scale factor give

$$\rho_{m0} = \frac{f_0}{3\kappa^2}(5 - 4n)(2n-1)^2 R_0^{n-1} a_0^{\frac{6n}{2n-1}}, \quad I_2 = 4n(4n-5)(2n-1)f_0 a_0^3 R_0^{n-1}, \quad (99)$$

where $n \neq \frac{1}{2}, \frac{5}{4}$ due to $\rho_{m0} \neq 0$. The power-law solution of this case works for $n = 2$, i.e., $w = 1/3$, and it gives negative energy density as $\rho_{m0} = -54 f_0 a_0^4 / \kappa^2$.

- $n = \frac{3}{2}, m = 1$: We will firstly consider the case $w = -2/3$ which requires that $\rho_{m0} = \alpha/2\kappa^2$, α is a constant. For this case, there are *three* Noether symmetries X_1, X_2 with $K = -2\alpha t$, and $X_3 = tX_2$ with $K = -9f_0 a\sqrt{R} - \alpha t^2$, where X_2 is the same as given in (38). Thus the constants of motion for the vector fields X_1, X_2 and X_3 are, respectively,

$$\frac{\dot{a}^2}{a^2} + \frac{\dot{a}\dot{R}}{2aR} - \frac{R}{18} = \frac{\alpha}{9f_0 a\sqrt{R}}, \quad (100)$$

and

$$I_2 = -9f_0 a \sqrt{R}\left(\frac{\dot{a}}{a} + \frac{\dot{R}}{2R}\right) + 2\alpha t, \quad I_3 = I_2 t - \alpha t^2 + 9 f_0 a \sqrt{R}. \quad (101)$$

Using above Noether constants, the scale factor for the case $w = -2/3$ yields

$$a(t) = \frac{1}{9f_0\sqrt{R}}(I_3 - I_2 t + \alpha t^2). \tag{102}$$

For $w = 0$, the Noether symmetries are identical to vector fields given in (38), but X_3 has a non-zero function $K = -6\kappa^2 \rho_{m0}$. Redefining the Noether constants such as $I_2 = -9f_0 \bar{I}_2$, $I_3 = -9f_0 \bar{I}_3$ and $I_4 = -9f_0 \bar{I}_4$, after some algebra, we find the scale factor

$$a(t) = \frac{1}{6\bar{I}_2}\sqrt{(\bar{I}_2 t - \bar{I}_4)^4 + \frac{16}{f_0}\kappa^2 \rho_{m0}(\bar{I}_2 t - \bar{I}_4) + 18\bar{I}_2 \bar{I}_3 + \frac{12}{f_0}\kappa^2 \rho_{m0} \bar{I}_4}, \tag{103}$$

and the Ricci scalar

$$R(t) = \frac{36 \bar{I}_2^2 (\bar{I}_2 t - \bar{I}_4)^2}{a(t)^2}. \tag{104}$$

- $n = \frac{7}{8}, m = 1$: If $w = 4/3$, there are *three* Noether symmetries, in which X_2 and X_3 are the same as (43). The first integral $I_1 = -E_{\mathcal{L}} = 0$ due to X_1 becomes

$$\frac{\dot{a}^2}{a^2} - \frac{\dot{a}\dot{R}}{8aR} + \frac{R}{42} = \frac{8}{21f_0}\kappa^2 \rho_{m0} a^{-7} R^{\frac{1}{8}}. \tag{105}$$

The scale factor for this case has the same form as (49), but now it is difficult to gain the explicit form of $a(t)$ using (105).

For the dust matter ($w = 0$), there are *three* Noether symmetries which are the same form as (43), but the function K is non-trivial such that $K = -8\kappa^2 \rho_{m0} t$ for X_2 and $K = -\frac{21}{4} f_0 a^3 R^{-1/8} - 4\kappa^2 \rho_{m0} t^2$ for X_3. Thus the first integrals for X_1, X_2 and X_3 are given by, respectively,

$$I_1 = -E_{\mathcal{L}} = 0 \quad \Leftrightarrow \quad \frac{\dot{a}^2}{a^2} - \frac{\dot{a}\dot{R}}{8aR} + \frac{R}{42} = \frac{8\kappa^2 \rho_{m0} R^{\frac{1}{8}}}{21 f_0 a^3}, \tag{106}$$

and

$$I_2 = \frac{21}{4} f_0 R^{-\frac{1}{8}}\left(-3a^2 \dot{a} + a^3 \frac{\dot{R}}{8R}\right) + 8\kappa^2 \rho_{m0} t, \quad I_3 = I_2 t + \frac{21}{4} f_0 a^3 R^{-\frac{1}{8}} + 4\kappa^2 \rho_{m0} t^2. \tag{107}$$

After redefining $I_2 = -\frac{21}{4} f_0 \bar{I}_2$ and $I_3 = -\frac{21}{4} f_0 \bar{I}_3$, the second relation in (107) implies the scale factor

$$a(t) = R^{\frac{1}{24}}\left(\alpha t^2 + \bar{I}_2 t - \bar{I}_3\right)^{\frac{1}{3}}, \tag{108}$$

where $\alpha \equiv \frac{16\kappa^2 \rho_{m0}}{21 f_0}$.

- $n = \frac{1}{2}, m = \frac{1}{4}$: In addition to $X_1 = \partial/\partial t$, the condition for existing extra Noether symmetry is that the EoS parameter should be $w = 0$. Thus, an additional Noether symmetry is obtained as follows

$$X_2 = t\frac{\partial}{\partial t} - 2R\frac{\partial}{\partial R} - 4G\frac{\partial}{\partial G} \quad \text{with} \quad K = -2\kappa^2 \rho_{m0} t. \tag{109}$$

Then the Noether constants for these vector fields yield

$$H^2 - \frac{\dot{R}}{2R}H + \frac{R}{6} + \frac{g_0}{4f_0}\sqrt{R}G^{\frac{1}{4}}\left(1 - 6H^3 \frac{\dot{G}}{G^2}\right) = \frac{2\kappa^2 \rho_{m0} \sqrt{R}}{3f_0 a^3}, \tag{110}$$

and
$$I_2 = -3a^3 \left(f_0 R^{-\frac{1}{2}} H + 2g_0 G^{-\frac{3}{4}} H^3\right) + 2\kappa^2 \rho_{m0} t, \tag{111}$$

which can be written as

$$H^3 + \frac{f_0}{2g_0}\left(\frac{G^3}{R^2}\right)^{\frac{1}{4}} H - \frac{(2\kappa^2 \rho_{m0} t - I_2)}{6g_0 a^3} G^{\frac{3}{4}} = 0. \tag{112}$$

This is a cubic equation for H.

- $n = 1, m = \frac{1}{2}$: For $w = 0$, it is found the Noether symmetries X_1 and X_2 which are the same as (58), but X_2 has the non-trivial function $K = -6\kappa^2 \rho_{m0} t$. Then the Noether constants for X_1 and X_2 take the following forms

$$H^2 + \frac{g_0}{f_0}\sqrt{G}\left(\frac{1}{12} - H^3 \frac{\dot{G}}{G^2}\right) = \frac{\kappa^2 \rho_{m0}}{3a^3}, \tag{113}$$

and

$$I_2 = 6a^3 \left[-2f_0 H + \frac{g_0 H^2}{\sqrt{G}}\left(\frac{\dot{G}}{G} - 4H\right) + \frac{\kappa^2 \rho_{m0} t}{a^3}\right]. \tag{114}$$

For $w = 1$, there are *two* Noether symmetries that are the same as (58), where $K = 0$ for both of symmetries. Therefore, the first integral for X_2 is the same as (60), and the first integral for X_1 becomes

$$H^2 + \frac{g_0}{f_0}\sqrt{G}\left(\frac{1}{12} - H^3 \frac{\dot{G}}{G^2}\right) = \frac{\kappa^2 \rho_{m0}}{3a^6}. \tag{115}$$

Case (ii): $F(R,G) = f_0 f(R) g(G)$. The functional forms $f(R) = R^n$ and $g(G) = G^m$ are also assumed in this section.

- n, m arbitrary: For this case, there exist *two* Noether symmetries which are the same as (64), and the EoS parameter becomes

$$w = \frac{1}{4m + 2n - 1}. \tag{116}$$

The first integral for X_2 is the same as (66), and it has the following form

$$H^2 + \left[(n-1)\frac{\dot{R}}{R} + m\frac{\dot{G}}{G}\right] H + \frac{12mR}{G}\left[\frac{\dot{R}}{R} + \frac{(m-1)}{n}\frac{\dot{G}}{G}\right] H^3 - (n+m-1)\frac{R}{6n} = \frac{\kappa^2 \rho_{m0} a^{\frac{6(2n+m)}{4m+2n-1}}}{3f_0 n R^{n-1} G^m}, \tag{117}$$

for X_1. Note that the Equation (116) includes important EoS parameters, for example $w = -1$ if $n = -2m$; $w = -1/3$ if $n = -(2m+1)$; $w = 1/3$ if $n = 2(1-m)$ and $w = 1$ if $n = 2, m = -1/2$. In the case of dust matter ($w = 0$), we have *two* Noether symmetries given by (64), but where the function K for X_2 is $K = -6\kappa^2 \rho_{m0} t$. Therefore, the first integrals for X_1 and X_2 are, respectively,

$$H^2 + \left[(n-1)\frac{\dot{R}}{R} + m\frac{\dot{G}}{G}\right] H + \frac{12mR}{G}\left[\frac{\dot{R}}{R} + \frac{(m-1)}{n}\frac{\dot{G}}{G}\right] H^3 - (n+m-1)\frac{R}{6n} = \frac{\kappa^2 \rho_{m0} R^{1-n}}{3f_0 n a^3 G^m}, \tag{118}$$

$$I_2 = 6f_0 R^n G^m a^3 \left\{2(n+2m-2)H\left(\frac{n}{R} + \frac{4mH^2}{G}\right)\right.$$
$$\left. -(4m+2n-1)\left(\frac{n}{R}\left[(n-1)\frac{\dot{R}}{R} + m\frac{\dot{G}}{G}\right] + \frac{4mH^2}{G}\left[n\frac{\dot{R}}{R} + (m-1)\frac{\dot{G}}{G}\right]\right)\right\} + 6\kappa^2 \rho_{m0} t. \tag{119}$$

- $m = 1 - n$: In this case, the EoS parameter takes the form $w = \frac{1}{3-2n}$, and this model admits *two* Noether symmetries, which are the same as (71). The first integral due to X_1 yields

$$H^2 + (n-1)H\left(1 - \frac{4R}{G}H^2\right)\left(\frac{\dot{R}}{R} - \frac{\dot{G}}{G}\right) = \frac{\kappa^2 \rho_{m0}}{3 f_0 n}\left(\frac{R}{G}\right)^{n-1} a^{\frac{6(n-1)}{3-2n}}, \qquad (120)$$

and the first integral for X_2 becomes

$$I_2 = 6 f_0 n \left(\frac{R}{G}\right)^{n-1} a^3 H \left\{ 2n - 3 + 2(1-n)\left(3 - \frac{4R}{G}H^2\right) + \frac{\kappa^2 \rho_{m0}}{3 f_0 n}\left(\frac{R}{G}\right)^{1-n} \frac{a^{\frac{6(n-1)}{3-2n}}}{H^2} \right\}, \qquad (121)$$

using (120).

5. Conclusions

In this work, we have considered both the vacuum and the non-vacuum theories of $F(R,G)$ gravity admitting Noether symmetries. First of all, we have obtained the dynamical field equations for those of gravity theories, which also come from the Lagrangian of $F(R,G)$ gravity in the background of spatially flat FLRW spacetime such that it gives rise to the dynamical field equations varying with respect to the configuration space variables. Afterwards we have used the point-like $F(R,G)$ Lagrangian (16) to write out the Noether symmetry equations, and solve them to get the Noether symmetries in both the vacuum and the non-vacuum cases. It has been appeared very rich cosmological structures from the Noether symmetries for the several functional form of the $F(R,G)$ functions in each of the cases.

The main results of this study can be summarized in the following. First of all, we can verify that all the $F(R,G)$ models studied here admit trivial first integral, namely $E_{\mathcal{L}} = 0$, as they should. Secondly, it is obtained the previous results choosing the $F(R,G)$ function, for example, the case (i) in the vacuum recovers the results of [30] on the Noether symmetries for $n = \frac{3}{2}, \frac{7}{8}$. Using the first integrals directly, we found the analytical solutions (42) and (51) for $n = \frac{3}{2}$ and $n = \frac{7}{8}$, respectively. These cases are also generalized to the non-vacuum and it is found analytical solutions (103) for $n = \frac{3}{2}$ related with the EoS parameter $w = -2/3$, and (108) for $n = \frac{7}{8}$ with the EoS parameter $w = 4/3$. For other values of n, the scale factor $a(t)$ is analytically calculated by (32) in the vacuum section of this study. In each of the cases (i) and (ii) for the vacuum and the non-vacuum, we found the first integrals of Noether symmetries which can be used to provide analytical solutions. As it is pointed out in [25], we also note that the classical Noether symmetry approach with a boundary term K constrains the $F(R,G)$ gravity as a selection criterion that can distinguish the $F(R,G)$ models to utilize the existence of non-trivial Noether symmetries. In this study, we found the maximum number of symmetries as five at the non-vacuum case, but it is four at the vacuum case [28].

This work not only plays complementary role to the previous two studies [19,25], but also includes the the non-vacuum case and it is explicitly found some scale factors in the vacuum case. It might be interesting to perform an analysis of the cosmological parameters for the obtained cosmological models in both of the cases. This will be an argument of future work.

Funding: This work was financially supported through Emergency and Career Fund of the Scholar at Risk (SAR).

Acknowledgments: I would like to thank unknown referees for their valuable comments and suggestions.

Conflicts of Interest: The author declares no conflict of interest.

References

1. Perlmutter, S.; Aldering, G.; Goldhaber, G.; Knop, R.A.; Nugent, P.; Castro, P.G.; Deustua, S.; Fabbro, S.; Goobar, A.; Groom, D.E.; et al. Measurements of Omega and Lambda from 42 high redshift supernovae. *Astrophys. J.* **1999**, *517*, 565. [CrossRef]
2. Riess, A.G.; Filippenko, A.V.; Challis, P.; Clocchiatti, A.; Diercks, A.; Garnavich, P.M.; Gilliland, R.L.; Hogan, C.J.; Jha, S.; Kirshneret, R.P.; et al. Observational evidence from supernovae for an accelerating universe and a cosmological constant. *Astron. J.* **1998**, 116, 1009. [CrossRef]
3. Riess, A.G.; Strolger, Lo.; Casertano, S.; Ferguson, H.C.; Mobasher, B.; Gold, B.; Challis, P.J.; Filippenko, A.V.; Jha, S.; Li, W.; et al. New Hubble space telescope discoveries of type Ia supernovae at $z \geq 1$: Narrowing constraints on the early behavior of dark energy. *Astrophys. J.* **2007**, *659*, 98. [CrossRef]
4. Ade, P.A.R.; Aghanim, N.; Arnaud, M.; Ashdown, M.; Aumont, J.; Baccigalupi, C.; Banday, A.J.; Barreiro, R.B.; Bartlett, J.G.; Bartolo, N.; et al. Planck 2015 results. XIII. Cosmological parameters. *Astron. Astrophys.* **2016**, *594*, A13.
5. Ade, P.A.R.; Aghanim, N.; Ahmed, Z.; Aikin, R.W.; Alexander, K.D.; Arnaud, M.; Aumont, J.; Baccigalupi, C.; Banday, A.J.; Barkats, D.; et al. A Joint Analysis of BICEP2/Keck Array and Planck Data. *Phys. Rev. Lett.* **2015**, *114*, 101301. [CrossRef] [PubMed]
6. Ade, P.A.R.; Ahmed, Z.; Aikin, R.W.; Alexander, K.D.; Barkats, D.; Benton, S.J.; Bischoff, C.A.; Bock, J.J.; Bowens-Rubin, R.; Brevik, J.A.; et al. Improved Constraints on Cosmology and Foregrounds from BICEP2 and Keck Array Cosmic Microwave Background Data with Inclusion of 95 GHz Band. *Phys. Rev. Lett.* **2016**, *116*, 031302. [CrossRef] [PubMed]
7. Komatsu, E.; Smith, K.M.; Dunkley, J.; Bennett, C.L.; Gold, B.; Hinshaw, G.; Jarosik, N.; Larson, D.; Nolta, M.R.; Page, L.; et al. Seven-Year Wilkinson Microwave Anisotropy Probe (WMAP) Observations: Cosmological Interpretation. *Astrophys. J. Suppl.* **2011**, *192*, 18. [CrossRef]
8. Hinshaw, G.; Larson, D.; Komatsu, E.; Spergel, D.N.; Bennett, C.; Dunkley, J.; Nolta, M.R.; Halpern, M.; Hill, R.S.; Odegard, N.; et al. Nine-Year Wilkinson Microwave Anisotropy Probe (WMAP) Observations: Cosmological Parameter Results. *Astrophys. J. Suppl.* **2013**, *208*, 19. [CrossRef]
9. Cognola, G.; Elizalde, E.; Nojiri, S.; Odintsov, S.D.; Zerbini, S. Dark energy in modified Gauss-Bonnet gravity: Late-time acceleration and the hierarchy problem. *Phys. Rev. D* **2006**, *73*, 084007. [CrossRef]
10. Nojiri, S.; Odintsov, S.D.; Gorbunova, O.G. Dark energy problem: From phantom theory to modified Gauss-Bonnet gravity. *J. Phys. A Math. Gen.* **2006**, 39, 6627–6634. [CrossRef]
11. Copeland, E.J.; Sami, M.; Shinji, T. Dynamics of dark energy. *Int. J. Mod. Phys. D* **2006**, *15*, 1753–1936. [CrossRef]
12. Nojiri, S.; Odintsov, S.D. Introduction to modified gravity and gravitational alternative for dark energy. *Int. J. Geom. Meth. Mod. Phys.* **2007**, *4*, 115-146. [CrossRef]
13. De Laurentis, M. Topological invariant quintessence. *Mod. Phys. Lett. A* **2015**, *30*, 1550069. [CrossRef]
14. Odintsov, S.D.; Oikonomou, V.K.; Banerjee, S. Dynamics of Inflation and Dark Energy from $F(R, G)$ Gravity. *Nuclear Phys. B* **2018**, in press.
15. Nojiri, S.; Odintsov, S.D.; Oikonomou, V.K. Modified Gravity Theories on a Nutshell: Inflation, Bounce and Late-time Evolution. *Phys. Rep.* **2017**, *692*, 1–104. [CrossRef]
16. Stephani, H. *Differential Equations: Their Solution Using Symmetries*; Cambridge University Press: Cambridge, UK, 1989; p. 99.
17. Ibragimov, N.H. *CRC Handbook of Lie Group Analysis of Differential Equations: Symmetries, Exact Solutions and Conservation Laws*; CRC Press: Boca Raton, FL, USA, 1994.
18. Capozziello, S.; Lambiase, G. Selection rules in minisuperspace quantum cosmology. *Gen. Relat. Gravit.* **2000**, *32*, 673–696. [CrossRef]
19. Capozziello, S.; De Laurentis, M.; Odintsov, S. D. Noether symmetry approach in Gauss-Bonnet cosmology. *Mod. Phys. Lett. A* **2014**, *29*, 1450164. [CrossRef]
20. Camci, U.; Kucukakca, Y. Noether symmetries of Bianchi I, Bianchi III, and Kantowski-Sachs spacetimes in scalar-coupled gravity theories. *Phys. Rev. D* **2007**, *76*, 084023. [CrossRef]
21. Kucukakca, Y.; Camci, U.; Semiz, I. LRS Bianchi type I universes exhibiting Noether symmetry in the scalar-tensor Brans-Dicke theory. *Gen. Relat. Gravit.* **2012**, *44*, 1893–1917. [CrossRef]

22. Tsamparlis, M.; Paliathanasis, A. Lie and Noether symmetries of geodesic equations and collineations. *Gen. Relat. Gravit.* **2010**, *42*, 2957–2980. [CrossRef]
23. Kucukakca, Y.; Camci, U. Noether gauge symmetry for $f(R)$ gravity in Palatini formalism. *Astrophys. Space Sci.* **2012**, *338*, 211–216. [CrossRef]
24. Camci, U.; Jamal, S.; Kara, A.H. Invariances and Conservation Laws Based on Some FRW Universes. *Int. J. Theor. Phys.* **2014**, *53*, 1483–1494. [CrossRef]
25. Dialektopoulos, K.F.; Capozziello, S. Noether Symmetries as a geometric criterion to select theories of gravity. *Int. J. Geom. Meth. Mod. Phys.* **2018**, *15*, 1840007. [CrossRef]
26. Noether, E. Invariante variation problemes. *Transp. Theory Statist. Phys.* **1971**, *1*, 186–207. [CrossRef]
27. Capozziello, S.; Stabile, A.; Troisi, A. Spherically symmetric solutions in $f(R)$-gravity via Noether Symmetry Approach. *Class. Quantum Gravity* **2007**, *24*, 2153–2166. [CrossRef]
28. Tsamparlis, M.; Paliathanasis, A. Symmetries of Differential Equations in Cosmology. *Symmetry* **2018**, *10*, 233. [CrossRef]
29. Capozziello, S.; Makarenko, A.N.; Odintsov, S.D. Gauss-Bonnet dark energy by Lagrange multipliers. *Phys. Rev. D* **2013**, *87*, 084037. [CrossRef]
30. Paliathanasis, A.; Tsamparlis, M.; Basilakos, S. Constraints and analytical solutions of $f(R)$ theories of gravity using Noether symmetries. *Phys. Rev. D* **2011**, *84*, 123514. [CrossRef]
31. Elizalde, E.; Myrzakulov, R.; Obukhov, V.V.; Sáez-Gómez, D. ΛCDM epoch reconstruction from $F(R,G)$ and modified Gauss-Bonnet gravities. *Class. Quantum Gravity* **2010**, *27*, 095007. [CrossRef]
32. Bamba, K.; Odintsov, S.D.; Sebastiani, L.; Zerbini, S. Finite-time future singularities in modified Gauss-Bonnet and $F(R,G)$ gravity and singularity avoidance. *Eur. Phys. J. C* **2010**, *67*, 295–310. [CrossRef]

 © 2018 by the author. Licensee MDPI, Basel, Switzerland. This article is an open access article distributed under the terms and conditions of the Creative Commons Attribution (CC BY) license (http://creativecommons.org/licenses/by/4.0/).

MDPI
St. Alban-Anlage 66
4052 Basel
Switzerland
Tel. +41 61 683 77 34
Fax +41 61 302 89 18
www.mdpi.com

Symmetry Editorial Office
E-mail: symmetry@mdpi.com
www.mdpi.com/journal/symmetry

www.ingramcontent.com/pod-product-compliance
Lightning Source LLC
LaVergne TN
LVHW071949080526
838202LV00064B/6712